PROCEEDINGS OF THE 30TH INTERNATIONAL GEOLOGICAL CONGRESS
VOLUME 22

HYDROGEOLOGY

Proceedings of the 30th International Geological Congress

Volume 1 : Origin and History of the Earth
Volume 2 : Geosciences and Human Survival, Environment, and Natural Hazards
Volume 3 : Global Changes and Future Environment
Volume 4 : Structure of the Lithosphere and Deep Processes
Volume 5 : Contemporary Lithospheric Motion / Seismic Geology
Volume 6 : Global Tectonic Zones / Supercontinent Formation and Disposal
Volume 7 : Orogenic Belts / Geological Mapping
Volume 8 : Basin Analysis / Global Sedimentary Geology / Sedimentology
Volume 9 : Energy and Mineral Resources for the 21st Century / Geology of Mineral Deposits / Mineral Economics
Volume 10 : New Technology for Geosciences
Volume 11 : Stratigraphy
Volume 12 : Paleontology and Historical Geology
Volume 13 : Marine Geology and Palaeoceanography
Volume 14 : Structural Geology and Geomechanics
Volume 15 : Igneous Petrology
Volume 16 : Mineralogy
Volume 17 : Precambrian Geology and Metamorphic Petrology
Volume 18.A : Geology of Fossil Fuels - Oil and Gas
Volume 18.B : Geology of Fossil Fuels - Coal
Volume 19 : Geochemistry
Volume 20 : Geophysics
Volume 21 : Quaternary Geology
Volume 22 : Hydrogeology
Volume 23 : Engineering Geology
Volume 24 : Environmental Geology
Volume 25 : Mathematical Geology and Geoinformatics
Volume 26 : Comparative Planetology / Geological Education / History of Geosciences

PROCEEDINGS OF THE
30TH INTERNATIONAL GEOLOGICAL CONGRESS

BEIJING, CHINA, 4 - 14 AUGUST 1996

VOLUME 22

HYDROGEOLOGY

EDITORS:

FEI JIN

INSTITUTE OF HYDROGEOLOGY AND ENGINEERING, MGMR, ZHENGDING, CHINA

N.C. KROTHE

DEPARTMENT OF GEOLOGICAL SCIENCES, INDIANA UNIVERSITY, BLOOMINGTON, USA

CRC Press
Taylor & Francis Group
Boca Raton London New York

CRC Press is an imprint of the
Taylor & Francis Group, an **informa** business

First published 1997 by VSP BV Publishing

Published 2019 by CRC Press
Taylor & Francis Group
6000 Broken Sound Parkway NW, Suite 300
Boca Raton, FL 33487-2742

© 1997 by Taylor & Francis Group, LLC
CRC Press is an imprint of Taylor & Francis Group, an Informa business

First issued in paperback 2019

No claim to original U.S. Government works

ISBN 13: 978-0-367-44819-6 (pbk)
ISBN 13: 978-90-6764-253-8 (hbk)

**Visit the Taylor & Francis Web site at
http://www.taylorandfrancis.com**

**and the CRC Press Web site at
http://www.crcpress.com**

CONTENTS

Application of a neural network to groundwater quality evaluation
Shu Longcang, Lin Xueyu and Zhu Xiaojun   1

A study on the water-soil interaction in the process of sea
water intrusion
Wu Jichun, Xue Yuqun and Xie Chunhong   5

Conjunctive use of groundwater and surface water,
Central Beijing, China
Lu Xiaojian   15

System analysis and reasonable use of karst groundwater in
Shentuo spring catching in Shanxi Province
Zhang Lai   22

The new understanding of water movement in saturated clay soil
Hou Jie, Su Qingshan, Pan Shusen and Zu Liaoyuan   29

Study on interchanges among atmospheric, surface unsaturated soil
and groundwater under the conditions of vegetation
Zhang Dezhen and Xu Shimin   35

Modeling exploration and development of deep water level confined
karst water in Zhungeer coal District
Yao Jikun, Duan Guiyin, Cao Yunquan and Lin Zengping   40

Combination of water drainage and supply to prevent disasters and
multi-purpose use of mine water in North China karst inundation
mining areas
Yu Pei and Ye Guijun   47

Coal mine water control in big-discharge coal regions in China
Wang Mengyu   53

Some problems on riverside groundwater
Han Zaisheng   59

The theory and practice on the conjunctive utilization of surface water
and groundwater for the urban water supply in Xi'an, China
Li Peicheng, Zhang Yiqian and Zheng Xilai   64

Study on utilization and reclamation of shallow saline groundwater
Fang Sheng and Chen Xiuling   71

Hydrogeochemical characteristics and genetic analysis of
Dongying Basin, China
Yang Tianxiao, Wang Min and Chen Liang   80

Palaeohydrogeology of the Tuhar Basin (North-West China):
Two stages of groundwater development and their impacts on
distribution of petroleum
Wang Jianrong, Zhang Dajing, Wan Li and Tian Kaiming   89

Realizing comprehensive control of multiple objectives for irrigation
districts by conjunctive management of surface and groundwater
Fang Sheng and Chen Xiuling   96

Groundwater development for agriculture in Northern China
Ji Chuanmao and Wang Zhaoxin   105

Artificially induced hydrogeological effects and their impact of
environments on karst of North and South China
Lu Yaoru and Duan Guangjie   113

Chinese karst and engineering protection
Yan Tongzhen, Tang Huiming and Luo Wenqian   121

Assessment of groundwater contamination from an industrial river by
time series-, flow modeling- and particle tracking methods
Hongbin Sun, M. Koch and Xinlan Liu   139

Karst and hydrogeology of Lebanon
H.S. Edgell 165

The flushing of saline groundwater from a Permo-Triassic sandstone
aquifer system in NW England
J.H. Tellam and J.W. Lloyd   177

A model for coupled fluid flow and multicomponent chemical
reactions with application to sediment diagenesis
Ming-Kuo Lee   185

The physical and economic impact of aquifer over-exploitation at
Hangu, China
T.R. Shearer, B. Adams, R. Kitching, R. Calow, X.D. Cui,
D.J. Chen, R. Grimble and Z.M. Yu 204

Optimization of pump-and-treat technology for aquifer remediation
Yunwei Sun and Bingchen Wang 213

A discussion of scientific issues in hydrogeology of low
permeability strata
Chin-Fu Tsang 224

The paleohydrogeology of the Siberian platform
E.V. Pinneker and A.A. Dzyuba 229

Exploiting ground water from sand rivers in Botswana using
collector wells
R. Herbert, J.A. Barker, J. Davies and O.T. Katai 235

The groundwater assessment for the Western Jamahiriya System
wellfield, Libya
J.W. Lloyd, A. Binsariti, O. Salem, A. El Sunni, A.S. Kwairi,
G. Pizzi and H. Moorwood 258

Optimal allocation of water resources with quality consideration in
Kaifeng City, Henan Province, China
M. Chitsazan, Lin Xueyu and Shu Longcang 270

Annual variation of $\delta^{13}C$ infiltrating water through a soil profile in
a mantled karst area in Southern Indiana
G.H. Yu and N.C. Krothe 281

Groundwater protection strategy, policy and management
J. Vrba 291

Implementation of groundwater protection strategies in the UK
Mengfang Chen and C. Soulsby 299

An investigation of salt contamination of groundwater in
Inner Mongolia
Yun-Sheng Yu, B.C. Wang, W.F. Sun, C.H. Dong,
M.Z. Wang and Ming-Shutsou 308

Proc. 30th Int'l Geol. Congr., Vol.22, pp. 1-4
Fei Jin and N.C. Krothe (Eds.)
© VSP 1997

Application of a Neural Network to Groundwater Quality Evaluation

SHU LONGCANG   LIN XUEYU
Institute of Applied Hydrogeology, Changchun University of Earth Sciences, Changchun, Jilin, 130026, P.R.. CHINA
ZHU XIAOJUN
Department of Applied Geophysics, Changchun University of Earth Sciences, Changchun, Jilin, 130026, P. R.. CHINA

Abstract

Artificial backpropagation (BP) neural network is widely used in robotics, automotive applications and real-time systems. Considering the pattern recognition function of BP network, groundwater quality of Lankao county, Henan province was evaluated by using BP network. BP network used in this application is composed of three layers, they are input, hidden and output layers. Thirty-seven groundwater quality samples from Lankao county, Henan province were chosen as training patterns. Then, thirty-nine groundwater quality samples were evaluated by the trained BP network and   the evaluation results are good.

Keywords: BP network, groundwater quality evaluation, Lankao county

INTRODUCTION

To accurately evaluate groundwater quality is very important for developing groundwater. Some traditional methods have shortcomings [Shu Longcang, 1988], which can be overcome by using an artificial neural network. In this paper, we chose an artificial backpropagation(BP) neural network, it is a good tool in the evaluation of groundwater quality. Backpropagation network operates similarly to linear regression in that a least-squares error criterion is used to determine the goodness of fit to the training data. Through use of a nonlinear transfer function, backpropagation neural network is able to determine complex nonlinear relationships between a set of inputs and a set of outputs. It is this property that makes the method especially well suited to evaluation of groundwater quality.

ARCHITECTURE OF A NEURAL NETWORK

The architecture of an artificial backpropagation neural network is illustrated in Figure 1. The network generally consists of an input layer, one hidden layer, and an output layer. Each layer consists of a number of artificial neurons (also called processing elements or nodes). Each neuron is connected to those in the layers above and below it through weights.

Artificial backpropagation neural network operates as follows. Each input to the network is multiplied by the connection weights connecting it to neurons in the hidden layer. A sum of these products is taken and passed through a nonlinear transfer function such as sigmoid function. The output of the transfer function is passed to the output layer, where it is multiplied by the connection weights between the output layer and the hidden layer, and again a sum of products is taken to generate the output for the network.

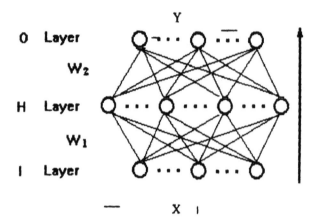

Figure 1. Architecture of an artificial backpropagation neural network (The direction in the figure is information flow direction.).

Notes: I   Layer -- Input layer;

H   Layer -- Hidden layer;

O   Layer -- Output layer;

W_1 -- Connection weights between the elements of the input and hidden layers;

W_2 -- Connection weights between the elements of the output and hidden layers;

X   -- Input vector;

Y   -- Output vector.

The network is trained by learning. During the network is trained, the outputs from the network is compared with the desired outputs. There are many learning methods, but, we used the gradient descent technique in this paper.

APPLICATION TO THE GROUNDWATER QUALITY EVALUATION

According to the demands of groundwater quality evaluation, an artificial backpropagation neural network with four input elements in the input layer, ten hidden elements in a single   hidden layer, and three elements in the output layer, which is shown in Figure 2.

The inputs, shown in the lowest layer, are contents of Sulfate(SO_4^{2-}), Chloride(Cl^-), Fluoride(F^-) and Hardness(H) of the groundwater in Lankao county, Henan province. The processing is performed in the single hidden layer (see Figure 2). The output elements (or nodes) represent different grades of groundwater quality. A sigmoid function was used as the transfer function in the hidden layer.

TRAINING AND PREDICTING

Thirty-seven groundwater quality samples(in September 1992) from Lankao county, Henan province of China are chosen as training patterns. The connection weights were modified until good match between the network's outputs and sample's outputs had been obtained. When the training times is 4000, the average errors of the network are smaller than 0.01. Then, thirty-nine groundwater quality samples(in May 1992) were evaluated by running the trained BP network. The prediction results are shown in Table 1.

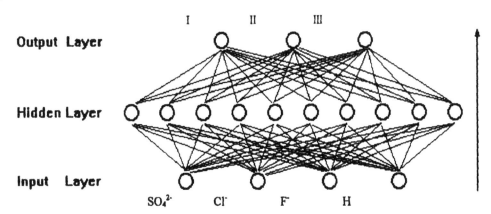

Figure 2. Architecture of the BP network used to evaluation of groundwater quality (The direction in the figure is information flow direction.).

Table 1. Results of groundwater quality evaluation by using the trained BP network

Grades	Number of Samples	Percentage
1	19	48.71
2	17	43.59
3	3	7.70

According to the results in the table 1, we know that over half of groundwater in the study area can not be used for drinking.

CONCLUSIONS

This paper illustrates the feasibility of using an artificial backpropagation network to evaluate groundwater quality. According to the application of the artificial backpropagation neural network to evaluate of groundwater quality in Lankao county, the following conclusions are obtained.

(1). As neural network shows some properties, like association, parallel searching, and adaptation to the changes to environment, which are analogous to human brain properties, these properties can improve the precision of groundwater quality evaluation results.

(2). It is very important to point up that the BP network has low requirements of statistics knowledge and data pre-processing, which makes it practicable.

(3). We are sure that artificial neural network will be widely used in hydrogeology.

REFERENCES

1. Jeffrey L. Baldwin, Richard M. Bateman and Charles L. Wheatley. Application of a neural network to the problem of mineral identification from well logs, *The Log Analyst*, 279-293 September-October, 1990.
2. Mary M. Poulton, Ben K. Sternberg and Charles E. Glass. Neural network pattern recognition of subsurface EM images, *Journal of Applied Geophysics*, 29 (1992) 21-36.
3. Robert G. Cubero. Neural networks for water demand time series forcasting, *Artificial Neural Networks*, Springer-Verlag, 453-460 (1991).

Proc. 30th Int'l Geol. Congr., Vol.22, pp. 5-14
Fei Jin and N.C. Krothe (Eds.)
© VSP 1997

A Study on the Water-Soil Interaction in the Process of Sea Water Intrusion

WU JICHUN, XUE YUQUN, XIE CHUNHONG

Dept. of Earth Sciences, Dept. of Mathematics, Nanjing University, 210093, P.R. China

Abstract

The field observed and the laboratory experimental results of water-soil interaction in the process of sea water intrusion are presented. Results of the field observed and the laboratory experiments show that there exist significant cations exchange of Na^+ and Ca^{2+}, and moderate cations exchange of Mg^{2+} and Ca^{2+} between water and soil in the process of sea water intrusion. On the basis of the above work, a new three-dimensional mathematical model for sea water intrusion in aquifer is presented. This model considers not only many important factors(Such as the existence of transition zone, the effect of variable density and viscosity on fluid flow, the effect of phreatic surface fluctuation on the process of sea water intrusion, etc.), but also the cations exchange of $Na^+ - Ca^{2+}$ and $Mg^{2+} - Ca^{2+}$ between water and soil during sea water intrusion. It can simulate simultaneously the transport behavior of Ca^{2+}, Na^+ and Mg^{2+} in aquifer in the process of sea water intrusion. This model is used to describe the transport behavior of Ca^{2+}, Na^+ and Mg^{2+} in the process of sea water intrusion in Huangheying of Longkou City, Shandong Province, China. The simulated values agree very well with the field data(e.g., the total mean values of the absolute errors of Ca^{2+}, Na^+ and Mg^{2+} concentrations are 6.48mg/l, 11.53mg/l and 2.72mg/l, respectively, and that of the hydraulic head is 0.19m).

Keywords: Water-soil interaction, Sea water intrusion, Mathematical model, Numerical
     simulation

INTRODUCTION

Sea water intrusion has attracted great attention recently. Much work has been published on this subject. Several researchers have noted that there exists water-soil interaction in the process of sea water intrusion[1,2], and a series of column laboratory experiments, together with geochemical calculations have been set up[3,4] , but they have no opportunity to do much more work on this subject. In the past few years, We had also done some research work on this subject, such as the consequent monitoring of the transport behavior of exchanging cations in the process of sea water intrusion in Huangheying of Longkou City, Shandong Province, a series of laboratory experiments of cation exchange between water and soil, and numerical modeling of sea water intrusion with multiple cation. Here we will present some positive results of our late research work.

FIELD OBSERVED RESULTS

The study area is in Huangheying region, a place located in the northeastern tip of Longkou City, Shandong Province (Figure 1).

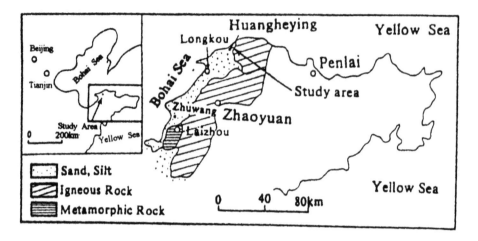

Fig. 1. Location of the study area and its feature map

In the study area, we had set up an observation network, in which there are 31 special observation wells, 12 ordinary wells and three pumping wells. Among these, 24 of the special observation wells are divided into eight groups, each group has three wells located at approximately the same site but at different depths--at the top, middle, and bottom of the aquifer. The screens of all the special observation wells are less than 0.5 m in length. Since January 1988, a study of the groundwater of the area has been consecutive conducted by means of these observation wells, with water levels measured every five days and water samples taken for chemical analysis every month. The field observed results along the main observed section in June 20, 1993 is shown in Figure 2.

Figure 2 shows that in sea water intrusion region, the concentrations of Cl^-, SO_4^{2-}, Mg^{2+}, $K^+ + Na^+$ and TDS decrease, while the concentration of Ca^{2+} increases rapidly and is greater than those in sea water and fresh water regions, which demonstrates that the water in sea water intrusion region should not be a simple mixture of local fresh water and sea water. Why is this phenomenon popular in sea water intrusion region of the study area? Figure 2 shows that the contents of SO_4^{2-} and HCO_3^- in groundwater of sea water intrusion region do not increase when the content of Ca^{2+} increases. In addition, no large amount of calcite, dolomite and gypsum precipitation was found in the study aquifer. For this reason, we think this phenomenon may be caused by the cation exchange between water and soil. The alluvial and diluvial sandy clays have a high calcium content due to their previous saturation of water from land and therefore provide good opportunity for cation exchange. Analyzing the field observed results thoroughly, we think there may exist cation exchange of $K^+ - Ca^{2+}$, $Na^+ - Ca^{2+}$, and $Mg^{2+} - Ca^{2+}$ between water and soil in sea water intrusion region of the study area.

LABORATORY EXPERIMENTAL RESULTS

In order to confirm our conclusion, we had carried out a series of laboratory experiments. Results of cations exchange experiments using the soil sample collected from fresh water region and the exchanging solution mixed by sea water and fresh groundwater at different ratios are listed in Table 1. It shows that during mass exchange between water and soil in the process of sea water intrusion, there exist significant cations exchange between Na^+ and Ca^{2+}, moderate cations exchange between Mg^{2+} and Ca^{2+}, and insignificant cations exchange between K^+ and Ca^{2+}.

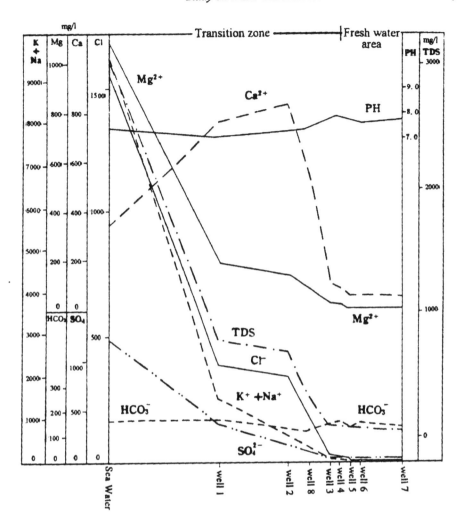

Fig. 2. The generalized hydrochemical section from sea to the fresh water region

Table 1. Changing result of composition of exchanging solution and exchange complex of soil sample(meq)

	Exchanging solution				Soil sample			
	K^+	Na^+	Ca^{2+}	Mg^{2+}	K^+	Na^+	Ca^{2+}	Mg^{2+}
1	-0.0740	-0.2244	+0.3771	-0.0615	+0.0250	+0.2181	-0.3691	+0.0580
2	-0.0723	-0.2200	+0.3627	-0.0599	+0.0230	+0.2182	-0.3503	+0.0531
3	-0.0711	-0.2130	+0.3543	-0.0548	+0.0220	+0.2080	-0.3414	+0.0481
4	-0.0507	-0.2106	+0.3408	-0.0535	+0.0160	+0.2031	-0.3242	+0.0440

Note: + stands for increase, and - decr-ease

Result of the soil column experiment is similar to that of Appelo, et. al.[4]. It also shows that during mass exchange between water and soil in the process of sea water intrusion, there exist significant cations exchange between Na^+ and Ca^{2+}, moderate cations exchange between Mg^{2+} and Ca^{2+}, and insignificant cations exchange between K^+ and Ca^{2+}.

In order to describe the transport behavior of exchangeable cations and reflect the important role of the cation exchange between water and soil in the forming and evolving of the main cations in the groundwater in sea water intrusion region, we have to develop a new miscible transport model, in which the cation exchange between groundwater and soil is considered.

MATHEMATICAL MODEL

Flow model
For the phreatic aquifer, the flow model can be written in the following form:

$$\frac{\partial}{\partial X_i}\left[K_{ij}\left(\frac{\partial H}{\partial X_j}+\eta c_T e_j\right)\right]=s_s\frac{\partial H}{\partial t}+\varphi\eta\frac{\partial c_T}{\partial t}-\frac{\rho_f}{\rho_0}q \tag{1}$$

initial condition: $H(X_i,0)=H_0(X_i)$ (2)

Dirichlet boundaries: $H(X_i,t)\big|_{\Gamma_1}=\overline{H}(X_i,t)$ (3)

Neumann boundaries: $v_i n_i\big|_{\Gamma_2}=-q_n(X_i,t)$ (4)

phreatic surface boundaries: $H^*(X_i,t)\big|_{\Gamma_{2-1}}=X_3(X_i)$ (5)

$$K_{ij}\left(\frac{\partial H}{\partial X_j}+\eta c_T e_j\right)n_i\big|_{\Gamma_{2-1}}=\left(W-\frac{\rho^*}{\rho_0}S_Y\frac{\partial H^*}{\partial t}\right)n_3 \tag{6}$$

Motion equation
The relevant motion equation can be expressed by

$$v_i=-\frac{K_{ij}}{\rho_r}\left(\frac{\partial H}{\partial X_j}+\eta c_T e_j\right) \tag{7}$$

In the equations above, H is the reference hydraulic head referred to as fresh water head, defined as $H=\dfrac{P}{\rho_0 g}+X_3$, P is the pressure of the fluid, ρ_0 is the reference density referred to as fresh water density, X_3 is elevation, and g is the gravitational acceleration; For the sake of contrast, the actual head is $h=\dfrac{P}{\rho_f g}+X_3$, ρ_f is the density of the mixed fluid; K_{ij} is the hydraulic conductivity tensor($i, j=1, 2, 3$); S_s is the specific storage; φ is the effective porosity; q is the volumetric flow rate of sources(or sinks) per unit volume of porous medium; $c_T=c_1+2c_2+c_3$, c_1, c_2 and c_3 are the concentrations of Ca^{2+}, Na^+ and Mg^{2+} in the mixed fluid, respectively; $\eta=\dfrac{\rho_s-\rho_0}{\rho_0 c_{Ts}}$ is the density coupling coefficient; c_{Ts} is the concentration of c_T corresponding to the maximum density   ;   is the jth component of the gravitational unit vector, i.e.,   ,   ; t is time; .   and .   (   ) are cartesian coordinate, which conform to the summation convention; .   is the initial head; .   is the prescribed head on the Direchlet boundaries Γ_1; n_i is the outward unit vector normal to Neumann boundaries Γ_2 and the phreatic surface Γ_{2-1}; q_n is the prescribed lateral flow rate per unit area on Γ_2 ; H^* is the reference hydraulic head on the phreatic surface; $\rho_r = \dfrac{\rho_f}{\rho_0}$ is the density ratio; ρ^* is the density of groundwater in the zone where the water level fluctuates, and it becomes ρ_f when the water level declines and approaches ρ_0 when the water level ascends; S_Y is specific yield when water level declines and is equal to $1 - S_w$ (S_w is degree of saturation) when water level ascends; v_i is the Darcian velocity tensor of the mixed fluid.

Solute Transport Model
Considering the fact that during mass exchange between water and soil in the process of sea water intrusion, significant cation exchange takes place between Na^+ and Ca^{2+}, moderate cation exchange between Mg^{2+} and Ca^{2+}, and insignificant cation exchange between K^+ and Ca^{2+}, the former two cation exchanges need to be considered in our model. Assuming C_1, C_2 and C_3 are the concentrations of Ca^{2+}, Na^+ and Mg^{2+} in groundwater respectively, the solute transport model for the description of the transport behavior of exchange cations Ca^{2+}, Na^+ and Mg^{2+} can be written as follows:

$$\left(1 - \frac{\eta c_T}{\rho_r}\right)\frac{\partial c_T}{\partial t} = \frac{\partial}{\partial X_i}\left(D_{ij}\frac{\partial c_T}{\partial X_j}\right) - \left[u_i + \frac{\eta c_T}{\rho_r \varphi}\frac{K_{ij}}{\rho_r}\left(\frac{\partial H}{\partial X_j} + \eta c_T e_j\right)\right]\frac{\partial c_T}{\partial X_i}$$

$$+ \frac{c_T S_s}{\rho_r \varphi}\frac{\partial H}{\partial t} + \frac{q}{\varphi}\left(c_T^* - c_T\right) \tag{8}$$

$$\frac{\partial c_1}{\partial t} + \frac{\rho_m S_T}{\varphi}\left(\frac{\partial F_1}{\partial c_1}\frac{\partial c_1}{\partial t} + \frac{\partial F_1}{\partial c_2}\frac{\partial c_2}{\partial t} + \frac{\partial F_1}{\partial c_T}\frac{\partial c_T}{\partial t}\right) = \frac{\partial}{\partial X_i}\left(D_{ij}\frac{\partial c_1}{\partial X_j}\right) - u_i\frac{\partial c_1}{\partial X_j}$$

$$- \frac{\eta c_1}{\rho_r \varphi}\frac{K_{ij}}{\rho_r}\left(\frac{\partial H}{\partial X_j} + \eta c_T e_j\right)\frac{\partial c_T}{\partial X_i} + \frac{c_1 S_s}{\rho_r \varphi}\frac{\partial H}{\partial t} + \frac{\eta c_1}{\rho_r}\frac{\partial c_T}{\partial t} + \frac{q}{\varphi}\left(c_1^* - c_1\right) \tag{9}$$

$$\frac{\partial c_2}{\partial t} + \frac{\rho_m S_T}{\varphi}\left(\frac{\partial F_2}{\partial c_1}\frac{\partial c_1}{\partial t} + \frac{\partial F_2}{\partial c_2}\frac{\partial c_2}{\partial t} + \frac{\partial F_2}{\partial c_T}\frac{\partial c_T}{\partial t}\right) = \frac{\partial}{\partial X_i}\left(D_{ij}\frac{\partial c_2}{\partial X_j}\right) - u_i\frac{\partial c_2}{\partial X_j}$$

$$- \frac{\eta c_2}{\rho_r \varphi}\frac{K_{ij}}{\rho_r}\left(\frac{\partial H}{\partial X_j} + \eta c_T e_j\right)\frac{\partial c_T}{\partial X_i} + \frac{c_2 S_s}{\rho_r \varphi}\frac{\partial H}{\partial t} + \frac{\eta c_2}{\rho_r}\frac{\partial c_T}{\partial t} + \frac{q}{\varphi}\left(c_2^* - c_2\right) \tag{10}$$

$$c_3 = c_T - c_1 - 2c_2 \tag{11}$$

where D_{ij} is the dispersion coefficient tensor; ρ_m is the soil bulk density; u_i is the pore velocity

vector $u_i = \dfrac{v_i}{\varphi}$; c_T^* is defined as $c_T^* = c_1^* + 2c_2^* + c_3^*$, c_1^*, c_2^* and c_3^* are the concentrations of

Ca^{2+}, Na^+ and Mg^{2+} in the pumped(or injected) fluid, respectively; S_T is the cation exchange capacity, defined as $S_T = s_1 + 2s_2 + s_3$, s_1, s_2 and s_3 are the sorbed phase concentrations of Ca^{2+}, Na^+ and Mg^{2+} in soil, respectively; F_1 and F_2 are nonlinear algebraic functions, which can be expressed as:

$$F_1(c_1,c_2,c_T) = \frac{r_1 K_{12} c_1}{4\psi^2}\left[-r_2 c_2 \pm \left(r_2^2 c_2^2 + 4\psi\right)^{\frac{1}{2}}\right]^2 \tag{12}$$

$$F_2(c_1,c_2,c_T) = \frac{r_2 c_2}{2\psi}\left[-r_2 c_2 \pm \left(r_2^2 c_2^2 + 4\psi\right)^{\frac{1}{2}}\right] \tag{13}$$

where $\psi = \dfrac{r_3 K_{12}}{K_{13}}(c_T - c_1 - 2c_2) + r_1 K_{12} c_1$; r_1 ,r_2 and r_3 are the activity coefficients of

Ca^{2+}, Na^+ and Mg^{2+}, respectively; K_{12} and K_{13} are the cation exchange selectivity coefficients of $Ca^{2+}-Na^+$ and $Ca^{2+}-Mg^{2+}$, respectively, which can be expressed as:

$$K_{12} = \left(\frac{s_1^-}{r_1 c_1}\right)\left(\frac{r_2 c_2}{s_2^-}\right)^2, \quad K_{13} = \left(\frac{s_1^-}{r_1 c_1}\right)\left(\frac{r_3 c_3}{s_3^-}\right) \tag{14}$$

where s_1^- , s_2^- and s_3^- are the equivalent fractions of Ca^{2+}, Na^+ and Mg^{2+} in the sorbed phase,

defined as $s_1^- = \dfrac{s_1}{s_T}$, $s_2^- = \dfrac{2s_2}{s_T}$, $s_3^- = \dfrac{s_3}{s_T}$, respectively.

The solute transport equations (9), (10) and (11) are taken to satisfy following initial and boundary conditions:

$$c_1(X_i,0) = c_{1_0}(X_i), \quad c_2(X_i,0) = c_{2_0}(X_i), \quad c_3(X_i,0) = c_{3_0}(X_i)$$

$$c_1(X_i,t)\big|_{B_1} = \overline{c_1}(X_i,t), \quad c_2(X_i,t)\big|_{B_1} = \overline{c_2}(X_i,t), \quad c_3(X_i,t)\big|_{B_1} = \overline{c_3}(X_i,t)$$

$$\left(-D_{ij}\frac{\partial c_1}{\partial X_j} + u_i c_1\right)\varphi n_i\Big|_{B_2} = J_{b_1} n_i \quad , \quad \left(-D_{ij}\frac{\partial c_2}{\partial X_j} + u_i c_2\right)\varphi n_i\Big|_{B_2} = J_{b_2} n_i$$

$$\left(-D_{ij}\frac{\partial c_3}{\partial X_j} + u_i c_3\right)\varphi n_i\Big|_{B_2} = J_{b_3} n_i$$

For the phreatic aquifer, the following boundary conditions must be satisfied along the free surface:

$$-D_{ij}\frac{\partial c_1}{\partial X_j} n_i\Big|_{\Gamma_{2-1}} = \left(1 - \frac{\rho^*}{\rho_f}\right)\frac{c_1 s_T}{\varphi}\frac{\partial H^*}{\partial t} n_3 + \left(\frac{\rho_0 c_1}{\rho_f} - c_1'\right)\frac{W}{\varphi} n_3$$

$$-D_{ij}\frac{\partial c_2}{\partial X_j}n_i\Big|_{\Gamma_{2-1}} = \left(1-\frac{\rho^*}{\rho_f}\right)\frac{c_2 S_\gamma}{\varphi}\frac{\partial H^*}{\partial t}n_3 + \left(\frac{\rho_0 c_2}{\rho_f}-c_2'\right)\frac{W}{\varphi}n_3$$

$$-D_{ij}\frac{\partial c_3}{\partial X_j}n_i\Big|_{\Gamma_{2-1}} = \left(1-\frac{\rho^*}{\rho_f}\right)\frac{c_3 S_\gamma}{\varphi}\frac{\partial H^*}{\partial t}n_3 + \left(\frac{\rho_0 c_3}{\rho_f}-c_3'\right)\frac{W}{\varphi}n_3$$

where c_{1_0}, c_{2_0} and c_{3_0} are the initial concentrations of Ca^{2+}, Na^+ and Mg^{2+}, respectively; $\overline{c_1}$, $\overline{c_2}$ and $\overline{c_3}$ are the prescribed concentrations of Ca^{2+}, Na^+ and Mg^{2+} on the Dirichlet boundaries B_1, respectively; J_{b_1}, J_{b_2} and J_{b_3} are the prescribed mass fluxes of Ca^{2+}, Na^+ and Mg^{2+} per unit area on the Neumann boundaries B_2, respectively.

Mathematical model for sea water intrusion with multiple cation

The above-mentioned flow model and solute transport model, together with the motion equation (7), are integrated to form a sophisticated mathematical model for the description of transport behavior of exchanging cations Ca^{2+}, Na^+ and Mg^{2+} in aquifer during sea water intrusion. It can reflect the cation exchange of $Na^+ - Ca^{2+}$ and $Mg^{2+} - Ca^{2+}$ between water and soil during sea water intrusion. It is obviously more advanced than sea water intrusion models now available[5,6].

Numerical method
Equation (1) can solved numerically by the Galerkin finite element method. Equation (7) can be solved by the approach suggested by Yeh[7]. Due to nonlinearity of equations (8), (9) and (10), the characteristic finite element method formerly used is not applicable. So we improved the method and adopted improved characteristic finite element method to solve (8), (9) and (10) [8].

NUMERICAL SIMULATION FOR EXCHANGEABLE CATIONS TRANSPORT

Numerical simulation
The developed model was applied to simulate the transport behavior of exchange cations Ca^{2+}, Na^+ and Mg^{2+} in the process of sea water intrusion from January 6, 1988 to June 19, 1990 in Huangheying of Longkou City, Shandong Province, China. The whole domain is divided into 2430 elements and 3592 nodes. The simulation time step is 1 month. In the simulation period of two and a half years, the simulated hydraulic head and the concentrations of Ca^{2+}, Na^+ and Mg^{2+} of all the 24 special observation wells are compared with the observed data. Figure 3 shows part of the results. It shows good agreement. The average absolute errors between the simulated and the observed are: 0.29 m for the hydraulic head, 6.48 mg/l for $[Ca^{2+}]$, 11.5 mg/l for $[Na^+]$, and 2.72 mg/l for $[Mg^{2+}]$. Good comparison between the simulated and the observed values demonstrates that the model is reliable and can be applied to solve practical problems.

Analysis of the results
Figure 4 is distribution map of some simulated results in cross section. It clearly illustrates the distribution and evolution of Ca^{2+}, Na^+ and Mg^{2+} in groundwater in sea water intrusion region. The evolutional pattern of concentrations of Na^+ and Mg^{2+} are similiar to that of Cl^-[6]: From shore to continent, concentrations of Na^+ and Mg^{2+} near shore decrease rapidly at the beginning, then decrease smoothly; Near pumping wells, they decrease rapidly again before reaching the fresh water region; There is a considerable wide transition zone between sea water and fresh

water. Moreover, there exists a relatively clear sea-water/fresh-water interface; In the transition zone, because of the cation exchange between groundwater and soil, the concentrations of Na^+ and Mg^{2+} decrease a little faster than that of Cl^-. However, the evolutional pattern of Ca^{2+} are different: From shore to continent, the concentration of Ca^{2+} increases rapidly at the beginning, then decreases gradually, and decreases rapidly near pumping wells. Compared with Na^+ and Mg^{2+} a considerable wide transition zone and a steep sea-water/fresh-water interface also exist. The above-mentioned cation exchange results in the accumulation of Ca^{2+} in the transition zone and the higher content of Ca^{2+} than in both sea water and fresh water. The results also show a good correspondence between the simulated and observed concentration field.

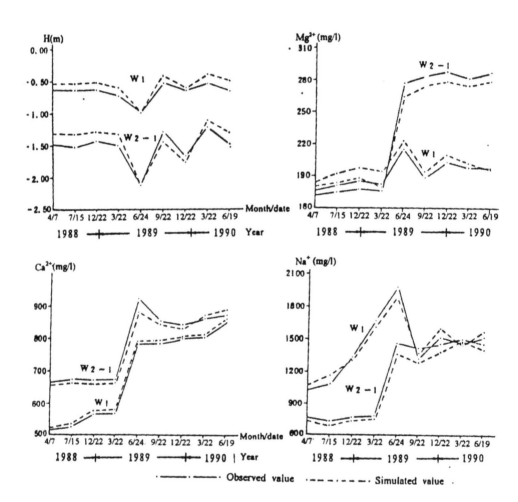

Fig. 3. Comparison between simulated and observed results

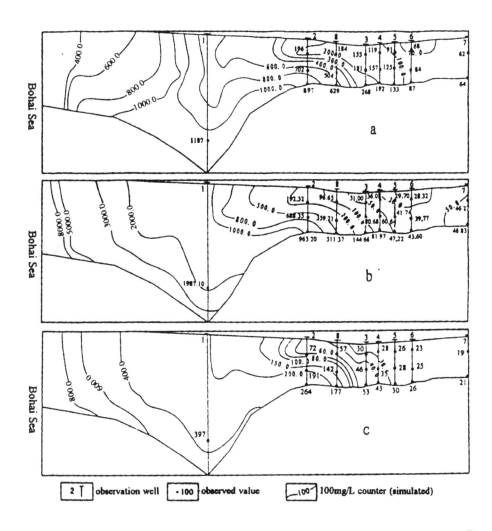

Fig.4. Distribution of simulated results in cross section (June 19, 1990) a concentration contour map of Ca^2; b concentration contour map of Na$^+$; c concentration contour map of Mg^2

CONCLUSIONS

(1) Results of field observed and laboratory experiments show that during mass exchange between water and soil in the process of sea water intrusion, there exists significant cation exchange between Na$^+$ and Ca^{2+}, moderate cation exchange between Mg^{2+} and Ca^{2+}, and insignificant cation exchange between K$^+$ and Ca^{2+}.

(2) The new three-dimensional sea water intrusion model established in the paper is reliable and more advanced than our former three-dimensional sea water intrusion model. Considering the main facts in the study area, the model not only can represent dynamics of groundwater level,

changes of flow velocity, transport pattern of sea-water/fresh-water interface and transition zone, but also can reflect the cation exchange between water and soil and the transport behavior of exchange cations Ca^{2+}, Na^+ and Mg^{2+} in aquifer. The simulated results of the model agree well with the field observed values.

(3) In order to study the environmental evolution and development in groundwater in sea water intrusion region, it is necessary to established new three-dimensional sea water intrusion mathematical model with multiple cation. The traditional solute transport model can not reflect the fact that the concentration of Ca^{2+} in groundwater increases abnormally in the transition zone.

REFERENCES

1. Howard, K. W. F. and J. W. Lloyd. Major ion characterization of coastal saline ground waters. *Ground Water* 21:4, 429-437(1983).
2. Custodio, E. Saline intrusion, hydrogeology in service of man. *Memories of the 18th Congress of the Int. Assoc. of Hydrogeologists*, Cambridge(1985).
3. Appelo, C. A. J. and A. Willemsen. Geochemical calculations and observations on salt water intrusions, I. A combined geochemical/mixing cell model, *Journal of Hydro.* 94, 313-330(1987).
4. Appelo, C. A. J., A. Willemsen, H.E. Beekman and J. Griffioen. Geochemical calculations and observations on salt water intrusions, II. Validation of a geochemical model with laboratory experiments. *Journal of Hydrology* 120, 225-250(1990).
5. Huyakorn, P. S., P. F. Anderson, et.al. Saltwater intrusion in aquifers: Development and testing of a three-dimensional finite element model. *Water Resour. Res* 23:2, 293-312(1987).
6. Xue, Yuqun, Chunhong Xie, Jichun Wu. A three-dimensional miscible transport model for seawater intrusion in China. *Water Resour. Res* 31:4, 903-912(1995).
7. Yeh, G.T. On the computation of Darcian velocity and mass balance: The finite element modeling of groundwater flow. *Water Resour. Res* 17:5, pp1529-1534(1981).
8. Wu, Jichun. Xue Yuqun. Improved characteristic finite element method in solving highly nonlinear sea water intrusion problem, *Chinese Journal of Comp. Physics* 13:2, 201-206(1996).

Proc. 30th Int'l Geol. Congr., *Vol.22*, pp. 15-21
Fei Jin and N.C. Krothe (Eds.)
© VSP 1997

Conjunctive Use of Groundwater and Surface Water, Central Beijing, China

LU XIAOJIAN
Beijing Prospecting Institute of Geological Engineering, No.38 Beiwa Road, Haidian District, Beijing, China

Abstract

Beijing is one of the biggest city which use groundwater as its main water supply source. With the rapid development of construction and the increase of population, water demand is also increasing rapidly. Owing to the lack of scientific management, groundwater has been over-exploited for a long time, which led to regional water level down. In the meantime, some ecological and environmental problem appeared, as groundwater pollution and land subsidence. By analyzing precipitation periods and establishing a most practical model of groundwater and surface water conjunctive use, this paper puts forward the optimal plans of conjunctive use of groundwater and surface water in Beijing proper and the suburbs.

Keywords: conjunctive use, model, groundwater, surface water

INTRODUCTION

The area discussed below ranges from Eastern Xishan Mountains of Beijing to the upper and middle part of Yongding River alluvial- proluvial fan, which is about 1,000 km². With the development of the urban's construction, industry, agriculture, and the increase of the population and economy, demands for water- resource grow rapidly.Before 1980s, groundwater yield had been 50% of the total water supply.Otherwise, in recent years surface-water supply has dropped and been rather expensive,so groundwater yield has grown to 60%- 70%. Continued fall in groundwater level, persistent drought climate and insufficient surface water, will greatly threaten the water supply in the Beijing central area. Therefore, to increase the effective utilization ratio of water resource for satisfying the water demand and to carry out the conjunctive use of groundwater and surface water are economical and rational measures, which can keep the balance and recycle between the water extraction and recharge, and can mitigate the pressure of water shortage and periodical contradictions of urban water supply .

CONJUNCTIVE USE MODELS OF GROUNDWATER AND SURFACE WATER

Scopes and methods

The water supply in the whole Beijing mainly consists of surface water, groundwater and water works, which the waterworks' sources come from both of surface-water and groundwater, and their water supply makes up 1/3 of the total urban water supply. The surface water supply system consists of two parts. One part is the network consisting of reserviors of Guanting, Miyun, Baihebu and Huairou and their accesory canals of Yongding River and Jingmi. The other part of water supply comes from the sources of scattered medium and small-sized reserviors and river gates. They form separate supply systems of their own and supply water for agriculture mainly. Nine waterworks locate in the central Beijing. No.1,2,3,4,5,7 and 8 are groundwater works; No.9 and Tiancun are surface water works. All the groundwater works abstract in the central area but No.8 which has its source from distant suburbs. If Water Work No.9 in the outer suburb which has its source from Miyun reservior is built and goes into operation, the water supply of waterworks in central Beijing will be more than 50% of the total. Therefore, the waterworks assiciated with the urban water supply is an important part of the conjunctive use. Besides, the western aquifers, i.e. groundwater reservoir, provide huge room for surplus surface water and discharge water, which can be stored by artificial recharge in wet years and used in dry years. Four artificial recharge sites locate in the western Beijing. They distribute in San Jia Dian, Xi Huang Cun, Lu Gou Qiao and Nan Han He, in which all sites recharge by river beds but Xi Huang Cun by sand-gravel pit(Fig.1). Furthermore, we choose the central Beijing as the area of the conjunctive use of groundwater and surface water.

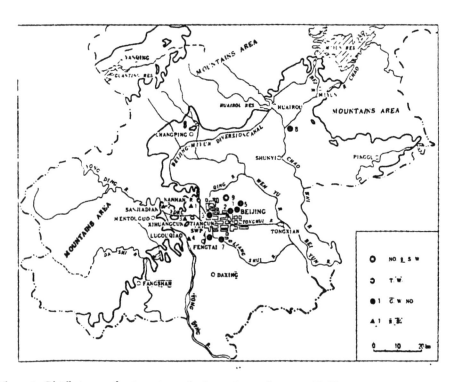

Figure 1. Distribute map of water system and urban water supply system of Beijing. Symbols: YRDC, Yongding River diversion Canal; No.9 S.W., No.9 surface water waterworks; T.W., Tiancun waterworks; G.W.No., groundwater waterworks No.; R.S., recharge site; 1, Nan Han He; 2, San Jia Dian; 3, Xi Huang Cun; 4, Lu Gou Qiao.

Objects of water supply are: 1) living usage; 2) the usage of industry and agriculture; 3) the specific usage of power generation and the artificial recharge.
With different rainfall in seasons, the methods and principles of water supply are different: in normal and wet years, surface water is used as more as possible, and groundwater pumpage is limited. In dry years,more groundwater is withdrawn to satisfy water demands. The waterworks in outer suburbs have priority to supply water for the central area over those water works in the suburbs and the urban which play a complement role.

Conjunctive use models of groundwater and surface water
The conjunctive use of groundwater and surface water can be formulated as a linear optimization model. Defining $f_i(Q_{gk},Q_{sk},Q_{rk})$ as the extraction function and cost function and assuming that the planning horizon(T^*) again consist of n planning periods of length T, the objective function of model is

$$Max(Min)Z = \sum_k \alpha_k \sum_i \mu_i f_i(Q_{gk}, Q_{sk}, Q_{rk})$$

where α_k is the discont rate during k period, and μ_i is the weights. Q_{gk}, Q_{sk}, Q_{rk} is groundwater, surface water and recharge water yield respectively at k period.

The policies of conjunctive use of groundwater and surface water are constrained in each period by:

The response equations of the groundwater system. The response equations may be written for each planning period using the matrix solution to the finite - difference equations. That is:

$$s_l(j,k) \le \sum_{k=1}^{n} \{ \sum_{i=1}^{N_g}[\beta(j,i,n-k+1)Q_g(i,k)] + \sum_{i=1}^{N_r}[\beta(j,i,n-k+1)Q_r(i,k)] \} \le s_h(j,k)$$

$$j=1,2,\cdots,N_g+N_r,$$

$k=1,2,\cdots,T.$

The well capacity restriction.

$$0 \le Q_g(i,k) \le Q_{gmax}(i,k)$$

where $Q_{gmax}(i,k)$ is the capacity of well i at k period.

The artificial recharge capacity restriction

$$0 \le Q_r(i,k) \le Q_{rmax}(i,k)$$

where $Q_{rmax}(i,k)$ is the capacity of recharge site i at k period.

The surface water supply capacity constriction

$$0 \le Q_s(i,k) \le Q_{smax}(k)$$

where $Q_{smax}(i,k)$ is the maximum surface water supply at k period.

The water target requirements. That is:

$$\sum_{i=1}^{N_g} Q_g(i,k) + \sum_{i=1}^{N_s} Q_s(i,k) \ge Q_d(k) .$$

The artificial recharge yield constriction.

$$\sum_{i=1}^{N_r} Q_r(i,k) \le SWR(k)$$

where SWR(k) is the usable recharge yield at k period.

The computed optimal dicisions maximize the total water supply from allocating the water supplies and recharge over a planning horizon which minimizing interference effects between the wells in the system, land subsidence, or other water quality problems.

COMPUTATION RESULTS OF THE CONJUNCTIVE USE OF GROUNDWATER AND SURFACE WATER

Analysis of precipitation period

Through computing and analyzing the implicated periods of precipetation records of 266 years by means of auto-corelation, windows spectrums, Fisher periodical map and variance analysis etc. , the results show that the apparent periods of precipitation have three stages: short period (2, 3 or 5 years); middle period (10-12 years); long period (85-118 years). In every 5 years, there's mostly a dry period of 2 or 3 years. Therefore, an optimized and implemented proposal of water supply, based on a dry period of 2 or 3 continuous years in the central area, has been submitted.

The optimal measures for satisfying water supply in a period with 2 continuous dry years

In the period with 2 continuous dry years (i. e. 5 years of wet, normal,dry, dry, wet), with a rule of storing water in 3 normal or wet years to make up for 2 dry years, the multi- purpose plans were computed (see Tab. 1).

Table 1.   (unit: million m³)

| | Groundwater Works NO. | | | | | | | Surface Water Works | | | Artificial Recharge Site | | | | |
	1	2	3	4	5	6	SUM	NO.9	T.C.	SUM	L.G.Q	X.H.C.	N.H.H.	S.J.D.	SUM
Objective: Minimum Cost															
W	0	0	101.08	87.6	0	0	182.5 371.18	136.88	58.95	195.83	64.34	30.6	73.0	43.65	211.59
N	0	0	129.08	87.6	0	0	182.5 399.18	136.88	58.95	195.83	0.71	0	0	35.29	36.0
D	0	0	289.85	87.6	0	0	182.5 599.95	0	62.05	62.05	0	0	0	0	0.0
D	0	0	282.06	87.6	0	35.79	182.5 587.95	0	62.05	62.05	0	0	0	0	0.0
W	10.57	36.04	47.43	0	4.93	3.62	182.5 285.08	330.87	62.05	392.92	0	30.6	73.0	0	103.6
S	10.57	36.04	849.50	350.4	4.93	39.41	912.5 2203.3	604.62	304.05	988.67	65.05	61.2	146.0	78.94	351.19
Objective: Maximum Water Supply															
W	0	0	350.4	87.6	0	5.78	182.5 626.28	273.75	62.05	335.80	64.53	30.6	73.0	44.74	12.87
N	0	0	0	87.6	0	0	182.5 270.1	273.75	62.05	335.80	0	0	0	36 0	36.0
D	0	0	279.85	87.6	0	0	182.5 549.95	10.0	62.05	72.05	0	0	0	0	0.0
D	0	0	313.73	81.73	0	0	182.5 577.95	10.0	62.05	72.05	0	0	0	0	0.0
W	9.98	34.52	13.94	0	3.17	6.84	182.5 250.95	365.0	62.05	427.05	78.70	30.6	73.0	63.78	46.08
S	9.98	34.52	957.92	344.53	3.17	12.62	912.5 2275.2	932.50	310.25	1242.8	143.24	61 2	146.0	144.52	94 95
Objective: Minimum Drawdown															
w	0	0	0	49.0	0	0	49.0				0	0	0	0	0 0
N	0	0	0	77.0	0	0	77.0				0	0	0	0	0.0
D	0	0	280.4	87.6	0	0	368.0				0	0	0	0	0.0
D	3.25	0	350.4	87.6	0	54.75	496.0				0	0	0	0	0.0
W	11.31	36.45	10.61	1.13	5.69	3.81	69.0				0	0	40.18	0	40.18
S	14.56	36.45	641.41	302.33	5.69	58.56	1059.0				0.0	0.0	40.18	0.0	40.18
Objective: Minimum Drawdown & Cost and Maximum Watersupply															
W	38.10	0	0	87.6	0	0	182.5 308.20	196.76	62.05	258.81	0	0	0	0	0.0
N	0	0	0	87.6	0	0	182.5 270.10	262.85	62.05	324.90	36.0	0	0	0	36.0
D	0	0	350.4	0	0	17.05	182.5 549.95	10.0	62.05	72.05	0	0	0	0	0.0
D	55.85	0	197.26	87.6	0	54.75	182.5 577.95	10.0	62.05	72.05	0	0	0	0	0.0
W	6.75	35.28	16.27	7.23	1.61	1.31	182.5 250.95	365.0	62.05	427.05	0	0	0	0	0.0
S	100.69	35.28	563.93	270.03	1.61	73.11	912.50 1957.2	844.61	310.25	1154.9	36.0	0.0	0.0	0.0	36.0

T.C., Tiancun; L.G.Q., Lu Gou Qiao; X.H.C., Xi Huang Cun; N.H.H., Nan Han He; S.J.D., San Jia Dian; W, wet year; N, normal year; D, dry year; S, sum.

The optimal measures for satisfying water supply in a period with 3 continuous dry years

When 3 continuous dry years (i.e. 9 years of dry, normal, wet, dry, dry, normal, wet, wet, normal) appear, the plans of storing water in 6 normal or wet years to make up water shortage in 3 dry years, were computed(see Tab. 2).

Table 2 .   (unit: million m³)

	1	2	3	4	5	7	6	SUM	NO.9	T.C.	SUM	L.G.Q.	X.H.C.	N.H.H.	S.J.D.	SUM
	Groundwater Works NO.								Surface Water Works			Artificial Recharge Site				
Objective: Minimum Cost																
D	0	0	350.4	25.12	0	0	182.5	558.0	5.0	62.05	67.1	0	0	0	0	0.0
N	0	0	77.2	0	0	0	182.5	259.7	273.8	62.05	335.8	0	0	0	36.0	36.0
W	0	0	13.0	0	0	0	182.5	195.5	365.0	62.05	427.1	73.18	30.6	73.0	51.05	227.83
D	0	0	342.5	58.46	0	0	182.5	583.5	5.0	62.05	67.1	0	0	0	0	0.0
D	0	64.10	248.8	87.6	28.5	0	182.5	611.5	5.0	62.05	67.1	0	0	0	0	0.0
N	0	0	100.0	0	0	0	182.5	282.5	365.0	62.05	427.1	0	0	0	36.0	36.0
W	0	0	51.0	0	0	0	182.5	233.5	438.0	62.05	500.1	86.38	30.6	73.0	69.19	259.17
W	0	0	79.0	0	0	0	182.5	261.5	438.0	62.05	500.1	64.42	30.6	73.0	59.27	227.30
N	0	0.06	72.2	35.70	0	0	182.5	290.5	438.0	62.05	500.1	0	0	0	36.0	36.0
S	0.0	64.16	1334.2	206.87	28.5	0.0	1642.5	3276.2	2332.8	558.45	2091.3	223.99	91.8	219.0	287.5	822.3
Objective: Maximum Water Supply																
D	0	0	323 60	0	0	0	182 5	506 10	0	62 05	62 05	0	0	0	0	0.0
N	0	0	125.54	0	0	0	182.5	308.04	225.46	62.05	287.51	0	0	0	36.0	36.0
W	0	0	13.0	0	0	0	182.5	195.5	365.0	62.05	427.05	71.83	30.6	73.0	51.1	26.51
D	0	0	324.14	76.86	0	0	182.5	583.5	5.0	62.05	67.05	0	0	0	0	0.0
D	0	60.14	252.49	87.60	28.5	0	182.5	611.5	5.0	62.05	67.05	0	0	0	0	0.0
N	0	0	100.0	0	0	0	182.5	282.5	365.0	62.05	427.05	0	0	0	36.0	36.0
W	0	0	51.0	0	0	0	182.5	233.5	438.0	62.05	500.05	86.26	30.6	73.0	69.1	259.0
W	0	0	79.0	0	0	0	182.5	261.5	438.0	62.05	500.05	64.37	30.6	73.0	59.2	227.2
N	0	0.57	71.76	35.67	0	0	182.5	290.5	438.0	62.05	500.05	0	0	0	0	0.0
S	0.0	60.98	1340.5	200.13	28.5	0.0	1642.5	3272.7	2279.4	558.45	2837.9	222.5	91.8	219.0	251.4	784.7
Objective: Minimum Drawdown																
D	0	0	270.41	53.18	0	0		323.60				0	0	0	0	0.0
N	0	0	0	78.0	0	0		78.0				0	0	0	36.0	36.0
W	0	0	0	13.0	0	0		13.0				74 66	30.6	73.0	46.8	225.04
D	0	0	350.4	55.6	0	0		406.0				0	0	0	0	0.0
D	0	0	302.83	79.34	0	0		434.0				0	0	0	0	0.0
N	0	0	100.0	0	0	0		0.0				0	0	0	36.0	36.0
W	0	0	51.0	0	0	0		51.0				87.13	30.6	73.0	65.8	256.48
W	0	0	79.0	0	0	0		79.0				65.0	30.6	73.0	56.7	225.27
N	0	0	88.89	19.11	0	0		108.0				0	0	0	36.0	36.0
S	0.0	0.0	1242.6	298.23	0.0	0.0		1592.6				226.8	91.8	219	277.2	814.8
Objectives: Minimum Cost & Drawdown and Maximum Water Supply																
D	0	0	318.6	0	0	0	182.5	501.1	5.0	62.05	67.05	0	0	0	0	0.0
N	0	0	77.2	0	0	0	182.5	259.7	273.8	62.05	335.85	36.0	0	0	0	36.0
W	0	0	0	13.0	0	0	182.5	195.5	365.0	62 05	427.05	4.73	30.6	73.0	55.4	163.7
D	0	0	327.06	73.94	0	0	182.5	583.5	5.0	62.05	67.05	0	0	0	0	0.0
D	55.8	0	274.51	87.6	0	11.09	182.5	611.5	5.0	62.05	67.05	0	0	0	0	0.0
N	0	0	100.0	0	0	0	182.5	282.5	365.0	62.05	427.05	0	0	0	36.0	36.0
W	0	0	51.0	0	0	0	182.5	233.5	438.0	62.05	500.05	90.24	30.6	73.0	69.7	263.6
W	0	0	79.0	0	0	0	182.5	261.5	438.0	62.05	500.05	66.64	30.6	73.0	59.4	229.7
N	0	0	75.39	32.61	0	0	182.5	290.5	438.0	62.05	500.05	0	0	0	36.0	36.0
S	55.8	0.0	1302.8	207.16	0.0	11.09	1642.5	3219.3	2332.8	558.45	500.05	197.61	91.8	219	256.6	764.9

T.C., Tiancun; L.G.Q., Lu Gou Qiao; X.H.C., Xi Huang Cun; N.H.H., Nan Han He; S.J.D., San Jia Dian; D, dry year; W, wet year; N, normal year; S, sum.

THE OPTIMAL PLANS OF CONJUNCTIVE USE OF GROUNDWATER AND SURFACE WATER

The least precipitations in short periods in Beijing area are dry, normal, wet, dry and dry years, which has 2 continued dry years, in which the shortage of water supply could cause great troubles according to the present water supply programme of Beijing city. Thus, the conjunctive use of groundwater and surface water, optimized distribution and comprehensive management will be accomplished by sufficient and rational utilizations of the water- supply system. In dry seasons, groundwater is exceedingly withdrawn to make up urban water supply. In rainy seasons, the groundwater loss is replenished by various measures.

In order to meet the water demand , the water supply ability of groundwater works in the central area should be reajusted

According to the water supply standards in urbans and the water supply probability 95%, the design and programme of conjunctive use are feasible. On the basis of the lithology, tectonism, groundwater runoff conditions , and the computation results of an optimal management model, the present capacity of exploitation equipments and water supply of waterworks in the central area is limited and needs to be enlarged. If the conjunctive use of groundwater and surface water is carried out with the current pumpage equipments and capacity of groundwater, the problems of water supply in continuous dry years could hardly be solved. Meanwhile, stronger regulated ability of the thick aquifers (underground reservoir) hasn't been completely utilized. Therefore, if wells in Water Work No.3, 4, are added to a certain amount (total water supply ability can be risen to 1.5-2 times), then the groundwater in dry years can be withdrawn a lot so as to make room of groundwater reservoir for the infiltration of artificial recharge water, rainfalls and floods, and to provide a static condition for the urban water supply .

The enviromental protection of water resources, especially the groundwater, should be strengthened

The total dissolved solids in groundwater in the western suburbs is low,and groundwater quality is good. Care must be taken on waste water from some factories and living areas along river beds and artificial recharge areas. The effective measures must be immedialtely taken, and the environmental protective area should be established for artificial recharge work.

The measures of conjunctive use of groundwater and surface water

During the driest years of 5 years' period, i.e. 3 dry years of 5 years,in which there're 2 continuous dry years. Water Work No. 9, as one of the main water supply factories in the urban, is out of water. Meanwhile, the most of water in Miyun Reservoir is used by Tiancun Water Work in the west suburb.Thus, except the water supply of Water Work No.8 in Shunyi County and Tiancun Water Work in the western suburb, there will be short of about 400 million m^3 water in the central Beijing every year, which can only be replenished by groundwater . In order to assure safe water supply in the central area and the safe operation of water works and to meet water requirement , the groundwater supply of water work should reach 3.8-4.3 billion m^3 per year in dry years. In order to recover the over-exploitated groundwater in dry years, it is necessary to use the later 2 normal or wet years while Water Work No.9 is operating on its full load, the important things are to conserve and store the groundwater resource in the central area, to carry out the groundwater artificial recharge, and to raise the groundwater level. The computation results show that the make- up measures are feasible.

The analysis and comparation of the conjunctive use of groundwater and surface water

In the computation of the planning problems, different optimal management objectives are selected. Obviously, Water Work No.9 plays a very important role in conservation and storage of groundwater in the central area. In the normal and wet years, Water Work No. 9 supplies a large amount of water for the central area, reduces greatly the preasure of groundwater works in the central area, and creates a better condition to conserve and store the groundwater source in the central area. Although its general supply quantity in the management periods, since the cost of water supply is higher, is slightly less than other management objectives, it still has great weight in the total water supply of urban area.

Artificial recharge of groundwater
In the conjunctive use, artificial recharge of groundwater is an important complementation. According to the computation results, the artificial recharge of groundwater is quite necessary when there are more rechargeable water sources. Especially in wet years (5% - 25%), annual artificial recharge quantity can reach 2 billion m^3, which speeds up the recover of groundwater .

The general optimal management of the conjunctive use of groundwater and surface water contributes to the rational exploitation and the utilization
The final optimal proposal of the conjunctive use of groundwater and surface water concerns not only the possibility of technology and the effection degrees of enviroments, but also physical conditions,economic benefits, etc. The water developed ways and the recharge proposal of the urban water supply system basically acorrd with the conjunctive use' principles, i.e. groundwater is withdrawn as more as possible in dry years. Surface- water is mostly used , groundwater recharge is increased in normal and wet years. In 2000, although new irrigation works and equipments haven't been set up or put into operation, the problems of water supply in continuous dry years will be solved to a certain extent, by means of raising the pumpage capacity of groundwater works, enlargeing and adding artificial recharge places.

Furthermore, it should be considered carefully that the water resource conservation and utilization in the conjunctive use of groundwater and surface water after projects that water is diverted from the south to the north have been completed in 2000.

REFERENCES

1. T.III.Maddock. *Algebraic technological function from a simulation model.* Water Resour. Res. V.18. No.1. p.129-134 (1972).
2. E.Aguado, N.Sitar, I.Remson. *Sensitivity analysis in aquifer studies.* J. Geophisical Res. V.13. No.14. p.733-737 (1977).
3. M.Meidari. *Application of linear system's theory and linear programming to groundwater management in Kansas.* Water Resour. Bulletin. V.18. No.6. p.1003-1012 (1982).
4. J.F.T.Houston. *Groundwater systems simulation by time series techniques.* Ground Water. V.21. No.3. p. 301-310 (1983).
5. S.M.Goreilick. *A review of distributed parameter groundwater management modeling methods.* Water Resour. Res. V.19. No.2. p.305-319 (1983).
6. R.Willis, P.Liu. *Optimization model for groundwater planning.* J. Water Resour. Plann. Manage. Div. 110. No.3. p.333-347 (1984).
7. D.R.Maidment, E.Parzen. *Casade model of monthly municipal water use.* Water Resour. Res. V.20. No.1. p. 15-23 (1984).
8. R.Willis. *A unified approach to regional groundwater management.* Water Resour. Monogr. V.9. p.393-407 (1984).
9. Beijing-Tianjin Study Team and East-West Center. *Water resources policy and management in Beijing-Tianjin region.* China Science and Technolohy Commision (1987).
10. Liu Jiaxiang et al. *Study on groundwater reservoir in western suburbs of Beijing.* China Geology Press (1988).
11. Lu Xiaojian. *Management model of groundwater resources in Beijing proper and the suburbs (unpublished).* Beijing Hydrogeology and Engineering Geology Company. Beijing (1990).
12. Liu Jiaxiang, Lu Xiaojian et al. *Water resources management study in Beijing.* Beijing Science and Technology Commison (1991).

Proc. 30th Int'l Geol. Congr., Vol.22, pp. 22-28
Fei Jin and N.C. Krothe (Eds.)
© VSP 1997

System Analysis and Reasonable Use of Karst Groundwater in Shentuo Spring Catching in Shanxi Province

ZHANG LAI

China National Administration Of Coal Geology, Hydrogeological Bureau, Handan,Hebei Province, P.R.China

Abstract

ShenTuo spring group in ShanXi province is the famouse and great karst spring group in the north of china, the average spring flow is 6.99 m^3/s, the water quality is good, it is the main water source of living industry and agriculture in this region. Because of the decreasing of the precipitation and increasing of development year by year, the spring flow declines year by year, if the trend develop continueously, it will affect the main industry and agriculture, at the same time, the whole karst water system will be destroyed. Based on the understanding to the karst water system, writer consider that the recharge capacity of the karst water is able to be increased by artificial storage which use the surface flow in flood season through constructing the flood-control dams and leaking reservoir. This idea is simulated by numerical model and the result is good, this way is practicable.

Key Words: Karst Water System, System Analysis, Artificial Storage

INTRODUCTION

ShenTuo spring group in ShanXi province is the famouse and great karst spring group in the north of china, the karst water is the main water source of living and industry and agriculture in SuoZhou, the even resource amount of the karst water is $27843 \times 10^4 m^3$/y in many years, the karst water is mainly consumed by ShenTuo power plant, Pinsuo open coal mine and agriculture irrigation. Based on the statistical data in 1992, the utilization factor of the karst water is 69.5 percent, the water quantity consumed by ShenTuo power plant at the spring mouth is $7961 \times 10^4 m^3$/y; the water quantity consumed by PinSuo open coal mine and industry is $1716.9 \times 10^4 m^3$/y; the water quantity consumed by agriculture irrigation is $9688 \times 10^4 m^3$/y. Recently, the water quantity consumed by pinsuo open coal mine and municipal water supply is planed to add by $2008 \times 10^4 m^3$/y (table 1). Because of the decreasing of precipitation and increasing of development year by year, if the trend develop continueously, it will affect the power plant and the agriculture irrigation, at the same time, the whole karst water system will be destroyed, for this reason, based on the full understanding to the karst water system, we should make use of the the favorable conditions of hydrogeoloy, geomorphology, geoloy and the method and techonology of artificial storage to increase the karst water rechage capacity, carry out both increasing development of the karst water and maintaining the stability of the spring flow and impel the karst water syttem to develop in the directiou of good cycle in addition to planing and using the limited karst water resource reasonablly.

HYDROGEOLOGY BACKGROUND

ShenTuo spring area is situated in SUOZHOU in ShanXi province, in this area, the weather is both cold and dry, the precipitation is 396mm and mainly concentrate in the three monthes of july, august, september in a year, ShenTuo spring group is situated in ShenTuo Town in the northeast of SuoZhu basin, which is the source of Shan Gang river, Based on the statistical data in 28 years (1966-1994), the even spring flow in many years is 6.99m³/s, the water quality is good.

Table 1. The statistic list of the development and use of the karst water   unit:1 × 10⁶m³/y

total resource		278.4	the way of using the karst water
water quantity consumed	power plant	79.6	spring mouth
	open coal mine	17.2	water supply pumping
	agriculture irrigation	96.9	flowing out of the spring
utilization ratio		69.5%	

The spring group appears at the sector of Yuan Zhi river valley in front of Hong Tao mountain. The spring water flows out of the alluvium of the river bed by Ordovician limestone, it is ascending spring in structure. It's cause of formation is that MaYi fault at downstream intercepts the karst fracture water flowing from west to east. It cause the ground water flowing horizontally to ascend on the west of the fault, the ascending water breaks through the thin and loose cover of Quaternary and the accumulation of the river shoal to go up to surface to form spring on the horst fault blook .The mumber of the great and small spring mouth is over 100 . the spring area is about 5 km².

The spring group is the main discharge site of karst fracture water in Cambrian-Ordovician limestone, the whole karst spring area is very large and about 5316km². the open recharge region is 3800km². The precipitation is the main recharge source of the ground water, there is no other water to flow into the spring area, under the control of the structure and the areal discharge datum plane, the ground water obtains recharge at open region, it converges and discharge from north, west and south to east .Because of the intercepting of MaYi fault, the ground water overflow to surface in the spring discharge and form the spring area which has very high closing degree.The karst aqufer mainly consists of ordovician limestone (O_1)and ordovicin limestone (O_2), which are mainly dolomite limestone and rock dolomite. the thickness is great and stable, the karst is very developed, the degree of the filling is low, the spaces of storing water mainly consist of the karst fracture, the karst hole is developed in the depth, because the karst fractures are mixed with the karst hole and karst cavity, the structure of the aqufer is of network pattern and has the feature of the typical north karst.

SYSTEM ANALYSIS OF KARST GROUND WATER IN SPRING AREA

Based on the viewpoint and the method of system theory, the whole spring area is known as a independent karst ground water system, the form of the exchange between the system and the outer environmeut is mainly that the system obtains the precipitation recharge and discharge in spring,There is no relationship with contiguous ground water system vertically except partial recharge of the karst water to Quaternary during ascending, the forms of the import and the export of the system are simple.

Based on the data obtained by various exploring methods, we make use of various math methods to analise the inner structure of the system, the feature of the ground water migration, the function of the storage and the exchanging machnisme with outer environment systematically, at last, a correct hydrogeology model is obtained.

Feature of Rock Structure and Storing Water Basin

Topographically, the spring area system is surrounded by mountains on the north , west and south, it becomes open and gently in the east, which leads to the whole trend that karst water flows from north, west and south to the basin, In this trend, Neocathaysian Xiao Jia Zhug -Ning Wu syncline basically controls the pattern of the spring area system, the secondary MaGuang syncline, Suoxhu syncline and ShenChi syncline form the basins which have relationship with each other and good storing water condition, the fault which extends from east to west has the feature of conducting water vertically and obsteructing water horizontally, It provides good flowing environment for the karst ground water.

Feature of Water Dynamic

Based on the distribution pattern of the unit rate of flow of the drilling on plane and the analysis of flow field, it is found that there are three strong flowing zone and two weak flowing zone in the system. we name them as S_1, S_2, S_3 an W_1, W_2 flowing zone respectively. After leaking in the recharge region, the precipitation flow to the spring region along the flowing zones, the feature is that the karst ground water exchange actively in the strong zone and it is able to be recharged and recover fast after the karst ground water is developed, so the great water supply source should to be constructed at the strong flowing zone, the karst water exchange slowly in the weak flowing zone, but the weak flowing zone has the ability of storing water, the great water supply source shouldn't to be constructed at the weak flowing zone.These above flowing features are verified by analysis of water chemistry further.

Featuere of Water Chemistry

By the analysis of six kinds of ions in the drilling water sample and the calculation of Ca^{2+} saturation deficiency, the result reflect the above feature alikely, the contour of Ca^{2+} saturation deficiency reflect the unsaturated status of Ca^{2+} on S_1, S_2 and S_3 flowing zone and the saturated status of Ca^{2+} on W_1, W_2 flowing zone.

Based on the temperature , hardness, mineralization degree of various spring site and the drilling water sample in the system and the result of cluster analysis of similarity of six kinds of ions. It is found that the whole system consist of three relatively independent subsystems which are Shi Ma Pu spring group subsystem; ShenTuo spring group subsystem and Xiao Pu spring group subsystem respectively, under the control of the condition of the recharge, flow, discharge of the karst water system, there are relatively independent recharge direction, flowing feature and discharge spring group in every subsystem (figure).

Numerical Model

Based on the above analysis, a two-dimensional numerical simulation model that reflects inhomogentity, anisotropy of the aqufer and linear flow of the ground water transport is founded, the changing law of the natural flow field is used to calibrate the numerical model, the law reflected by the parameter of the model which is calibrated is the same as the above analysis result. Based on the analizing great result, we simulate to construct three great water supply source wells which provide $1 \times 10^4 \sim 5 \times 10^4 \, m^3$ water quantity per day in Shi Ma Pu spring group subsystem early or late, two of the three water supply source wells are simulated to develop for several years. the result indicated that the water quantity is ample, the water level is stable, they are good water supply source wells.

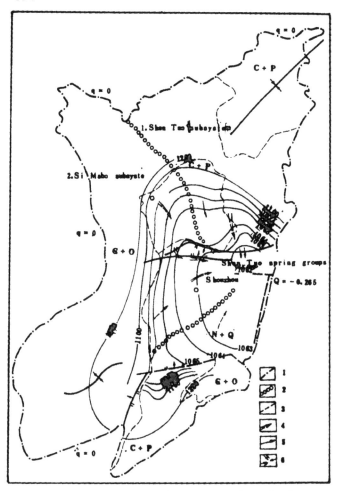

Figure 1. Shentuo spring area concept model sketch figure legend: 1. the border of system; 2. the border of subsystem; 3. ground water leave isogram ; 4. strong flowing zone; 5. week flowing zone; 6. disharge area

REASONABLE DEVELOPMENT OF KARST GROUND WATER AND ARTIFICAL STORAGE

Based on the above analysis and the experience obtained from water supply source wells construction before, the strong flowing zone should be choosen to develop the karst water, because the strong flowing zone is favorable to extracting water and recharge. The pumping test of the well group indicate that if the karst ground water is pumped at strong flowing zone, the recover of karst ground water is fast, the recharge channel is unblocked, which is very favorable to the adjusting of the ground water.

Additionally the puming water at the discharge region should be avoided, in case the spring flow decline fast, which will affect the consumer using the spring water and destroy the environment.

Because the precipitation decrease in the karst water system year by year, the recharge capacity decrease, furthermore, the ground water exploitation increase continueously, the system is in a negative state, which causes the spring flow to decrease and the ground water level to decline widely.

The method to solve the problem is to decrease the development to the karst water, or to increase the recharge capacity to the karst water, the former is a negative method, the latter is a positive measure.

Based on the feature of the karst water of Shen Tuo spring area, we may use the surface flow in the flood season, look for the suitable place, construct the flood-control dams and the leaking reservoir to increase the recharge capacity.

Analysis of Leaking Condition
The recharge region is suited in the uncovered area of cambrian and Ordovician limestone in the north, west and south , there is little vegetation in this area, because of the tectonic disturbance many time and the weathering erosion for a long time, the karst fractures are very developed, these fracture are both open and vertical. They are good channel to precipitation leaking. There are two ways to recharge; one is areal leaking, the other is linear leaking (river leaking), the river are seasonal, the river valleys are very developed, MaYin river and MaGuan river are leaking river,   paticularly, MaYin river is strongly infiltrating, the drainage area is 1489 km^2 , the even width of the river channel is 19.5m , the leaking part is located at the end of the river, it is about 15km, the leaking capacity is about 0.6 m^3/s.

There is water in MaGuan river throughout a year, the flow changes seasonally, the drainage area is 151km^2 , the leaking part is located in Zjo jia kuo at down river, it is about 500m long, the leaking capacity is abuot 0.024m^3/s.

Flow Analysis of Leaking River
Based on exploration and hydrology analysis, the even flow in a year is 10810×10^4 m^3/y at the down reach of MaYin river, the even flow is 3510×10^4 m^3/y at the down reach of MaGuan river(table 2), the leaking rates are 0.41 and 0.10 respectively, the flows are seasonal and concentrate in june, july, august, september, these surface flow that discharge concentratively and strong leaking of the river provide the beneficial conditions for increasing the recharge capacity.

Table 2. The statistic list of surface runoff capacity   unit: $1 \times 10^6 m^3$/y

month	6	7	8	9	10	11	12	1	2	3	4	5
Mayi river	8.2	32.1	28.5	23.2	2.5	0	0	3.4	0	0	0	0.6
Magan river	3.9	10.7	12.4	8.5	0	0	0	0	0	0	0	0
Total capacity	143.5											

Method of Increasing Recharge Capacity and Numerical Simulation
To intercept the surface flow effcetively, we assume to construct the flood-control dam and the leaking reservoir to increas the leaking capacity of the rivers artificially, carry out to increase ground water reservior capacity and maintain the stability of the spring flow.

Using the above numerical model of the karst water system, we simulate trend of the water level and the spring flow in the system after constructing the flood-control dam and the leaking reservoir, the original source and sink in the numerical model remain unchanged, based on the leaking rate, location and flow patten, we distribute the intercepted flood quautity to every time step in source and sink.

Based on the relationship between the spring flow and ground water level at the spring head, we use the numerical model and the feed-back mechanism of the model to determine the change law of the spring flow caused by increasing recharge capacity, the period of operation is one year, the result of the operation of numerical model indicate that the rising range of the water level at the down reach of the leaking part is about 0.5m after constructing the multi-step flood-control dams, the rising of the water level at the down reach of the leaking part is very obvious, the declining of the water level at the upper river is very slow, the spring flow increase by 3311.3×10^4 m³/y over the simulation initiation, the effect is very obvious(table 3).

Table 3. The statistic list of increment of the spring flow   unit: m³/s

month	6	7	8	9	10	11	12	1	2	3	4	5	average spring flow
real flow	4.56	4.33	4.25	4.20	4.36	3.88	5.08	4.30	4.01	4.06	4.52	4.58	4.33
simulation flow	4.56	4.37	5.09	8.89	5.99	5.99	5.88	5.79	5.39	5.31	5.31	5.06	5.38
increment of spring flow	1.05m³/s=33.1 × 10⁶m³/y						capacity	increment of stortage				10.3 × 10⁶ m³/y	

CONCLUSIONS

1) By system analysis, there are three relatively independent subsystem and five flowing zone of various strength in the karst ground water system in ShenTuo spring area, every subsystem has relatively independent recharge direction flowing feature and discharge spring group, Using the result, we successfully construct three great water supply wells at S_1, S_2 strong flowing zone and verify the reliability of analizing result further.

2) Because of the decreasing of the recharge capacity and the increasing of the development of the karst water year by year, the spring flow declines continueously, there is a lot of risk to pumping water at the spring head, so it is important to carry out artificial storage at the recharge region.based on the system analysis and the leaking condition analysis of the rivers, we found the numerical model of artifical recharge, operate the model and obtain the following conclusion.

a. After one year of artifical recharge, the storage capacity increases by 1028.1×10^4 m³/y in ShenTuo spring group subsystem.
b. After one year of artifical recharge, the spring flow increases by 3311.3×10^4 m³/y.
c. After artificial recharge, although the exploitation of the ground water increases by 5.5×10^4 m³/d, there is no problem of lacking water to the consumer using spring water.

d. The method to use natural leaking condition and few surface projects and the transitting flow of the flood is practicable.

It is a new idea of developing karst water resource reasonablly to carry out artificial storage to karst water system, from the self-optimization distribution of karst water resource to the transform of "three water" in nature, it is important to improve the utilization rate of water resource, it can alleviate the shortage of water resource in the north of China.

Proc. 30th Int'l Geol. Congr., Vol.22, pp. 29-34
Fei Jin and N.C. Krothe (Eds.)
© VSP 1997

The New Understanding of Water Movement in Saturated Clay Soil

HOU JIE[①] , SU QINGSHAN[①], PAN SHUSEN[②] and   ZU LIAOYUAN[①]

①Department of Hydrogeology and Engineering Geology, Changchun University of Earth Sciences, Changchun, 130026, P.R.C.
② Water Conservancy Bureau of Hei LongJiang Province, Harbin, 150030, P.R.C.

Abstract

In saturated clay soil, there are large pores and micro-pores, and the total micro-pore space is always greater than the total large pore space. The large pores are mainly filled with gravitational water and capillary water, but The micro-pores are filled with bound water. Through studying the groundwater movement in saturated clay aquitard which is distributed between two aquifers in Daqing area (in P.R.C), we have found that the water movement in saturated clay soil has different stages, and the movement rules depend upon both the transformation and the reaction of different forms of water during the filtration.

Keywords: large pore, micro-pore, water movement, saturated clay soil, Darcy's law, bound
        water, filtration

INTRODUCTION

Since the 1940s, there are mainly three different understandings of the characteristics of water movement in saturated clay soil: (1) Darcy's law can be used, (2) Darcy's law can not be used, and (3) Darcy's law is true for gravitational water, but not true for bound water. In 1964, a Chinese professor, Zhang Zhongyin, proposed that the bound water is the main existing form for the saturated clay water; therefore the water movement rule in clay soil is actually the movement rule of bound water [1]. In this paper, through studying the water movement in saturated clay aquitard which is distributed between two aquifers in Daqing area (in P.R.C), we have found the types and sizes of   the   soil   pores, and the water existing forms, and set up an equation about the relation between the filtration velocity and the hydraulic gradient. The new understanding of water movement in saturated clay soil was achieved.

THE STRUCTURE AND THE POROSITY OF   THE CLAY SOIL

By means of SEM (scanning electron microscope),   we have found that in natural condition there is granular structure and coagulation structure in saturated clay soil. The granular structure and coagulation structure are favourable for moisture migration, storage and preservation, and the coagulation structure has a   property of water-releasing due to deformation. The pores in clay soil can be classified as the pores between and inside group particles, and the pores at contact points (Fig. 1). The pores between group particles and inside some group particles are large pores. Their diameter is greater than 2 times of the thickness of the bound water film (i.e. $d>0.07\mu m$).

According to the measurement of mercury compressing, the large pores are continuously distributed in the clay soil profile, and the total pore space has a decreasing tendency from up to down. The large pores are mainly filled with gravitational water and capillary water. The pores inside the group particles and at the contact points are micro-pores. Their diameter is less than 2 times of the   thickness of   the bound water film ($d > 0.07\mu m$). This kind of pores is also continuously distributed, and the total pore space has an increasing tendency from up to down in the clay soil profile. The micro-pores are filled with bound water. We found that the total micro-pore space is always greater than the total large pore space (Table 1).

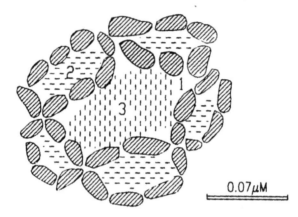

0.07μM

Figure 1. Types of the pores in clay soil: 1—contact point pore between group particles,   2—the pore inside group particle and   3—the pore between group particles

EXPERIMENT AND ANALYSIS

Experiment Result
Putting the saturated in-situ clay sample into the test container, and doing the compressed filtration test under different hydraulic gradients [2], we obtained a V-I curve, which is curve OABC (Fig. 2), and the equations as follows.

Line OA:

$$V_1 = K_1 I,$$
(1a)

$$K_1 = tg\alpha_1.$$
(1b)

   Initial condition:

$$I(0) = 0$$
(1c)

      and

$$V(0) = 0.$$
(1d)

   Boundary condition:

$$0 < I \le I_{01}.$$
(1e)

Table 1. The statistical table of pore types and volume tested by mercury compressing in clay soil body

Number of the i-n-situ clay soil sample	1	2	3	4	5	6
Depth of the in-situ clay soil sample (m)	15	20	28	42	58	67
Porosity of the in-situ clay soil sample	0.45	0.41	0.42	0.39	0.38	0.40
Total volume of all pores (cm^3)	9.5426	8.6943	8.9064	8.2702	8.0582	8.4823
Volume of large pores (cm^3)	3.4267	3.0308	2.2533	2.3768	2.0765	1.8822
Proportion of large pores in all pores (%)	35.91	34.86	25.30	28.74	25.77	22.19
Volume of micro-pores (cm^3)	6.1159	5.6635	6.6531	5.8934	5.9817	6.6001
Proportion of micro-pores in all pores (%)	64.09	65.12	74.70	71.26	74.23	77.81

Line AB:

$$V_2 = V_{01} + K_2 (I - I_{01}),$$

(2a)

$$V_{01} = K_1 I_{01},$$

(2b)

$$K_2 = tg\alpha_2.$$

(2c)

　　Initial condition:

$$I(0) = I_{01},$$

(2d)

Houjie et al

$$V(0) \quad = \quad V_{01}.$$

(2e)
   Boundary condition:

$$I_{01} < I \leq \quad I_{02}.$$

(2f)

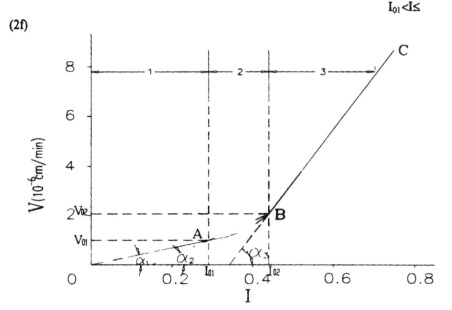

Figure 2. The V-I curve of the water filtration in saturated clayey soil. There are three stages: 1–the stage of large pore channels filtration, 2–the stage of large pore channels filtration combined with the latent filtration in micropore channels and 3–the stage of large pore channels filtration combined with the evident filtration in micro-pore channels.

Line BC:

$$V_3 \quad = \quad V_{02} \quad + \quad K_3 \ (\ I \ - \ I_{02} \ ),$$

(3a)

$$V_{02} \quad = \quad K_2 \quad I_{02} \quad ,$$

(3b)

$$K_3 \quad = \quad tg\alpha_3.$$

(3c)
   Initial condition:

$$I(0) \quad = \quad I_{02} \quad ,$$

(3d)

$$V(0) \quad = \quad V_{02}.$$

(3e)
   Boundary condition:

$$I \quad > \quad I_{02}.$$

(3f)

where I is the hydraulic gradient, I_{01} and I_{02} are respectively the initial hydraulic gradient of the latent filtration and the evident filtration, V_{01} and V_{02} are respectively the initial filtration velocity of the latent filtration and the evident filtration, α_1, α_2 and α_3 are respectively the inclined angles of lines OA, AB and BC, K_1 is the hydraulic conductivity of large pore channels (cm/min), K_2 is the hydraulic conductivity of large pore channels and micro-pore chann+ls with the latent filtration (cm/min), and K_3 is the hydraulic conductivity of large pore channels and micro-pore channels with the evident filtration (cm/min).

Test Result Analysis

Although the micro-pore space has been fully filled with bound water, not all the bound water is migrated   when the filtration occurs, because the anti-shearing strength of the bound water depends upon the distance   from the walls of micro-pore channels. The bound water in the center axis of the channels has the smallest anti-shearing strength; therefore it is easy to be migrated under some hydraulic gradient. However, the   bound water near the channel walls has a big anti-shearing strength, and therefore it is difficult to be moved. The amount of bound water movable is related to the hydraulic gradient [1]. We defined the space occupied by the bound water movable as "filtration pores". Their diameter is called "filtration diameter"(d). It is a   function   of hydraulic gradient.

Under the small gradient I ($0<I\leq I_{01}$), only does the gravitational water filtrate in large pore channels, and Darcy's   law is satisfied. The filtration can be described by line OA (Fig. 2, Equation 1).

When I increases from I_{01} to I_{02}, Darcy's law can't be used directly. Under the condition of $I_{01}<I\leq I_{02}$, in large pore channels the migration of gravitational and capillary water is strengthened; furthermore, in micro-pore channels some of the bound water that has a small anti-shearing strength will migrate. In this stage, the hydraulic conductivity changes very fast relatively, but this stage is short. The relation between V and I is actually described by curve AB. Because the degree of curve AB is small, we can let line AB approximately represent curve AB in this V-I curve. So Darcy's law can approximately apply to the water movement in this stage (Fig. 2, Equation 2). This movement of bound water can be called latent filtration in micro-pores.

Under the large gradient I ($I>I_{02}$), not only is the filtration in large pore channels increased, but more bound water in micro-pore channels   is also migrated. During this stage, the bound water that has a small anti-shearing strength has moved, and it is difficult to move the bound water that has a big anti-shearing strength though the hydraulic gradient is very big. In this stage, the hydraulic conductivity increases very slowly and trends towards a constant with the increase of hydraulic gradient (I), and the variance of the hydraulic conductivity (the slope of line BC) is very little. So the relation between V and I can approximately be described by line BC, and Darcy's law can be used in this stage (Fig. 2, Equation 3). In this stage the movement of bound water is evident. This can be called evident filtration in micro-pores.

VERIFICATION OF THE TEST RESULT

The test result has been used in the calculation of   water   leakage from unconfined aquifer to confined aquifer in Daqing area (in P.R.C). The ground water in the confined aquifer was actually in the equilibrium of recharge and discharge during calculation period. The water balance analysis indicates that the variance of the storage of the ground water in the confined aquifer is very small, and the recharge is less than the discharge because the yield of the aquitard was not considered. On the whole, the result is effective, and the filtration rules from the test are practical (Table 2).

CONCLUSIONS

(1)Both large pore and micro-pore exist in clay soil. In the large pore channels, gravitational water exists, and Darcy's law can be used directly. In micro-pore channels, bound water exists, and the water movement can't be described by Darcy's law directly.

Table 2. The results of the water balance analysis in the confined aquifer in Daqing area

Year	1984	1985	1986	1987
Leakage from the upper unconfined aquifer (10^4t/d)	31.59	33.53	32.70	**34.90**
Lateral recharge (10^4t/d)	12.90	12.80	13.00	13.50
Lateral discharge and pumpage (10^4t/d)	46.12	48.33	47.20	50.30
Variance of storage (10^4t/d) ("-" indicates the decrease of storage)	-1.63	-2.00	-1.50	-1.90

(2)Different stages can be divided for the water filtration in clay soil. These stages can be described by three lines in a V-I curve (OA, AB and BC). *First stage*: $I \leq I_{01}$, only gravitational water moves in the large pore channels according to the rule of $V_1 = K_1 I$. *Second stage*: $I_{01} < I \leq I_{02}$, the filtration in large pore channels combined with the latent filtration in micro-pore channels. *Third stage*: $I > I_{02}$, the filtration in large pore channels combined with the evident filtration in micro-pore channels. In these three stages, if the soil structure is not changed, then K_1 is a constant, while K_2 and K_3 will increase with the increase of I. The variance of K_3 is great, but the variance of K_3 is very small. Here is a rule: $K_1 < K_2 < K_3$.

(3)Under ordinary condition, compared with the large pore spaces, the micro-pore spaces are very well developed, and the total volume of micro pore spaces is larger than that of large pore spaces. The micro-pore channel is the main path of groundwater movement, and the large pore channel is the important filtration passage and has a function of rapid transmitting of water pressure. So the migration of bound water in micro-pore channels is the main form of groundwater movement, and the transmitting of the water pressure by the large pore channels is the driven force of groundwater movement in saturated clay soil.

REFERENCES

1. Zhang Zhongyin. *The Problems on Bound Water Dynamics*. Geological Press, Beijing (1980).
2. Su Qingshan, Liu Li and others. *Modern Experimental Hydrogeology*. Jilin Science and Technological Press, Changchun, P.R.C. (1991).

Proc. 30th Int'l Geol. Congr., Vol.22, pp. 35-39
Fei Jin and N.C. Krothe (Eds.)
© VSP 1997

Study on Interchanges among Atmospheric, Surface Unsaturated Soil and Groundwater under the Conditions of Vegetation

ZHANG DEZHEN
Division of Geological Environmental Management, Department of geology and Mineral Resources of Henan Province, Zhengzhou, 450007

XU SHIMIN
General Station of Environmental Hydrogeology, of Henan Province, Zhengzhou, 450006

Abstract

The article has studied the relations and interchange quantities among atmospheric, surface, unsaturated soil and underground water. The infiltration and evaporation differences with and without plants are specially em phasized. The circulating, storage and regulating functions of unsaturated soil water to rainfall and agriculture are also expounded. A new method to mitigate the water shortage in the east plain of Henan Province is given in this case study.

Key words: plant conditions,   four types of water,   interchanges,   experimental result

The four-types of water in the study is referred to atmospheric, surface, unsaturated soil and underground water. They are in one water resource s ystem. On one hand, they are relat ively independent and on the other hand, they are also interrelated and co existed. There is a clear interchanges in good order among them.

It should be stressed that the interchange of four types of water without plant is relatively simple,while interchange with plant is rather complicated. The case study shows that there is a large and obviously difference.

EXPERIMENTAL SITE AND FACILITIES

An experimental site was selected in Shangqiu in the east part of Henan Province to realize the research purpose. It includes meteorological observatory station,underground observation room,experimental field and groundwater balance area. It is described briefly as follows.

Meteorological observatory station.   Observation items include rainfall, evaporation, wind speed, sunshine time, underground temperature, humidity, temperature etc.

Underground observation room. This is a cemented underground room installed with 36 sets of ground osmometer which are filled with four types of soil samples (sandy clay, sandy clay and clay sand interlayes, clay sand as well as fine sand) respectively and 104 negative pressure recorders are inserted in them. All samples are kept undisturbed. The water heads are automatically adjusted and fixed to 9 level from 0.5 m to 4.5 m with an equal interval of 0.5 m.

Experimental field. Situated in west site of water balance area, the test field   consists of two parts based on the water supply way. The north part is traditional and the south is a water saving supply area and a separation zone just in between. There is a nagetive pressure recorder observation well and a neutron moisture meter observation well respectively in both part. The observation well of negative pressure recorder is 4.5 m deep, groundwater level depth was about 5 m, the exposed soils are sandy clay and fine sand. There is a WM-1 type mercury negative pressure recorder installed inside the well. The observation well with IH- II neutron moisture meter is 4.0 meters deep. The plants were wheat and maize.

The water balance area. The total area is 5.3 km^2, the cultivated land is about 3 km^2. The plants during experiments were wheat and maize, legume and cotton. The groundwater regime, groundwater exploitation and surface water runoff were carefully measured and registered. The study has considered the situation like point and area, surface and ground,   with and without plant, it was actually formed a 3-D experimental observation system.

CASE STUDY

By using the facilities mentioned above,the experiment in the obsence of crops for 4 kinds of different soils, 9 different groundwater heads had been conducted for 7 years, and a lot of parameters such as infiltration coefficient α, evaporation coefficient C   were obtained. These data can be used for analysing the rainfall infiltration and groundwater evaporation. Another point ought to be mentioned is that the condition of 4.0 m water level of sand clay and clay sand interlayes is similar to   the condition in the experimental field,thus it is reasonable that the result from both experiments can be compared each other and analysed later.

Using experimental field to carry out the study on the four   types of water interchange with plant, the basic principle is the potential theory of soil and water. The water   movement is driven by the potential difference. The potential at any point is the work from that point change to the standard or reference level. The total potential of the water content,therefore,is summed by 5 factors,which are formulated as follows:

$$\psi = \psi_z + \psi_v + \psi_m + \psi_s + \psi_t \tag{1}$$

where:

    ψ :total potential ψ_z: gravity potential

    ψ_v:pressure potential ψ_m: ground state potential

    ψ_s:solute potential ψ_t:temperature potential

The water contents in the soil were measured by using neutron moisture meters:

$$\theta_r = \alpha \cdot R/R_w + b \tag{2}$$

where:

    θ_r: soil water content   R: neutron counting rate in soil

    R_w: standard counting rate in water   b: coefficient

In the case study, on one hand, soil water were recharged by rain, then changed to groundwater later. On the other hand, soil water was recharged by groundwater, and then changed into atmospheric water by evapotranspiration. All of these were calculated by using the zero flux method when the zero flux plane was measured and fixed flux method when the zero flux plane was absent.

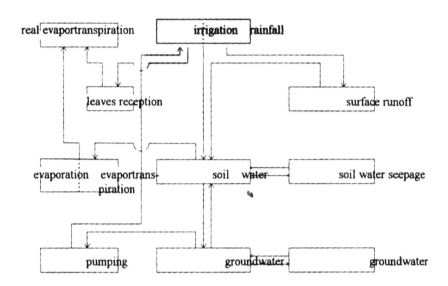

Fig.1 Interchange relation diagram among four-types of water

The formula is written as

$$q = - k \ (\overline{\psi}_m) \cdot \frac{\partial \varphi}{\partial z} \qquad (3)$$

where:

q: soil water flux at any section plane

$K(\overline{\psi}_{\bullet})$: unsaturated water conductivity

$\frac{\partial \psi}{\partial z}$: gradient of total soil and water potential

For different test fields, different water supply ways were used. Economized water supply field was irrigated based on the lower limit of soil water content, traditional field was irrigated as usual. The relation of four-types of water interchanges with plant is shown in the figure below.

All these factors above were taken into account for the water balance calculation in the presence of crops in the area of 5.3km^2, with formular (4) and the experimental and calculated result were considered satisfactory.

$$P+W=R+\Delta\theta+q+ET_D+PE \qquad\qquad (4)$$

where:

P: rainfall

W: irrigation

R: surface runoff

$\Delta\theta$ soil water storage

q: groundwater recharged by rainfall

ET_D: evapotranspiration

PE: retained by leaves and evaporation during rain time

RESULT ANALYSIS

In the experimental site, the annual rainfall was 700 mm, unsaturated soil was sandy clay and fine sand with thickness of 5 m, the production of wheat and maize was about 300kg/mu (1 mu $\approx$ 667m^2). This harvest was retatively good in the area.

The result of four types of water interchanges in different water supply way were listed in table. From table it can be analysed as below.

Table 1   Interchange result of four types of water in different water supply conditions   unit:mm

plant	summer maize			winter wheat		
test methods	method 1	method 2	difference	method 1	method 2	difference
water supply way	traditi-onal	water-saving		traditi-onal	water-saving	
negetive pressure wells	FY1	FY2		FY1	FY4	
neutron meter wells	ZZ2	ZZ4		ZZ1	ZZ3	
total rain	412.6	412.6	0.0	288.1	288.1	0.0
total irrigation	165.0	70.0	95.0	70.0	0.0	70.0
soil water change (0-100cm)	123.4	100.0	23.4	-72.2	-77.3	5.1
soil water upward flux (100cm)	1.9	15.3	13.4	5.1	5.7	0.6
soil water infiltration (100cm)	87.5	55.3	32.2	16.4	13.0	3.4
net infiltration	85.6	40.0	45.6	11.3	7.3	4.0
real evaportranspiration	368.6	342.6	26.0	419.0	358.1	50.9

Test results showed that the rainfall on one day reached 50 mm, 83 mm within two days, and continuously raining for six days with the total rainfall reached to 147 mm, but no surface runoff occurred, which demonstrates that most of the rain was infiltrated to unsaturated zone except a little leave reception. Because this unsaturated zone had a large storage space, the recharged soil water is proportional to the correlated rainfall. This can just explain the reason why the surface runoff is decreased in recent decades. But the decreased surface water and groundwater do not mean total water resources is decreased, in other words, the decreased part is changed to soil water and storaged in unsaturated zone.

With plant, the growth of plant stimulated and intensified the circulation of soil water. Based on the flux measurement, the 80% of rainfall were changed to soil water, only 10% to groundwater, while the traditional infiltration parameter to groundwater without plant was about 20%. On the other hand, the function of root zone also intensified the evaporation and evapotranspiration. Table 1 also shows that the evapotranspiration in the maize fields was 370 mm for traditional and 340 mm for water saving supply field, while the evapotranspiration in wheat field was 420 mm for traditional and 360 mm for water saving supply field. These results are higher than that without plant. Traditionally, the limited depth of groundwater evaporation is about 3-4 m, but on the condition with plant, there should be a limited depth of water absorption of root system, and the traditional conception of limited depth of groundwater evaporation, in fact, has become senseless.

Based on the fixed flux method of soil section plane (100 cm below surface), 90% of evapotranspiration is from soil water. Minimum limit of soil water content is 350 mm. The soil water contents could be adjusted to 220 mm one time, then it could be circulated many times. This again explains soil water is a very important key to agriculture.

CONCLUSIONS

The conclusions can be drawn from the case study as follows:
(1) The relation and quantity of four types of water interchange with and without plant are quite different. The infiltration and evaportation under two conditions are extremely different.

(2) With plant, in the interchange of four types of water, 80% of rainfall is changed into soil water, less than 10% into groundwater and 8% into surface water, 90% of plant evaportranspiration is from soil water, which is identified by soil section test of 100cm in depth.

(3) With plant, the groundwater recharge action of rainfall is obviously weakened, the coefficient of infieltration of rainfall decreased 10%, while evapirtransporation quantity is enlarged greatly. Therefore, some parameters obtained from ground osmometer without plant can't be popularized in plain plant area. In water balance calculation, it is important to distinguish the plantation conditions and reasonably select parameters, and regard the soil water as an important balance factor.

(4) With plant, when unsaturated zone reaches a certain thickness, the soil water forms a soil water reservoir. It is more functional than groundwater reservoir in the aspects of rainfall circulating and regulating. It is most important for increasing agriculture product and improving agriculture ecologic setting. But, the improtance of groundwater might not be neglected. The key is to correctly evaluate groundwater quantity of recharge and the quantity that can be used in a more reasonable way according to the hydrogeological condition and the influence of vegetation and plant.

(5) The basic strategy to mitigate the shortage of water resource in the east plain of Henan Province is to effectively protect and fully utilize the soil water resource, properly develop the groundwater resource and make rational use of surface water resource.

Proc. 30th Int'l Geol. Congr., Vol.22, pp. 40-46
Fei Jin and N.C. Krothe (Eds.)
© VSP 1997

Modeling Exploration and Development of Deep Water Level Confined Karst Water in Zhungeer Coal District

YAO JIKUN, DUAN GUIYIN, CAO YUNQUAN
Hydrogeological No.2 Team China National Administration of Coal Geology
LIN ZENGPING
Hrogeological Bureau, China National Administration of Coal Geology

Abstract

This paper studies the condition about the occurrence and movement of   deep-water-level confined karst water,on the basis of analysing and studying the condition about recharge, runoff, discharge and movement regulation, we applied the method of synthetical exploration and the theory of karst water system, established karst water system and set up a karst water model for the transformation between three types of water, and multistage discharge in Yellow River valley area.Applying the theory of karst water strong runoff zone and storage structure,found out a strong runoff zone affectted by   silt filling zone far from the Yellow River Adopting the exploration method combined extraction and principle of relatively concentrated water extraction at several separated regions and progressive expanding,found out abundant deep-water-level confined karst water within the area, and two big-middle karst water sources,Chenjiagoumen and Tanggongta water source whose mining yield is really 74930 .m^3/d by the end of 1995, was set up in 1990 and 1993. The two water sources production proved that the water supply source are stable, good quality with less drawdown of water level. there are no environmental problems such as surface collapse ets. The explored result have achieved remarkable economic and social benefits. These have great guiding significance and practical applicability for developing lower Paleozoic Erathem deep-water-level confined Karst water, and relaxing tight situation in short of supplying water and solving development urgent need of water in district of north China.

Keywords: modeling, Deep-water-level, confined karst water, District

INTRODUCTION

1.The backgroud of the natural geography and geology

Zhungeer coal District is situated in the northeastern Zhungeer county, yikezhaoer prefecture at Inner Mongolia autonomous region of China.It is a big coal field with high reserves abundance and its construction has been affirmed as one of the major projects during the period of China's eighth five-year program.

The District is located in the west side of the Yellow River valley area,the contiguous area of Shanxi,Shaanxi,Inner Mongolia,the surface elevation of the loess plateau is higher at north and west and lower at south and east, it is an arid District with great shortage of water.the average annual precipitation over year is 395mm.,while the averaged annual evaporation discharge is 2054mm..The water from the Yellow River which in 15km.from the center of the District,is a silt water and with bad quality and thus cannot immediately be used.

The outcrops of the strata within this area are Cretaceous,Jurassic, Triassic systems of Mesozoic

Erathem, clastic rocks of Permian, Carboniferous systems,upper Paleozoic Erathem, carbonate rocks of Ordoician and Cambrian systems,lower Paleozoic Erathem,and the rocks of Proterozoic and Archeozoic Erathems , Cenozoic loose deposits are widely dispersed,the coal- bearing strata are Shanxi group of lower Permian series and Taiyuan group of upper Carboniferous series,the structure of the strata as a whole is monoclinal,striking N-S and dipping to south-southwest,The crust has been uplifting slowly by strong Neotectonic movements.

The thickness of Ordovician and Cambrian strata under the coal formation is 355-725m.,average 593.71m., bearing karst water.The coal District is at the Shanxi,Shaanxi and Inner Mongolia.The buried depth of the karst aquifers is more than 400m.in general and the karst water level is more than 200m..

2. Tianqiao Karst Water System

The System runs 200km from south to north and extends 70km from west to east ,the area is 13974sq.km(Fig.1).At the east bank of the Yellow River, carbonate rocks of ordovician and cambrian strata are widely exposed ,the area is 4404sq.km.,while at the west bank they are deeply buried under Carboniferous and Permian strata,the system bears the following features:

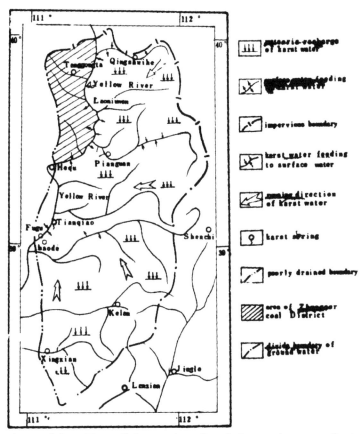

Fig.1 The sketch map of the multistage discharge and transformationbetween three types of water .on the east bank,the karst water is feeded by precipitated water and stream water .The Yellow River are both discharge and recharge zone of karst water .It form two local drainage areas-Laoniuwan and Yushuwan spring groups and one main drainage area-Tianqiao spring groups along the river.

2.1.The boundary conditions

North: Impervious boundary

East: Relative impervious boundary at the north part, divide boundary of groundwater
     at the south part .

South: Divide boundary of ground water and of surface water.

West: Relative impervious boundary,the elevation is 300-200m,that is a buried depth of
     the apical plate of karst aquifers about 800 m bellow the surface.

2.2.The karst aquifers of the system

The aquifers can be divided into three groups according to rock characters, karst development
features and difference of water abundance,they are:

   (1).Majiagou group,average thickness 136m

   (2).Liangjiashan-Yeli-Fengshan group ,average thickness 178m.

   (3).Gushan-Zhangxia group,average thickness 155m.

The main target aquifers for water supply group in the District is Gushan-Zhangxia group

2.3.The Karst Water System   possesses a   model of   multistage discharge and
     transformation between three types of water.

From North of qianfangzi, 983m. in elevation,Inner Mongolia, the Yellow River flows southwards
through the middle and   west   part   of the System,then goes out of the System at south part of
fugu county 850m.In elevation ,Shannxi province,running   across a Distance of 163 km. The
annual mean flow from 1952 to 1987 is 787-823 cub. m/s. The maximum on flood peak is 13800
cub.m./s. (fugu station, 1954.8), mean   silt   content is 5.6kg./cub.m., mean load-discharge ratio is
4.47t./s..

There're many streams on both sides of the Yellow River   within the system.,but most of the
streams are seasonal .On the east bank,in rainy   seasons, the karst water is feeded by stream
water through seepage.

The water level of the Yellow River is lower than that of karst water on east bank and higher
than that of buried karst water on west bank.The precipitated water on east bank seeps
into the carbonate rocks and becomes groundwater,running in the   direction of dipping of the
strata, i.e from east to west. Some   karst aquifers have been cut by the Yellow River On the
east bank, three   spring groups,i. e. Laoniuwan, Yushuwan and Tianqiao spring groups
from north to south,drain part of the karst water to   the   Yellow   River. Transforming
groundwater into surface water,while most of   the   karst water runs laterally and
westwards,under the   Yellow   River bed, to supply the buried karst water on west bank.
There're two local drainage   areas and one main drainage area along the   Yellow River
(Fig.2), the elevations (bottom) are +920 m (Laoniuwan), +863 m. (Yushuwan) and + 816 m
(Tianqiao), the height differences between the spring groups and the elevation of the karst water in
recharge area (+1450 m) are 530 m, 587 m and   634 m respectively. A calculation based on
hydrochemical and static isotopic δ D testing shows that on the west side of the Yellow River, the
ground water, which is a mixed water   of karst water and infiltrated Yellow River water,contains
about 40-50% of the Yellow River water within a short distance off the river. while at a distance of
7km westwards, it is 20-25%.

The multistage discharge of the karst water and the   complicated transformation relationship
between the Karst Water,the precipitation,and the Yellow River   water   within the System
control the storage and movement of the confined karst water in the District.The notable
difference in groundwater level between recharge and discharge areas provides some favourable
factors for deep karst development in the District.

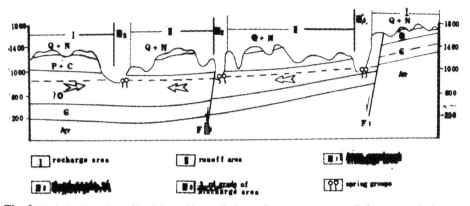

Fig. 2 The hydrodynamic profile of the multistage discharge, fromrecharge area to discharge area, the karst water have a characteristic of grade-3 discharge. The main drainage area is Tianqiao spring groups (III 3); other two local drainage areas, III 2: Yushuwan spring groups; III 1, Laoniuwan spring groups.

2.4.There is a strong runoff zone far from the Yellow River valley and affected by silt filling zone within the System

On the west side of the Yellow River ,about 7- 8km. from the River and within the District,there is a strong runoff zone(area) extending roughly N-S(Fig.3).The length from north to south is about 70 km. ,the width from west to east is about 3-4 km..They have been developed under the control of the particular model of Tianqiao Karst Water System and the boundary condition of the District.

The north boundary is an impervious boundary. The east is a recharge boundary, receiving the Yellow River water and the karst water from the east bank which flows westwards, and after a long geological period, the silt carried by the Yellow River water fills in the karst fissures more or less,formming a silt filling zone with varied width along the west bank of the Yellow River,thus the Yellow River water recharging to the karst water will first be filtered and precipitated.The south boundary, also the local discharge region of Yushuwan spring groups , but the karst water is mainly drained in the form of underground runoff, the spring groups is subsidiary. West is a poorly drained boundary, at a depth of 800m.below the surface, and buried under the clastic rock layers of the Carboniferous and Permian systems, the karst water runs very slowly, the water quality is the sodium chloride type.

To sum up,under the control of Yushuwan local discharge area and Tianqiao main discharge area in Tianqiao Karst Water System,and with the existence of boundary conditions and silt filling zone, the karst water in the buried runoff zone runs from northeast, north- northeast and then from north to south in a concentrated flow, formming a strong runoff zone which are affected by the Yellow River silt filling zone and far away from the River valley. This recognition provides a foundation for our theory,and guides us in looking for deep level confined karst water in this District.

3. The exploration and development of deep water level confined karst water

Since the begining of our work in 1980, several methods have been used such as remote sensing,surface geophysical exploration,drilling and logging,water pumping test, water behavior observation and hydrochemical and isotopic testing etc., the target strata being the Cambrian and Ordovician carebonate rock layers, in hydrogeological exploration for water supply. Two water sources, Chenjiagomen and Tanggongta,have been set up in 1990 and 1993. 24 big diameter

borholes for water supply have been completed,the total mining yield is 74973 cub. m/d, the
construction of the District thus can be going on successfully.

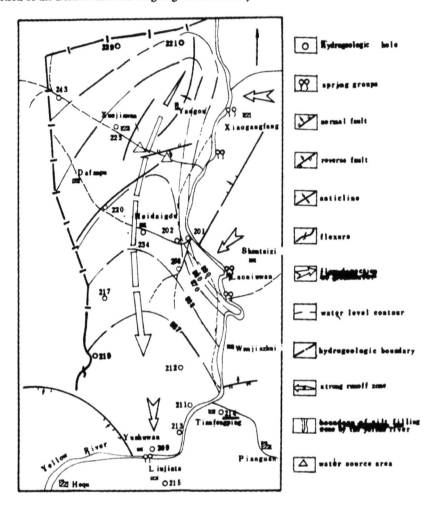

Fig. 3 The sketch map of strong runoff zone far from the Yellow River valley and affected by silt filling zone in the
District .The karstwater ,recharging by Yellow River water and groundwater from east bank which flows
westwards,runs from northeast to southwest.After across thesilt filling zone.it converge running fromnorth to south in
the central section of the District.

The following are our main experiences:

3.1. Applying the theory of Karst Water System,to study the features of the storage   and movement
of deep water level confined karst water, especially the features of strong runoff zone (area) in the
District.

Applying   the theory of Karst Water System, we set up the Tianqiao Karst Water System and
established a karst water model for the transformation between three types of water and the
multistage discharge in Yellow River Valley area. The District is   located between two local
discharge areas, i.e. Xiaogangfang, Laoniuwan spring   groups   and Yushuwan spring groups.The
recharge area at the east side of the Yellow River is far   away   from the   discharge   areas

(Xiaogangfang,Laoniuwan and Yushuwan areas), the distances are 40-60km. In addition, the discharge areas have the character of multistage discharge controled by Tianqiao main discharge area,the karstification in deep buried rock layers at the west bank of the Yellow River is rather strong and there is a poor water content silt filling zone along the west bank,Combining a further research on the boundary conditions, Karst water behaviors, hydrodynamic condition, hydrochemical and isotopic features and the character of the runoff field on pumping test,we have found,in the midst of the District, a strong runoff zone which is affected by the silt filling zone and far away from the River valley, and finally completed Tanggongta and Chenjiagoumen water sources on the strong runoff zone.

3.2. Applying the theory of karst water storage structures in the arrangement of holes

The development of the karst fissures is very heterogeneous in the District. In order to seek for the best water supply area in the strong runoff zone, Consulting geophysical data, we selected some favourable positions on water storage structures for water supply holes,for example, at chenjiagoumen water source area the structural compound part of F2, F82, F83 (FIG.4) on a monoclinal faulting water storage structure was selected and at Tanggongta water source area the holes were arranged at the south part of Yaogou pitching anticline water storage structure ,near the pitching end. All of the 24 big diameter boreholes were successful and water yield for each one is 2500-4000cub.m./d..

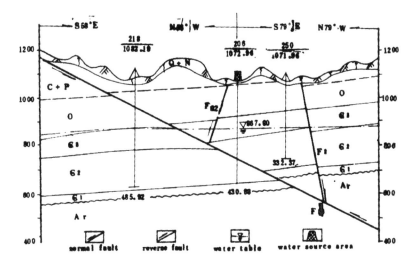

Fig. 4 The monoclinal water storage structure,showing the Tanggongta water source area.it was set up in the structural compound part of three fault

3.3. Adopting the exploration method by combined extraction to cut short working period and to ascertain the productivity of the water source.

The method of exploration by combined extraction is :On the basis of a thorough investigation of the Karst Water System,the hydrogeological condition of the coal District and a preliminary evaluation on the water supply capability of the water resources in the system, considering the demand for water supply at water sources,to drill big diameter borehloes for water supply at planned water source spot,then, the hydrogeological conditions of the coal District can be revealed after group well pumping tests in big diameter boreholes for water supply, so we can get correct and detailed hydrogeological data and reliable mining capability for the construction of water supply sites and on the other hand, we can cut short working period.In the construction of

Chenjiagomen and Tanggongta water supply sites,adopting this method,we found out that the recharge source for the two water supply sites is mainly (75-80%) the karst water from the exposed area east of the Yellow River, secondly (25-20%) the laterally recharge of the Yellow River water, confirmed the existence of a strong runoff zone(area)in the midst of the District,proved a silt filling zone and determined the production capabilities of the two water supply sources.

3.4.Adopting the principle of relatively concentrated water extraction at several separated regions and progressive expanding in developing deep water level confined karst water.

The first project for water supply begins in 1985-1986, at Chenjiagoumen water supply site,6 holes for water supply were drilled with the pump output of 12500cub.m./d..The maximum draw down of the water level for each hole was less than 5m..Further,to expand the water production capacity,we drilled another 8 holes from 1989-1990,the total 14 holes get a practical water supply of 40833cub. m/d. ,the average draw down of the water   level of the holes is still no more than 5 m..After some years of water supply, it is clear that the water source is stable, the draw down of the water level small and the water quality good.Thus, in 1993 we set up another water supply site, Tanggongta water source, 3.4 km west of Chenjiagoumen, 10 holes drilled with water production of 34140cub.m./d. and the average drawdown of water level was 1.64 m.

On the whole,adopting the principle of relatively concentrated water extraction at several separated regions and progressive expanding,, the total water production of two karst water sources is 74973 cub.m/d, and the production activities in 1994 has proved that the water supply sources are stable,good quality, with less draw down of water level,and there has been no environmental problems such as surface collapse etc.

Proc. 30th Int'l Geol. Congr., Vol.22, pp. 47-52
Fei Jin and N.C. Krothe (Eds.)
© VSP 1997

Combination of Water Drainage and Supply to Prevent Disasters and Multi-purpose Use of Mine Water in North China Karst Inundation Mining Areas

YU PEI
Department of Geological Environment, Ministry of Geology and Mineral Resources, Beijing, 100812, P.R.China
YE GUIJUN
China National Administration of Coal Geology, Zhouzhou,Hebei Province,072752 P.R. China

Abstract

This paper gives a brief introduction about the distribution of karst inundation mining areas in north China, the working tactics in preventing and control of groundwater disasters, the main patterns and mathematic models for combination of water drainage and supply, and the multipurpose use of the mine water.

Keywords: Combination, Prevent Disasters, Multipurpose Use, Mine Water, Karst, Mining Areas, Ordovician, North China

INTRODUCTION

The karst features of the carbonate rock layers are often seen from Proterozoic Erathem to upper Palaeozoic Erathem in north China, especially from the lower Palaeozoic Erathem. The distribution area of carbonate rock layers within Hebei, Shandong, Shanxi, Henan provinces, southeast of Liaoning province, Weibei mining area, Huainan and Huaibei mining area, Xuzhou district as well as Beijing, Tianjin etc.-- the ten concentrated distribution regions in north China -- is about 469 397 km^2, in which 73 403 km^2 are exposed karst rocks and the rest are covered of buried.

GENERAL DISTRIBUTION FEATURES OF KARST INUNDATION AREAS IN NORTH CHINA

North China is rich in coal and iron resources, the karst inundation ore deposits are also widely dispersed. About 70 karst inundation coal mines of carboniferous-permian and late Paleozoic are concentrated in Shanxi, Shaanxi, Henan, Hebei, Shandong, Liaoning, Inner Mongolia, Niningxia, Gansu, Jilin Provinces or rigions and Huainan and Huaibei Mining area, Xuzhou district, most of them have been developed.

The Skarn type karst inundation iron mines in north China are mainly distributed in Han-Xing iron ore field in Hebei province, Laiwu mine in Shandong province, Zhangmatun mine, Jinan district in Shandong province, Liguo mine in Jiangsu province, Anlin mine in Henan province.

THE WORKING TACTICS IN PREVENTING AND CONTROL OF GROUNDWATER DISASTERS IN KARST INUNDATION MINING AREAS

Groundwater hazards are a serious problem in karst inundation coal or iron mining areas in north China. There is a great discharge of mine water from coal and iron mines, the total is 2 900.25m^3/min., including 1 658.70m^3/min. of karst water, taking up 57.19% of the total. In coal mines, the water mainly comes from the karst limestone water bearing rock layers of middle Ordovician system beneath coal beds. According to incomplete statistics on the main coal mines in north China from 1956 to 1988, there have been 1 220 times of water bursting, including more then 200 times of shaft submergence. In iron mines, the roof and floor of the ore deposits are often the middle ordovician water bearing limestone rock layers, they are also a main source of inundation.

On the other hand, there is a sharp contradiction of water demand and the water supply in north China, where there has always been in a state of water shortage. Investigation and statistics show that in 1988 there have been a water shortage of 698 600m^3/d in the state owned mining areas in north China, in which, domestic shortage was 336 000m^3/d, industrial shortage was 362 600m^3/d. With the expansion of population and the development of mining activities the demand for water supply will be also on the increase.

The karst water is a very important groundwater resources for water supply in north China. According to the 1990 research result on karst in north China, the quantity of the natural karst groundwater resources is 127.45 × 10^8 m^3/a, the recoverable is 103.61 × 10^8 m^3/a, present mining yield is 68.62 × 10^8 m^3/a, the residual recoverable is 34.99 × 10^8 m^3/a. With good water quality and small temperature variation, it can be immediately put into use. So the best technical method in such mining areas is to apply the principle of combining mining activity with the development, utilization and protection of groundwater resources to meet the increasing needs for water   supply. It's an inexorable trend in karst inundation mining areas to combine drainage with water supply, and make multipurpose use so that to turn the groundwater from being a cause of hazards to being useful resources.

THE GENERAL MODELS FOR COMBINATION OF WATER DRAINAGE AND SUPPLY IN NORTH CHINA KARST INUNDATION MINING AREAS

A. Direct usage drained water
This pattern only remains at the level of a direct usage of groundwater and it is adopted by most of the karst inundation mines in north China.

The total discharge from Jincheng coal mines, Shanxi provines, is 19.03m^3/min., including 2.99m^3/min. of karst water, about 15.71% of the whole. To make full use of the discharged water, 275 projects for groundwater usage, 270 water pools and 30 km of seepage preventing ditches as well as other necessary facilities have been completed in Jincheng city.

B. Combination of water drainage and supply both on the surface and underground. The control of hazards and the developing and utilization of the ground water resources are considered simultaneously . Here are some main models:

(a) Drainage under ground and water supply on the surface
The model may be applied in locations where the exposure of ordovician limestone is rather limited so that the karst aquifer can receive very little precipitation. With a good premeability, when dewatering within a small region in mine, a similar dropping will be seen of the water levels

within the whole mine.

Since 1986 Zhaogezhuang shaft of kailuan coal mine, Hebei province, has been using this model. Through concentrated dewatering boreholes underground, the ordovician karst water is led to a collecting gallery which connects the surface extracting boreholes and then the water is pumping into high water tower(water store house) on the surface for direct use . That provides an additional water supply, the average is $0.285 \times 10^8 m^3$/a and $1\,756.56 \times 10^4$ tons of coal formerlly under the threat of groundwater have turned to be recoverable. By the end of 1994, 233.37×10^4 tons of such coal have been produced safely.

(b) Discharging the karst water of upper ordovician aquitard under ground and extracting the karst water of lower ordovician aquifer on the surface man-made water source.
This model may be applied in locations where, beneath the coal seams, the water yield property of the upper part of ordovician aquifer is poor and that of the lower part is good.

In Hancheng mines, Shaanxi province, for instance (Fig.1), the good aquifer (part V)
and the aquitard (part VI)of the middle ordovician pose a considerable threat on mining activities of coal No.11, which is located at the lower part of Taiyuan Series (Carboniferous system). In order to ensure safety in production and to meet the needs for water supply, while dewatering the aquitart (prat VI) under ground, boreholes for extraction of karst water from the good aquifer (part V) were drilled on the surface, at man-made water source, thus the leakage recharge from part V to part VI has been cut down and the hydrolic pressure has been reduced. The mining activeties are now less dangerous and water supply is more satisfied.

(c) Discharging shallow level karst water at man-made water source on the surface and draining deeper level karst water under ground
This model may be applied in such mines where indirect water bursting is a big problem. There is

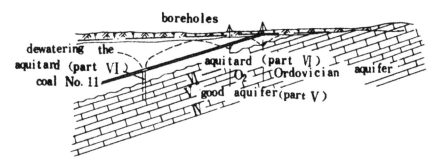

Figure 1. Model of discharging the karst water of upper ordovician aquitard underground and extracting the karst water of lower ordovician aquifer on the surface man-made water source

is a hydrolic relationship between the karst water of ordovician system beneath the coal seams and that of linestone rock layers sandwiched in between coal layers, the former often flooded indirectly into mines. Artificial water source can be set up on the surface to store the extracted karst water from ordovician system at shallow level of the mines, the water can then be put into reasonable use . This will cut down the rechargement to the interbeded aquifer and enhance the dewatering effect. As a result, the hydrolic pressure is released and the water supply is satisfied. The model is adopted by Yanmazhuang shaft, Jiaozuo coal mine in Henan province (Fig .2)

C. Intercepting or curtain grouting
This can be applied in karst inundation Skarn type iron mines in north China. By artificial

grouting, a heavy curtain is set up around the mine or the infllow mouth, that will stop the rechargement of ordovician karst water into the mine and greatly relieve it from drainage working. This leads to a safety production and a protection of the water sources nearby. The model has been used in Zhangmatun iron mine, Jinan City in Shandong province.

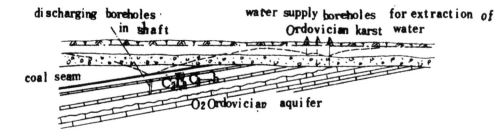

Figure 2. Model of discharging shallow level karst water at man-made water source on the surface and draining deeper level karst water underground

THE MATHEMATIC MODEL FOR THE COMBINATION OF WATER DRAINAGE AND SUPPLY

The optimal mathematic model here is only for above (b) and (c) items.
Simulated model for groundwater
The mathematic model for 2-Dimension unsteady flow is shown below:

$$
\begin{cases}
T_X \dfrac{\partial^2 H}{\partial X^2} + T_Y \dfrac{\partial^2 H}{\partial Y^2} + K_2(H_1 - H) + \varepsilon(t) + \sum_{i=1}^{N} \delta_i Q_i(t) = S \dfrac{\partial H}{\partial t} \\
(X,Y) \in \Omega \quad t > 0 \\
H(X,Y,t)\big|_{t=t_0} = H_0(X,Y) \quad (X,Y) \in \Omega \quad t=0 \\
H(X,Y,t)\big|_{\Gamma=\Gamma_1} = H_1(X,Y) \quad (X,Y) \in \Omega \quad t>0 \\
\dfrac{\partial H}{\partial n}\big|_{\Gamma=\Gamma_2} = q
\end{cases}
$$

where : X,Y,t - variables both space and time
       T_x, T_y - tramsmissivitys
       $\varepsilon(t)$ - quantity of valid infiltration and discharge at time t

$$
\delta_i(X,Y) = \begin{cases} 1 & X=X_i \quad Y=Y_i \\ i & \text{others} \end{cases}
$$

    Q_i - quantuity of pumping water at point (X_i, Y_i) and time t
    H,S - water level and storativity of the target aquifer (the aquifer being discharged under ground)
    H_0 - initial water level
    K_2, H_1 - the leakage coefficient of the leaky aquifer and the water level

of fist kind boundary

Γ_1, Γ_2 - first and second kind boundaries

Ω, n - outward normal vector of the leakage area and second kind
boundary

q - unit yield

The solution to the managemant model (Fig.3)

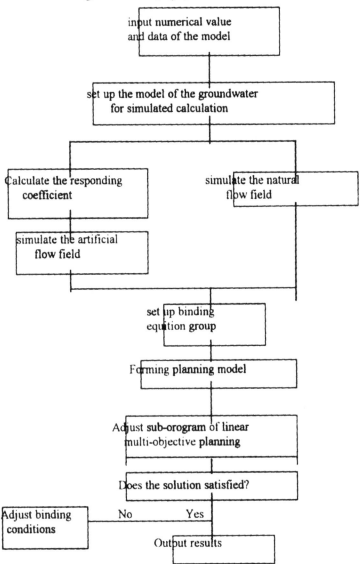

The diagram of the flow chart for solution

The optimal model

The multi-objective optimal model contains more then 2 or 3 objective functions and binding conditions.

The objective functions: In the combination of water drainage and supply, some requirements like the water level and water quantity must be satisfied, and some objectives must be optimized. Thus the problem of the combination of water drainage and supply is turned to be that of optimization with the discharging as the strategic variable and with binding conditions. This means reasonable extraction of   karst water both on the surface and under ground to achieve the optimal solution i.e. safety production and a satisfied water supply.

MULTIPURPOSE USE OF MINE WATER

The processing of multipurpose use of mine water in north China must be closely connected with the comprehensive treatment of groundwater hazards. First, we must   make full use of the mine water. Second, we must adopting the method of drainage of groundwater by combined water supply in the process of groundwater hazards control. In addition to free the coal resources from groundwater hazards, the water resources must be fully used to meet the growthing needs of water in production. Generally speaking, the combination of water drainage and supply in north China karst inundation coal mining areas has an immediate influnce upon the output of coal in eastern part of north China and the strategy for developing of coal resources in our country.

CONCLUSIONS

1.The combination of water drainage and supply, and the multipurpose use of groundwater, must be based on and satisfy the concrete conditions and demanding. So we must take an overall consideration and strengthen administration in order to obtain the best advantage.

2.To achieve a good combination of drainage with water supply and multipurpose use of mine water, and to guarantee safety in production, we must search for a technical settlement to the following problems: the optimal models, the method to predict and prevent the disasters of water bursting and flooding in mine within the funnel of groundwater surface on discharging, to prevent and control the pollution by mine water and to deal with a series of environmental geological problems due to the discharging of mine water etc.

3.The combination of water drainage and supply to make full use of mine water in order to raise the utilization rate of the water has a practical significance in prevention of groundwater hazards and relaxing the tension of water shortage, and also is an effective mesure to solve the problem of water supply for domestic and industrial use.

REFERECES

1.Xin Kuide, Yu Pei and Ye Guijun, Feb.,1993, The characteristics of water-filling zines in northern karst terrains of China, problems of conjunction of water supply and drainage, karst and karst water in north China, Guanxi Normal University Press, PP.141-148

2.Yu Pei, Ye Guijun, oct., 1994, Comprehensive utilization of mine water in karst water-impregnated mining areas of north China, Hunan activity and karst environment, Beijing Science Technical press, PP.233-237

3.Ye Guijun,Du Bingjian, Feb.,1993, Prevention and management of catastrophic karstic water in coal mining areas, souuthern Hancheng, Weibei coalfield, Karst and karst water in north China ,Guangxi Normal University Press,PP.154-161

4.Ye Guijun, May, 1993, Conjunction of water drainage and supply is the working tactics in preventing and control of coal mines karst groundwater disasters, The fourth national mining congress proceeding, PP.123-128

Proc. 30th Int'l Geol. Congr., Vol.22, pp. 53-58
Fei Jin and N.C. Krothe (Eds.)
© VSP 1997

Coal Mine Water Control In Big-discharge Coal Regions In China

WANG MENGYU

Xian Branch, CCMRI,44, Yanta Rd(N), Xian 710054, P. R. China

Abstract

The big-discharge coal regions are distributed widely over the Permo-Carboniferous coalfields in North China and Late Permian coalfields in South China. Coal mines suffered from Karst water disaster in these regions.

Since the founding of P.R.China, a great deal of research work on protecting against mine water has been done under the leadership of the Ministry of Coal Industry. Successful results have been obtained in respects of studies on mine water inrush mechanism, dewatering and depression of aquifer, sealing water inrush spots by grouting and cutting-off water flow by cement curtains, protection against water at surface and underground as well as dewatering combined with water supply. Many problems in coal mining have been solved and theory and technology of mine water control with typical Chinese feature have been gradually formed.

Keywords: Water control, Big-discharge, Coal regions, China

INTRODUCTION

Coalfields in China can be divided into six hydrogeological type regions (Fig. 1). The big-discharge coal regions are distributed widely over the Permo-Carboniferous coalfields in North China and Late Permian coalfields in South China. Coal mines suffered from Karst water hazards in these regions.

According to investigation on 417 pairs of state-owed shafts, 120 pairs, about 28.7 percent, are belonged to type of complex or extremely complex hydrogeological conditions. The discharge in many coal regions is over $1.0 \times 10^5 m^3/d$, Such as Jiaozuo $(8.91 \times 10^5 m^3/d)$, Kailuan $(4.18 \times 10^5 m^3/d)$, Fengfeng $(3.07 \times 10^5 m^3/d)$, Suzhou $(1.78 \times 10^5 m^3/d)$, Pingdingshan $(1.27 \times 10^5 m^3/d)$, Zibo $(1.11 \times 10^5 m^3/d)$, Lianshao $(1.05 \times 10^5 m^3/d)$ etc. (Fig. 2).

Since the founding of the P.R.China, a great deal of research work on protecting against mine water has been done under the leadership of the Ministry of Coal Industry. Successful results has been obtained in respects of studies on mine water inrush mechanism, dewatering and depression of aquifer, sealing water inrush spots by grouting and cutting-off water flow by cement curtains, protecting against water at surface and underground, as well as dewatering combined with water supply. Many practical problems in coal production have been solved and theory and technology of mine water control with typical Chinese feature have been gradually formed.

WATER INRUSH PREDICATION AND PROTECTION

Water inrush happened in big-discharge coal mines is the most widespread phenomenon. It is referred to as a phenomenon that the water from aquifers of roof or floor of a seam overcomes the intensity of relative impervious rock body between aquifers and resistance of the structural planes of faults and joints, and enters the mine galleries under the action of the water pressure and the mining pressure. More than thousands water inrushes happened in these coal mines. The inrush

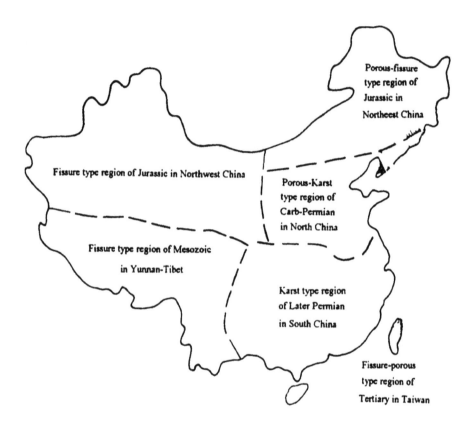

Fig.1 Sketch of hydrogeological type regions

inflow could be very large. Such as Kailuan (2053m³ / min), Lianshao (450m³ / min), Zhaozhuang (299m³ / min), Jiaozhao (243m³ / min), Huobi (225m³ / min), Laiwu (177 m³ / min), Fengcheng (166m³ / min), Fengfeng (150m³ / min). The enormous water inflow usually resulted in flooding mines.

As early as in 1950s, it was realized that water inrush had been associated with the water pressure in aquifers, the destruction of roof or floor caused by coal mining, the relative thickness of water-resisting layers and the existence of faults and joints. In 1961, we considered the factors mentioned above as a comprehensive phenomenon and on the basis of the statistic data, developed a concept of water-inrush coefficient, which is defined as the critical value of water pressure borne by the unit thickness of a water-resisting Layers. In 1966, a site test in which water was pressed into faults and water-resisting layers through wells was carried out to study the processe of water inrush. The limit intensity of different rock beds in per-meter thickness obtained from the test. In 1979, the site observation to the destructive depth of the floor caused by coal mining was made in Handan mine area. By means of the observation to wells with different depth, the destructive depth of the floor under mining pressure was determined to be from 8 to 12 meters. In 1982, the rupture and deformation of the floor beds arising from coal mine were tested and observed by using comprehensive methods with different observation instrument, such as the water injection through boreholes, ultra-sonic detection and borehole radio-frequency detection, steel-string instrument and pressure pillow, etc.. Other site tests were made in Fengfeng, Zibo, Hancheng coal mine areas later on, appropriate software was worked out. The research work on water inrush mechanism and regularity had guided mine water predication and protection. Numerous amounts of coal has been safely exploited in big-discharge coal regions in China.

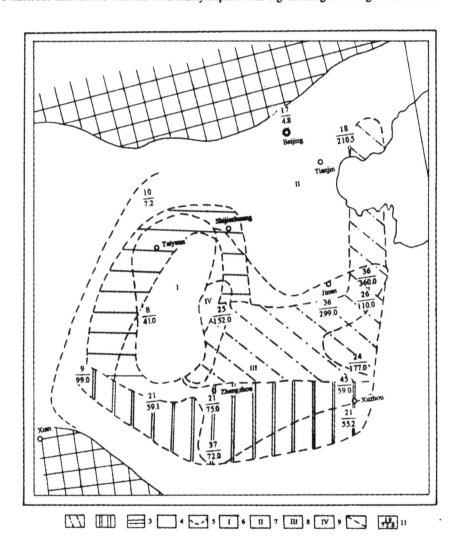

Fig. 2 The partitioned line of the number of water outrushesand water outrush rate in mine areas
1. water outrush rate>100, 2. water outrush rate=50-100, 3. water outrush rate=25-50, 4. water outrush rate<25, 5. The parting line of zones with different water outrush rate, 6. 0-5, 7. 5-30, 8. 30-50, 9. >50, 10.   The parting line of zones with different number of water outrush, 11.the uplifted zone.

Now, The methods of inrush predication in big-discharge coal mines include statistic methods such as inrush coefficient, inrush discriminate, structural mechanics method such as thin plate model, geophysical methods such as gallery radio-frequency detection, gallery seismic exploration, and geochemical method such as radioisotope analysis, etc. (Fig. 3). On the basis of predicating inrush, some measures for protecting inrush have been made in Fengfeng and Feicheng coal mines, such as pre-grouting cement to the seam floor, pre-dewatering with   submerge pump, etc.

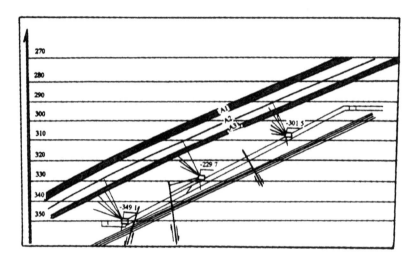

Fig. 3 Sketch of a comprehensive test for the destruction of the floor by mineing in Huainan mine

DEWATERING AND DEPRESSION

Dewatering and depression refer to dewatering of aquifers as roof of a seam and depression of aquifers as floor of a seam in mining area. In 1950s, dewatering was carried out in Fengfeng. Xuzhou mine areas, and depression in Zibo mine area. In 1960s, the investigation of the procedure and the exploration method of dewatering were made in Handan mine area. In 1970s, the control of mine water inflow was studied to regulate the cone of depression in Lianshao mine area, and the numerical method was developed to calculate water inflow at the same time. In the early 1980s, the overall study was carried out on dewatering and depression at 9 mines in Huainan mine area.

In China, the dewatering exploration is accomplished by pumping tests, discharging, hydrochemistry, hydrogeophysics, and laboratory modeling. In the old mine area, the test of discharging is commonly adopted, for the example, the dewatering of Xujiazhuang limestone in Zibo mine area. The discharge test was performed in the order of single borehole, double boreholes and multi-boreholes, to determine the quantity of the recharge and interference, the cone of depression, the residual heads and the hydraulic connection between aquifers.

In recent years, with the use of submerged pump with high-lift and large-discharge in mine in China, the swift development of pre-dewatering and depression will come into being.

GROUTING FOR SEALING WATER INRUSH SPOTS AND FORMING CEMENT CURTAINS FOR CUTTING OFF GROUNDWATER FLOW

Sealing water inrush spots and cutting off groundwater flow are important measures of mine groundwater control in coal mines in China. From more than 30 years' practice, the mature experience is obtained in respects of standing-water grouting and moving-water grouting to seal water-inrush spots, grouting and forming cement curtains outside the mine area to cut off groundwater flow. Standing-water grouting is used for sealing the water inrush spots when mine water level is in a still state after the mine is completely flood. The technique of sealing the water inrush spots by grouting can be illustrated by the case of Beidajing mine in Zibo Coal Mine Bureau. The gallery was cut into a normal fault with a throw of 30 meters, resulted in karst water from Ordovician limestone rushing into the mine with a inflow of $443 \mathrm{m}^3 / \mathrm{min}$. The mine was completely flooded 78 hours later. This accident happened in May 1935. Thirty-seven years later, the location of water inrush spots and the paths of water into mine were determined by means of borehole radio-frequency detection technique developed by Xian Branch, CCMRI. Borehole grouting from surface was used to seal the spots and twenty boreholes were laid out in three rows with a total drilling footage of 5000 meters. The water inrush spots were completely sealed by using 9100 tons cent, $115 \mathrm{m}^3$ pebble, and the effectiveness reached 100%.

Moving-water grouting is used for sealing water spots under the conditions of moving water when a mine is not completely flooded. In 1984, grouting was carried out at Fangezhuang Mine of Kai Luan Coal Mine Bureau to seal the water inrush spots where the maximum inflow is $2053 \mathrm{m}^3 / \mathrm{min}$. The groundwater came from karst subsidence column in Ordovician limestone (Fig. 4). The comprehensive grouting for blocking up galleries under the condition of running water of high velocity was adopted. Some instruments were used, such as the well horizontal fluid-velocity meter, the borehole radio-frequency detector, etc. In this grouting engineering, 75 boreholes were used and the total drilling footage was 32700 meters and stone of $47100 \mathrm{m}^3$, dreg of $63100 \mathrm{m}^3$, sand of $79100 \mathrm{m}^3$ and water glass of $9000 \mathrm{m}^3$ were used. The effectiveness reached more than 99% in this work that is the biggest and the most complicated one in the history of coal mining in the world.

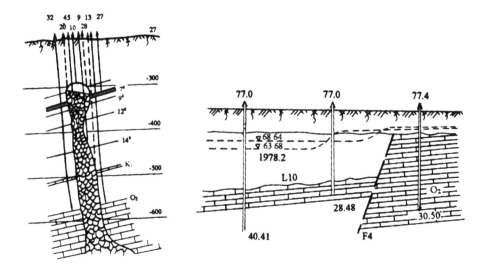

Fig. 4 Sketch of moving water grouting at Kailuan mine

Fig. 5 Sketch of grouting and forming cement curtains at Huangbei mine

The cement curtains were applied for cutting off water flows in the early 1960s. In thin limestone, cement curtains suitable for four types of different hydrogeological conditions were built up. The four types are, recharging from thin limestone to thin limestone, from alluviums to thin limestone, from Ordovician to thin limestone, from rivers to thin limestone. The cement curtain for cutting off water flow in Huangbei Mine of Zhaozhuang Coal Mine Bureau was built (Fig. 5). In this mine, Carboniferous coal seam is excavated and water flow comes directly from L_{10} limestone with a thickness from 5 to 8 meters. The L_{10} is being recharged by karst water from Ordovician limestone. After an investigation of F_4 fault, the main recharging section is determined to be 460 meters long and the recharge to be $32 m^3$ / min . The cement curtain was built in L_{10} limestone by using 77 boreholes with intervals of 10-20 meters. Quantities of injected cement and waste are 1017 tons and 948 tons, respectively. After completion of the curtain, the difference of piezometric levels between both sides of F_4 was 12 meters and the inflow reduced to 20% of the original one, the reduced cost of water drainage was 150 thousand RMB yuan each year. Furthermore, the problem of drinking water is solved in the mine area because the springs in Ordovician limestone dried up come to life.

REFERENCES

1. Wang Mengyu. Hydrogeologic characteristics of coal-fields for late Permian in South China and the problem of controlling mine discharge, Coal Science and Technology, 11, 33-36, (1977)
2. Wang Mengyu, Mechanism of Water inrush from Coal Floor and Investigation of Predictive Methods, Coal Science and Technology,9, (1978)
3. Wang Mengyu. Hydrogeological problems of karst coalfields in China, Coal Science and Technology, 12, 32-35, (1981)
4. Wang Mengyu. The Elemental Hydrogeological Characteristics of Coalfields in China and the Problems of protecting against MineWater, Mine water, Granada, Spain, 41-51 (1985)
5. Li Dean, Wang Mengyu, Technology of Coal Mine Water Control in China, proceedings of the 22nd International Conference of Safety in Mines, Beijing, 876-888, (1987)
6. Wang Mengyu, Zhang Zijai, Karst Water Systems and Coal Deposit Water Charge in the North of China, IAH XXIII International congress, volume 1, 247-250, Spain, (1991)

Proc. 30th Int'l Geol. Congr., Vol.22, pp. 59-63
Fei Jin and N.C. Krothe (Eds.)
© VSP 1997

Some Problems on Riverside Groundwater

HAN ZAISHENG

*Administration of Mineral Resources, Ministry of Geology and Mineral Resources,
Beijing, 100812, P.R.China*

Abstract

The characteristics of riverside groundwater and the main points in its exploration are discussed in this paper. For evaluation of riverside groundwater, the key point is the recharge from river water. The hydrogeologic concept model and mathematics model must take account of groundwater combined with surface water synthetically. Some problems must be paid attention in the exploitation of riverside groundwater. The riverside groundwater sources are classified herein with practical examples listed.

Keywords: riverside groundwater, river water seepage, allowable withdrawal, practical example

INTRODUCTION

Most cities in the world are situated by rivers. So the riverside groundwater is important for urban and industry water supply. Well field is the region where the groundwater resources could exploit assembly. There are about 300 riverside well fields in China. The complete theory and methods for the exploration of groundwater have formed in P.R.China.

The Yellow river valley and Haihe river valley in northern China are as two examples. Figure 1 is the map of Yellow river valley. There are eight capitals of Province situated along Yellow River and it's tributary. All the cities use groundwater for their urban water supply. There are also some industrial bases exploited groundwater for their water supply. There are more than 50 riverside well fields situated in this valley. Figure 2 is the map of Haihe river valley and Luanhe river valley. There are 16 large and middle cities in the basin include Beijing and Tianjin. Most cities and industrial bases in this area make use of groundwater for their water supply. Among them, the riverside groundwater is more assured of sources than other groundwater. The riverside groundwater is widely explored and exploited in suitable area. Some aspects on riverside groundwater are discussed as follows.

1. THE CHARACTERISTICS OF RIVERSIDE GROUNDWATER WELL FIELD

1.1 The riverside well field is close to a perennial river. It has the assurance of water supply. The situation of recharge from the river to the aquifers could be continuous or discontinuous. The constant river with abundant flow and the fine seepage capacity make the continuous recharge. Otherwise, the recharge is cyclical or intermittent. In that case, the storage of riverside aquifers must be used for regulating in dry seasons.

1.2 As a rule general, the riverside aquifers are consists of the Pleistocene and Holocene series of the Quaternary System. The lithological characters of riverside aquifers are gravel or sand in valley and basin. It is sands or sandy soil in alluvial and diluvial fan. It is clayey soil or clay in alluvial lacustrine plain and coastal plain. The riverside aquifer has a certain amount of storage and regulation ability. It has a close hydrodynamic relationship with the surface river water.

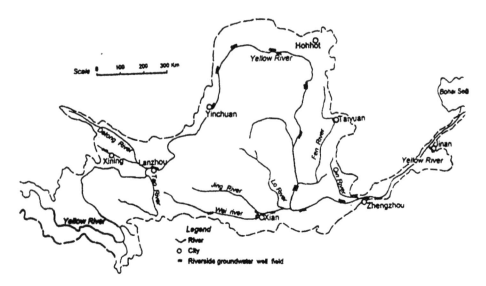

Fig. 1. The Map of Yellow River Valley

1.3 The groundwater recharge of riverside aquifers may consist of the infiltration from the rainfall, irrigation and the runoff from the upside flow. Among them, the infiltration from surface water is the most important recharge. The infiltration from river to the aquifer would increase after the well field exploited. The permissible yield is mainly composed of the infiltration from surface water.

1.4 The wells would be generally parallel to the river. The wells are relatively concentrate. They can arrange in single line or multi-lines as circumstances. Under condition of recharge per unit wide and the yield of single well, the distance between wells and wells' line could be designed. The riverside aquifers are shallow depth.

2. THE FOCAL POINTS OF EXPLORATION AND EVALUATION OF RIVERSIDE GROUNDWATER

2.1 The buried condition, distribution and seepage capacity of the riverside aquifer must be explored clearly. The type, distribution, contributing factors and laminated construction of the river terrace should be investigated. The adjacent aquifers, aquiclude and aquifue must be investedalso. The conventional exploration means are hydrogeological mapping, geophysical prospecting, hydrogeological drilling, pumping tests and groundwater observation, etc. Some new technique and methods such as remotes sensing technique, isotopic technique, shallow seismic prospecting and velocity logging method are applied in exploration.

2.2 The rock characters and seepage capacity of river bed must be studied with the field test in various positions and conditions for obtain the seepage coefficient.

2.3 The hydrologic analysis and budget must be taken for evaluating the river outflow. The wide of river surface and depth of river water in various seasons must be measured.

2.4 The ground water resources evaluation includes the calculation of the withdrawal of the well field and analysis of the influence of groundwater development on the environment. The concept model and mathematics model must conjoin the surface water with groundwater. It could be solved by analysis or numerical methods. The storage and regulation ability of riverside aquifers must be considered. The environment problems emerged after the groundwater exploited must be evaluated during the exploration.

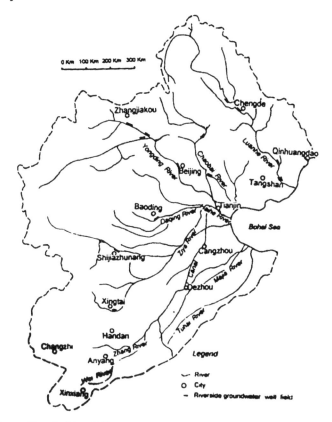

Fig. 2. The Map of Haihe River and Luanhe River Valley

2.5 For evaluating the allowable withdrawal of a well field, the key point is the amount of recharge from the river. There are some methods of deal with the river in the model. For two-dimensional model, a vertical is conjoined with the plane model. The three-dimensional model has the advantage in dealing with the river water recharge. For establishing three-dimensional model, the relevant data must be obtained in the exploration.

2.6 It must take account of groundwater combined with precipitation and surface water synthetically. Theirs transmute relationship must be known. The equilibrium and its alteration among the groundwater replenishment, storage and discharge should be appraised rationally. On the basis of the reasonable hydrogeological concept model, the groundwater mathematics model is set up for determining the allowable withdrawal of the well field. The restraints of the economy, technique and environment can fix as the constraint of groundwater level in the model.

3. SOME PROBLEMS FOR ATTENTION IN THE EXPLOITATION OF RIVERSIDE GROUNDWATER

3.1 It is very important to protect the river water quality for it concerned with the groundwater quality of riverside well field. The pollution resource to the river must be controlled and governed.

3.2 The wells' structure design must prevent the river flooding and adjust the rock characters.

3.3 The groundwater level, yield, temperature and quality must be sensed through the exploited period. It will provide the basis of regulation of the groundwater exploitation of the well field.

3.4 The groundwater resources management would be strengthened on the well field and the around area. A suitable management model is useful for it.

3.5 The groundwater resources development of riverside well field would be operated with the surface water in urban and industry water supply. It can overcome one's weaknesses by acquiring other's strong points.

4. THE CATALOG OF RIVERSIDE WELL FIELDS AND SOME PRACTICAL EXAMPLES IN P.R.CHINA

The riverside well field can classified into four types according their situation. Some of the following practical examples display in Figure 1 and Figure 2.

4.1 Riverside well field in valley. The Tar and Shijiazhuang well field in Qinghai Province, The Daguhe well field in Qingdao, Shandong Province, The Xilipu well field in Hebei Province. The aquifers are consist of coarse sand or gravel. The yield of single well is large. Because the riverside aquifers along the river are limited in scope, The allowable withdrawal of the well field ought less the minimum river flow in the dry season.

4.2 Riverside well field in basin. The Shalingzi power plant well field in Hebei province, The Hanzhong well field in Shaanxi province, The Yumenkou well field in Shanxi province. The aquifers are consist of medium sand and coarse sand. At the exit of basin the river flow in dry season can indicate the remind groundwater resources. The withdrawal of the well fields in the basin is ought to less the amount of the flow.

4.3 Riverside well field in alluvial and diluvial fan. The Hunhe well field in Shenyang, Liaoning province, The Matan Cuitan and Yingmentan well field in Lanzhou, Gansu province. The storage and regulation ability of the aquifer could take function of regulation. The riverside pumping wells can situate in a larger region. For evaluating the exploitable yield of the wells, several technical, economical and environmental restrain must satisfy.

4.4 Riverside well field in alluvial lacustrine plain and coastal plain. The Jiuwutan and Beijiao well field in Zhengzhou, Henan province, The Xibeijiao well field in Xian, Shaanxi province. The aquifers are consisted of fine or moderate sands. Besides of the alluviation of modern rivers, the ancient river bed could be rich of groundwater. The ancient river bed would be find out and ascertained in exploration. The special technical methods in drilling and complete well are needed in the silt and fine sand aquifers.

REFERENCES

1. Jacob. Bear. Hydraulics Of Groundwater, *McGraw - Hill*, New York (1979)
2. Fei Jin. Groundwater Resources in the North China Plain, *Environmental Geology and Water Sciences*, New York, Springier International, 12(1), 63 – 67 (1988)
3. He Weicheng and Xiao Yuquan. Principles and Methodology for Comprehensive Evaluation of Surface Water and Groundwater, *The Hydrological Basis for Water Resources Management*, IAHS Publication No. 197, 31 – 38 (1990)
4. Yin Changping, Sun Tingfang, Jin Liangyu, Wen Tingzuo and Long Shaodu. Exploration and Evaluation Of Groundwater Well Field, *Geological Publishing House*, Beijing, China (1993)
5. Han Zaisheng. Classification of Groundwater Resources and Categorization of Exploration Level, *Hydrogeology and Engineering Geology*, Beijing, China, Vol. 21, No. 2, 31 – 33 (1994)
6. Ji Chuanmao, Hou Jingyan and Wang Zhaoxin. Groundwater Development in the World and Guidelines for International Cooperation, *Seismological Press*, Beijing, China (1996)

Proc. 30th Int'l Geol. Congr., Vol.22, pp. 64-70
Fei Jin and N.C. Krothe (Eds.)
© VSP 1997

The Theory and Practice on the Conjunctive Utilization of Surface Water and Groundwater for the Urban Water Supply in Xi'an, China

LI PEICHENG , ZHANG YIQIAN , ZHENG XILAI
Xi'an University of Geology, Xi'an 710054, P.R. China

Abstract

Water shortage is a new problem caused by urban development,so this paper takes the water supply in Xi'an for example, and points out the conjuctive use of surface water and ground water is an important way to mitigate the water shortage and to impove the geological environment in Xi'an. After reviewing the history of water supply and the existing state of water shortage, the authors indicate the over-exploitaion of groundwater is a direct reason for the rapid drawdown of groundwater table, deterioration of exploitation conditions, water shorage ,land subsidence and ground fissures. Based on the analysis of various water sources, the studies of successful experience overseas, as well as the water cycling law ,a strategy--conjuctive use of surface water and ground water to relax the water shortage in Xi'an has been suggested, and then the water volumes from different rivers or brooks used for water supply in Xi'an and other purposes in the countryside are calculated. As a result, extra 0.3 billion m 3 water a year can be supplied to Xi'an, showing the strategy is effective to the solution of water shortage in short or middle term, and it gives directions for the future.

Keywords: conjuctive utilization, over-exploitation, water supply

URBAN DEVELOPMENT AND WATER SHORTAGE

Rapid Urbanization

Since the 20th century, the settlements are urbanizing, city number and scale are expanding, and urban population is growing in the world. It is reported that there are about 0.36 billion urban people in 1920 and will be 2.9 billion by the end of this century [6] .In other words, half of total people in the world will live in cities . The trend of urbanization will keep growing in the next century. In China , city number doubled from 1983 to 1993, and the urban population increased from 0.17 billion to 0.38 billion in the same period [5] .At the same time, the population in Xi'an also increastd from 0.4 million in 1950 to 3 millon in 1996.

Water Shortage in Xi'an

With the urban development, water consumption increases quickly, especially in the developing countries. For example, the water consumption in Xi'an is 10,000m^3 a day in 1952 and 800,000m^3 a day in 1994 (table 1).

Originally, ground water pumped from wells on the river banks was used for water supply in Xi'an. With the increment of water consumption, not only the water supply facilities can't meet the needs of growing requirement but also the over-exploitation results in a series of environmental problems, such as rapid drawdown of ground water level, deterioration of exploitation conditions, higher cost and water shortage (table 2).Every summer, the water supply has to be interrupted or stopped in part, which causes the resident people to be bothered and some factories to stop their production. Furthermore, the water shortage demages the imagery of Xi'an as the world-famous tourism city and limits the development of the biggest city and industrial base in the northwest of China. Comparison between the contours of confined water table in 1960 and 1988 (Fig.1) [1] indicates that the water table drawdown seriously affects and changes the vadose field in Xi'an area. In fact, the drawdown is continuing after 1988.

Table 1. Water consumption in Xi'an

Periods	Consumption per day ($\times 10^4 m^3$)	Population ($\times 10^4$)	Population rate (%)	Person's consumption per day (liters)
1952--53	0.2--0.4	1.6--15.4	6--23	9--11
1954--58	0.4--15.1	15.4--91.5	23--79	11--50
1959--72	16.6--16.9	92.9--126.0	80--95	24---140
1973--82	34.5--53.0	127.0--120.0	90--96	142--151
1982--89	49.0--61.0	147.0--177.0	92--96	114--134
1990--94	63.8--71.0	186.0--205.0	97--99	123--124

Table 2   The ratioes of water supply to water requirment in summer

Time	1982	1986	1987	1988	1989	1990	1991	1992	1993
The ratioes	0.71	0.78	0.70	0.65	0.60	0.70	0.66	0.63	0.56

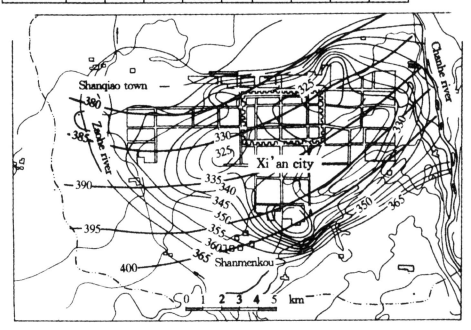

Figure 1. The contours of confined water table in Augest, 1960(thick lines) and Augest, 1988(thin lines) in Xi'an area, showing the variation of vadose field caused by the over-exploitation. The square is the old downtown area encircled by the city wall.

THEORETICAL BASE AND BASIC MODELS

Theoretical Base for Joint Use of the Two Waters

In order to solve the problems of urban water shortage, most scholars over the world agree to simultaniously use various water sources, especially both ground water and surface water in the areas with better conditions.The theoretical base of this understanding or recognition should be the water cycling law from which "the strategy of unified moniter and management of three waters" has been obtained [3] .

Water cycling law is well known to everyone, whose inportance, particularly with its importance for guiding the activities of human water affairs is still worth further probing into.We all   know that there exists a close relationship among the atmospheric water'surface water and ground water. Under the action of solar energy and gravity, they are in the continuous motion and conversion as well as in the dynamicbalance. Their existing states and spaces can be changed, but the amount of water remains no change.

Nowadays, it is difficult for human beings to control water in the big cycle between ocean and land, but human beings can exert the great impact upon the small cycle within the land areas. For this reason, various kinds of water engineering constructions including water supply works have been constructed by human beings with their objectives,whose working principles lie in changing the process and states of water cycle. If human beings can follow and use water cycling law, study and analyse the effect of new balance process upon human beings in combining with local conditions, human beings are bound to benefit from the good cycle of water purification by itself in the regions with water shortage where human induction should be strengthened. In contrast to the above mentioned, if three waters can not be used scientifically in water affair activities, such as only single expoitation of surface water or underground water resources,the irrational blockage or worsening of water cycling process and the damage of turning relationship between water and surroundings are bound to lead to the bad consequences of water cycle.

In order to make the basic natural water cycling law have the definite guiding importance, the first author of this paper [2,3] suggested the strategy on the unified monitor and management of three waters, comprehensive regulation and water control in time and space.This strategy has played an important role in solving the problem of farmland irrigation. Also, it can be concreted into the realization of combination of wells with canals. Meanwhile, surface water and underground water can be regulated in time and space so as to achieve the principle of "wells to supplement the canals and canals to feed wells", which can not only expand irrigation water resources but also prevent soil from becoming alkalized, thus, making it possible for the sustained development of irrigation causes [4].

The Optimal Model for Mimultaneous Utiliztion of the Two Waters
Various water bodies have different delay times and different advantages or disadvanages.Surface water receives the supplementation by rain water fast and recycles fast, whose disadvantages is short duration of water stay. If rainfall water can not be used , it will disappear immediately or run away fast. Even if surface water is stored in reserviors, it will loss seriously through evaporation and seepage in the arid and semiarid areas in particular, which may last over two years in fighting against droughts. The chief advantages of underground water are the wide distribution, great storage volume, less evaporation, longer delay duration, strong drought resistance and easy to be pumped for use.

(a)River discharge curve and low flow rate, (b)Dynamic curve of groundwater table and exploiting volume,(c)River discharge curve and usable flow rate for water supply by constructing reservoirs and (d)Recovery of ground water table by reducing exploitation in rainy season.

Therefore, in the practice of water supply, like the relation between hydro-power and thermal power,surface water is used to bear the basic water supply load, while underground water is used to bear the peak load and the drought resistant load during the low water period of rivers so as to form the joint water use system with wells to supplement rivers and with rivers to wordinate wells, as shown in Fig. 2.This system can be the optimal model for solving the water shortage in Xi'an and other similar cities.

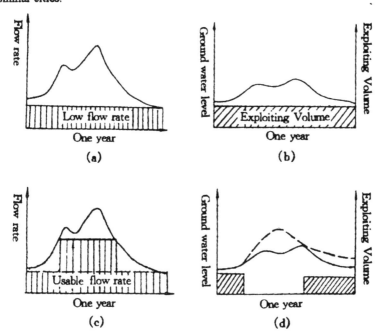

Figure 2. Diagrammatic sketch of surface water and groundwater joint use

THE MODEL OF JOINT UTILIZATION

The Objective Conditions in Xi'an
Although per capita water in Xi'an is much lower than that of national and international level with only 400m 3/person/year, water environment in Xi'an has its advantages. Precipitation in Xi'an is not very low ,about 600mm in the plain area and 800-1000mm in the nearby Qinling mountain area. Total rainfall in the mountain areas is about 4.2 billion m³, of which nearly half of water, about 2 billion m³, flows out of the mountains as river run-off through the gullies. These gullies in the mountain areas are locally known as the "valleys". The distribution of these valleys is shown in fig. 3, being high in the south and low in the north in landform, which is favourable for the water diversion to Xian through the trunk canal and branch canals.

In Xian, there spread rich aquifers consisting of alluvial sands and gravels. For this reason, there are not only plenty of underground water, but also underground reservoirs being favouerable for time and spacial regulation of various water resources.

The evaluted amount of exploitable underground water in Xian has a widely variable range from 0.258 million tons/day to 1.375 million tons/day. The volume of undergrond water exploited has reached 300 million m³ a year or so for 15 years, i.e.0.8 million tons/day, which has caused rapid drawdown of underground water table, land subsidence and fissures. In discussing the simultaneous utilization of surface water and undergrond water in this paper, the amount of water exploitation should be controlled at 80% of existing exploitation in the intermediate drought year(P=75%) and special big drought year (P=95%), 240 million m³/ year or so, but in the most years, the underground water exploitation should be less than this figure and the artificial water recharge must be strengthened so as to nurture or conserve underground water.

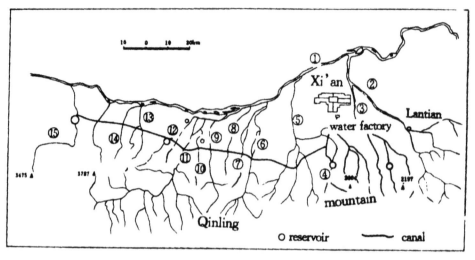

Figure 3. Distribution of surface water system and canals for the water supply in the northern Qinling mountain. Surface water coming from brooks or rivers and conrtolled with or without reservoirs can be diverted to Xi'an through the canals. 1.Weihe river, 2. Bahe river, 3. Chanhe river, 4. Shibianyu river, 5. Fenghe river, 6. Laohe river, 7. Ganhe river, 8. Gengyu river, 9. Tianyu River, 10. Jiuyu river, 11. Heihe river, 12. Luoyu river, 13. Niyu river, 14. Tangyu river, 15. Shitouhe river.

Engineering Layout of the Conjuctive   Utilization in Xi'an

In considering the favourable conditions for water supply and the simultaneous water use in the original benificiary regions, water coming from the river valleys (rivers)in the northern part of the Qinling mountains and controlled with or without reservoirs can be diverted through the trunk canal and branch canals into Xi'an as a surface water source combined with rational use of underground water, thus forming the water supply system of river valleys coordinated with wells to bear the task of water supply in Xi'an.

Hydrological Calculation for the Joint Disposal

In the process of calculation,the disposal principles should be followed. At first, the gravity flow in the rivers without any control should be fully used, and the exploitation of underground water should be as small as possible in order to conserve underground.And then, underground water and water in the reservoirs are mainly used to fight against droughts and to regulate water resouces.Thirdly, as previously described, in the intermediate drought year and the special drought year, 240 million m³/year or so should be limited, and in the rest of years, the amount of underground water exploitation should be lower than that figure, and also, artificial recharge must be carried out actively. Moreover, in order to complete the disposal calculation ,it is necessary to collect and sort out the following data or information:(1) Data of daily flow rate of the rivers without reservoir should be collected, and daily flow table in the years (P=75% and 95%) and the flow process curves should be worked out day by day.

(2) Rational water use system and irrigation water requirments of original beneficiary regions in the corresponding hydrological years (P=75%, P=95%) should be determined, and water demanding table for local city and town water supply and the flow rate table of irrigation water should be worked out day by day.

(3) The corresponding water use in the above (2) will be excluded from the incoming water in the above (1),thereby obtaining   total daily water supply flow rate from all the rivers, shown in Fig.4.

(4) In considering the urban and rural water supply, as well as old and new water supply, the amount of water supply from rivers and reserviors should be confirmed, and also, the initial system of water disposal must be established.(5) Based on the previously-described principle, the amount of underground water exploitation must be controlled.

(6) Daily water requirements should be determined based on the required planning year 2000 and 2010 in accordance with the hydrological years(P=75% and 95%).

According to the principles described above, canal water-carrying capacity and underground water exploitatoon capacity, as well as the data or information obtained from(3),(4),(5)and (6) should be optimized in fitting in order to obtain the corresponding charts for commanding water supply practice(Fig.4).

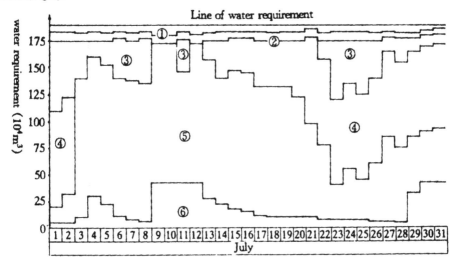

Figure 4. The matching diagram of joint water supply system in Xi'an(P=95%), showing the water volumes from gorges (rivers) and wells can be used in July as a example.① — Shibianyu river, ② — Chanhe river,③ — Shitouhe river, ④ — Groundwater, ⑤ — The gorges, ⑥ — wangchuanhe river.

OBSERVED EFFECT

The year 1994 is severe droughty in Xi'an so that water supply became a big problem. The research results of river valleys coordinated with wells began to be adopted in 1995 and 1996. As a result, the runoff water in the Fenghe river and the Hihe river was diverted into Xi'an, and the water from the Shitouhe river was diverted for the disposal of water supply, thus, achieving the joint and regulated use of surface water and underground water. In the water use peak stage in summer, there are 0.4–0.6 million tons/day of river water supplied to Xian,which has greatly improved water supply condition. Meanwhile, there are 15%~ 20% wells being pumped intermittently, thus, conserving underground water. This can only indicate the advantage and effectiveness of the joint water supply system. If this water supply system can get the overall implementation, water shortage in Xi'an will be overcome, and the bright future will be before us.

REFERENCES

1. Dong Fakai, Sun Nazheng, Studies on the management model of ground water resources in Xi'an city, [unpublished report] , First Hydrogeological Team in Shaanxi, Xi'an (1990).
2. Li peicheng, The construction of an undersurface reservoir by artifical recharge, Shaanxi Water Conservancy Technique, NO.3, 60-68(1973).
3. Li peicheng, A disccussion on the water resources problems and their solution, Irrigation Technique, NO.3,10-19(1975).
4. Li pecheng, On the water sources in the agricultural modernization of the northwest China, Proceedings on the symposium of agricuitural modernization in China, 54-61(1980).
5. Liu Guoguang, The ten years of Chinese city development, China Statistics Publishing House, Beijing(1994).
6. Mary E Pader, World Resources (1988-89), Basic Books Inc, New York (1990).

Proc. 30th Int'l Geol. Congr., Vol.22, pp. 71-79
Fei Jin and N.C. Krothe (Eds.)
© VSP 1997

Study on Utilization and Reclamation of Shallow Saline Groundwater

FANG SHENG   CHEN XIULING
Hebei Institute of Hydrotechnics, 28 Lingyuan Street, Shijiazhuang, P. R. of China

Abstract

Drought and fresh water shortage are the main limiting factors for sustainable development of agriculture in North China Plain. Using shallow saline groundwater to irrigating double cropping of wheat and summer corn enables to increasing yields 1.2-1.6 times than that of non-irrigation. Exploiting and utilizing shallow groundwater include slightly saline water for irrigation enables to regulate the groundwater table at a sufficient depth to transform the variable natural rainfall into a more reliable water resource for the comprehensive control of drought, waterlogging, salinity and saline water. This paper reviews the research and practice of utilization and reclamation of shallow saline groundwater: technology of using saline water for irrigation, salt tolerance index of crops, regulation of soil salt-water regime and reclamation of groundwater quality. That indicated the effective approaches for rational development and utilization of local water resources.

Keywords: Shallow Saline Groundwater, Irrigation, Utilization, Reclamation, Accumulation, Desalinization, North China Plain.

INTRODUCTION

There is a scarcity of fresh water resources in the North China Plain. The fresh surface water flow in the Haihe River catchment has been controlled and utilized for 96% (Fang Sheng et al,1993). Since the beginning of 1970's the groundwater has been exploited and utilized in large scale. Owing to serious overdrafting, there are cone depressions of groundwater table in large area. In order to overcome the serious constraints to agriculture by drought and water shortage, under the condition of Chao soil with irrigation and drainage systems, not only the fresh groundwater but also the Shallow saline groundwater to be used for irrigation, it not only increasing crops yield by considerable margin, but also enables to regulate the " four water ", (Fang Sheng and Wolter, 1990), reduce the evaporation of phreatic water, increase the rainfall infiltration, reduce the losses of runoff and enhance the function of waterlogging prevention, soil salt leaching and freshening of groundwater quality. Since 1974, Hebei Institute of Hydrotechnics had established the Nanpi Pilot area in the east of South Great Canal in Hebei Plain, engaged in the research on utilization and reclamation of shallow saline groundwater and comprehensive control of drought, waterlogging, salinity and saline water, and achieved remarkable research results. Applying the results to Nanpi Agricultural Development Programme Area(13333 ha) by IFAD loan from 1983 to 1987, the saline-alkali land has reduced 57% than that before controlled, the groundwater quality has certain freshened, the total output value of agriculture and income per capita all increased 3 folds. The utilization and reclamation of shallow saline groundwater with great economic, social and ecological benefits.

HYDROGEOLOGICAL BACKGROUND

The Hebei Plain belongs to a part of North China subsidence zone. The thickness of Quaternary system deposit was 500-600 m in downwarping area, and 300-400 m in upwarping area. In the Quaternary system, according to the mineralization less than 2 g/L as the fresh water, the boundary of fresh and saline groundwater approximately at the line of piedmont plain meet with alluvial plain, the fresh groundwater region in the west and the region with fresh and saline groundwater in the east. The area of fresh groundwater and shallow fresh groundwater which in the region with saline groundwater account for 67.7% and the saline groundwater occupies 32.3% in Hebei Plain. The severest shortage of fresh groundwater are in the Heilonggong region and the region east of South Great Canal. Their area and amount of exploitable fresh water only account for 18% and 14% respectively, but the exploitable slightly saline groundwater (2-3 g/L) occupies one half in Hebei Plain (Table 1)(Hebei Academy 1990).

Table 1. Resources and exploitable resources of fresh and slightly saline groundwater in Hebei Plain

region	fresh water (2 g/L)			salghtly saline water (2-3 g/L)		
	area (km^2)	resources (10^{8}m^3/a)	exp. res. (10^{8}m^3/a)	area (km^2)	resources (10^{8}m^3/a)	exp. res. (10^{8}m^3/a)
Hebei Plain in which:	49486.6	90.45	91.68	14118.1	21.04	15.36
Helonggang region	7244.2	11.45	8.95	5616.4	7.90	5.38
east of south Great Canal	1821.4	2.8	2.03	2042.7	3.1	2.00

In Nanpi Pilot Area, the Quaternary period stratum is located on the structure unit of Huanghua depression area, it is controlled with different degrees by bed rock structure from Lower Pleistocene (Q_1) to Recent pleistocene (Q_4). There are four aquifers, their hydrogeological characteristics are listed on table 2. The saline water was spread over in a depth about 100 m underground and most of shallow groundwater was saline water. The percentage of area of different qualities of shallow groundwater were: fresh water (<2 g/L) 18%, slightly saline water (2-3g/L) 33%, semi saline water (3-5 g/L) 24%, and saline water (>5 g/L) 25%. There was all saline water bellow the depth of 20 m. The highest salty occurred in 32-38 m, it may be up to 30 g/L. Its major chemical composition is Cl.SO-Na.Mg. The deep sodic fresh water pertains to HCO_3.Cl-Na or HCO_3-Na type. There was a grey or grey-black mud deposit as a weak permeable layer at a depth of 7-18 m, its coefficient of permeability was 1.5×10^{-3}-2.8×10^{-5} m^3/day. The shallow groundwater aquifer was a thin layer, its thickness about 0.8-3 m, with a coefficient of permeability 4.9 m/day, and specific yield 0.04.

Table 2. Aquifer characteristics of Quaternary system

aquifers	depth of top and bottom (m)	thickness of aquifer (m)	water qualities (m)	output of single well (m³/h.m)
Q₄	bottom 36	0.8-3	tf,ss,sw	2-6
Q₃	36-160, upper portion-saline water, then fresh water underneath, but the later is unvaluable to exploit alone.			
Q₂	160-360	40-82	af	6-11, max.17
Q₁	360-500	21-32	af	3.4-5.6

Note: tf - thin layer fresh water; ss - sligtltly saline water; sw - saline water; af - sodic fresh water

USING SALINE WATER FOR IRRIGATION

When the crops are irrigated with saline water in dry season, the soil moisture increases and the concentration and osmotic pressure of soil solution decreases, which is beneficial to the moisture and nutrient absorption by crops. During the seedling stage of wheat, soil evaporation is strong and the soil solution becomes concentrated (it may increase from 6 g /L to 14 g/L). However, after the seedling stage, when the wheat is irrigated with saline water, the concentration of soil solution may fall to 6-10 g/L (Chen Xiuling et al. 1988). According to experimental results, the physiological tolerance of soil solution should not exceed 10 g/L in seedling stage and 20 g/L in jointing stage(Beijing University of Agriculture 1976). Because the concentration of soil solution is kept below their limit, the wheat can grow well. Although the salt content of non-irrigated soil is lower, its moisture content is lower too, and therefore the concentration of soil solution is higher (30-80 g/L). This makes it difficult for the crops to absorb moisture and results in damage by drought. According to the research in Nanpi Pilot Area during the 1980-1989 decade.,the yield of double cropping of wheat and summer corn irrigated with saline water of 4-6 g/L and 2-4 g/L were 6961.5 kg/ha and 8355 kg/ha which were an increase of 1.2 and 1.6 times respectively as compared with non-irrigation (Table 3).

Table 3. Crops yields irrigated with various quality of water

crops	<1g/L kg/ha	2-4g/L kg/ha	4-6g/L kg/ha	non-irrigated kg/ha
(1) wheat	4848.0	3630.0	2925.0	840.0
(2) summer corn	5542.2	4725.0	4036.5	2340.0
(1)+(2)	10390.5	8355.0	6961.5	3180.0
(3) spring corn	5883.0	5332.5	4797.0	4882.5
(4) soybean	1704.0	1252.5	960.0	915.0

Fang Sheng   Chen Xiuling

The alternative use of saline water and fresh water, according to the salt tolerance in different growth stage, allows to optimise the role and benefits of saline water and fresh water respectively. Saline water is used when irrigating salt tolerance crops in the rotation or when irrigating a salt sensitive crop during a salt tolerance growth stage. The fresh water is used at all other times. Whatever salt build-up occurs in the soil from irrigating with saline water, is leached in a subsequnt cropping period when fresh water is applied.(Hoffman et al. 1990). In Nanpi Pilot Area, fresh water (<1 g/L) was used during the seedling stage and saline water (5-6 g/L) was used after the jointing stage, which achieved good harvest. The yield of wheat was 4549.5 kg/ha, only 2.2% less than when irrigating with freshwater (Table 4).

Table 4. Experimental results of cyclic irrigation with saline and fresh water in Nanpi Pilot Area (1989 - 1990)

growth stages				yield of wheat	
before seeding	turn green	jointing	booting	yield (kg/ha)	relative yield(%)
fresh water	fresh water	fresh water	fresh water	4650	100
fresh water	fresh water	saline water	saline water	4550	97.8
saline water	saline water	saline water	saline water	3660	73.7
non-irrigated				2258	48.4

The blending of saline water with sodic fresh water decreases the salinity and sodicity due to the mutual dilution of the two types water, and decreases the residual sodium carbonate (RSC) by the chemical combination of ions in the two types water (Table 5). When the sodic water which contains more Na^+ is blended with saline water which contains more Ca^{++}, Mg^{++}, the chemical combination of CO_3^{--}, HCO_3^- and Ca^{++}. Mg^{++} forms harmless salts ($CaCO_3$, $MgCO_3$ $Ca(HCO_3)_2$ and $Mg(HCO_3)_2$. Also, Na^+ is chemical combined with Cl^-, SO_4^{--} to form $NaCl$ and $NaSO_4$ which are far less harmfull than Na_2CO_3 and $NaHCO_3$ and are easy leached by the concentrated rainfall or by irrigating with fresh water. In Nanpi Pilot Area, double cropped wheat and corn was irrigated with shallow saline groundwater (5-6 g/L) blended with deep sodic fresh water (<1g/L, pH=8.5). The average yield during 1980-1989 was 8355 kg/ha, which was an increase of respectively 163% and 20% over non-irrigated and irrigated with saline water of 4-6 g/L (Table 3).

Table 5. Chemical properties of saline water, sodic water and mixture in Nanpi Pilot Area (October 10,1982)

types	pH	mineral -ization (g/L)	total- irons meq/L	C+ HC meq/L	RSC meq/L	SSP	N/ C+M	SAR	L+S meq/L
sedic water	8.45	0.88	12.85	6.44	5.34	31.44	10.68	15.84	6.21
mixture	7.73	2.46	35.69	3.49	-10.71	67.81	1.72	9.16	30.21
saiine water	7.25	6.95	115.2	5.85	-49.55	51.93	1.08	11.37	87.18

Note: C+HC is CO_3^{--} + HCO_3^-, N/C+M is Na^+/Ca^{++} + Mg^{++}, L+S is Cl^- + SO_4^{--}

SOIL SALINITY CONTROL IN ROOT ZONE AND CROPS SALT TOLERANCE

Using saline water for irrigation differs from irrigation with fresh water, because it not only should meet the requirement of moisture of the crops but also control the damage by salt. The principles of salinity control for irrigation with saline water are:

1. The salt accumulated in the soil should not exceed the crops salt tolerance limits.

2. The salt added to the soil by irrigating with saline water should be leached by rain or irrigation water so that the long term balance of soil salinity is maintained.

3. The salt accumulation should not occurred in root zone of soil. The salt balance relationship can be expressed as follow (Chen Xiuling et al. 1988):

So+Si-Sd≤Tc

where So= original salt content in the soil before irrigation
Si= addition of salt to the soil due to irrigation with saline water
Sd= salt leached from soil by rain water or irrigation water
Tc= threshold value of crops salt tolerance

Research results indicate that soil salinity does not reduce crop yield measurably until a threshold level is exceeded. Beyond the threshold yield decrease approximately linearly as salinity increases. The salt tolerance index of crops is expressed as the ratio in yield under saline and nonsaline condition. For the Nanpi Pilot Area where saline water is used for irrigation (taking the yields of non-saline soil irrigated with fresh water as 100) the relation between the salinity of root zone (0-40cm) and the relative yield of crops can be expressed by the following equation (see also Figure 1):

Ywheat = 100-25.038(C-1.20)
Ysummer corn= 100-16.232(C-1.38)
Yspring corn = 100-11.87(C-1.72)
Ybarley = 100-11.975(C-2.02)
Ysoybean = 100-37.722(C-1.80)
Ycotton = 100-10.3(C-4.2)

Where Y= relative crop yield

C= electric conductivity (mS/m) of saturation extract of root zone (0-40cm) in the growth period of crops

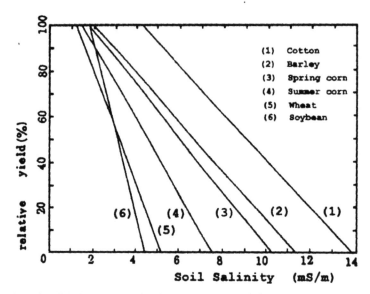

Figure 1. Relationship between soil salinity and crops yield

In Nanpi Pilot Area the regime of the soil salinity in root zone (0-40cm) is such that long term salt accumulation does not occur and that the soil salinity remains controlled within the limits of crop salt tolerance (Figure 2).

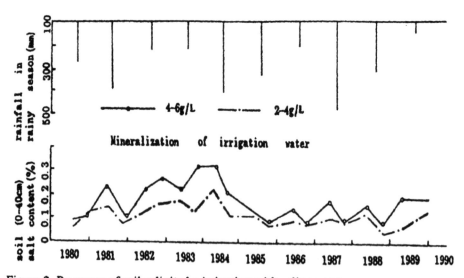

Figure 2. Processes of soil salinity by irrigation with saline water

RECLAMATION OF SHALLOW SALINE GROUNDWATER

The use of shallow saline groundwater for irrigation in the region with saline water can regulate the groundwater depth, reduce the evaporation of phreatic water, increase the rainfall infiltration, reduce the loss of runoff and strengthen the functions of waterlogging prevention and salt leaching, thus overall promoting the reclamation of saline-alkali land and the freshening of groundwater quality. In the Nanpi Pilot Area, the shallow saline groundwater has been exploited and utilized for irrigation, using well irrigation and drainage combined with canal irrigation and ditch drainage. The saline-alkali land reduced to 77 ha in 1983 from 931 ha in 1974, a reduction by 92%. Applying these result to Nanpi Agricultural Development Programme (IFAD programme of 13333 ha),the area of saline alkali land reduced 57% from 1983 to 1987. The exploitation of shallow groundwater has thoroughly changed the condition of groundwater depth. The groundwater depth was 1-2 m in 1970's but in 1980's it had lowered below 3-5 m. In the Agricultural Development Programme Area, the irrigated area increased to 4614 ha (1987) from 2000 ha (1982), the amount of exploited groundwater reached 19.58 million m^3 from 6 million m^3, in 1986-1988, the groundwater depth was 3.3-3.7 m in March, 4.5-5.0 m in June before and 1.1-2.95 m after rainy season. Control of the groundwater at this regime created the good condition for rain water storage and supplementation of fresh water. From 1974 to 1977 there were wet years with average precipitation of 750 mm and the groundwater depth was too shallow to store rain water. The precipitation of the July 23, 1977 rain storm, including some preceding rain was only 156mm, but since the groundwater depth merely 1.1 m in Nanpi Pilot Area, there occurred waterlogging and runoff (47 mm) and much fresh water drained off. The rain was 236.8 mm in August 26, 1987, but the groundwater depth was 4.6 m, all the rainwater infiltrated into subground and no runoff and waterlogging appeared (Table 6). This helped the salt leaching and water quality freshening. The net fresh groundwater resources increase was 84,000 m^3/km^2. From 1974 to 1988, the groundwater had freshened significantly. The area with fresh shallow groundwater (2 g/L) increased from 20% to 50% in the pilot area. In some areas, the originally saline groundwater has become fresh water from surface to 8 m depth (Figure 3). The fresh water body from depth 4 to 8 m increased 639,000 m^3 in the monitoring area and the modulus of exploitable amount of fresh groundwater has increased 15,000 m^3/km^2 /a.

Table 6. Influence of water table depth on infiltration and runoff of rainfall

observed section	area (km²)	date	H (m)	P (mm)	Pa (mm)	P+Pa (mm)	R (mm)	ground water depth 2 days after the rain (m)
south branch ditch	4.51	23/7/1977	1.10	102.0	54.4	156.4	47.2	0.01
south branch ditch	4.53	5/8/1977	2.05	123.0	121.0	244.0	44.7	0.15
north branch ditch	15.05	29/7/1975	2.36	186.3	18.2	194.5	24.5	0.65
south branch ditch	4.51	8/3/1987	3.20	149.0	12.0	171.0	4.0	1.73
north branch ditch	15.05	26/8/1987	4.62	189.1	47.7	236.8	0	1.92

note: H= groundwater depth, P= individual rain storm, Pa= preceding rainfall, R= depth of runoff

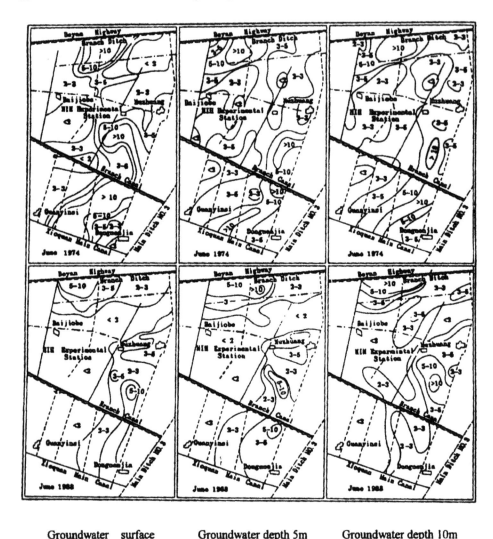

Groundwater   surface   Groundwater depth 5m   Groundwater depth 10m

Figure 3. Map of mineralization (g/L) of groundwater in monitoring area in Nanpi Pilot Area

CONCLUSIONS

Using shallow aline groundwater for irrigation can help to increase yields by providing some key Waterings for wheat which is the crop that requires most water in dry season. The seedling of crops may reach full stand by irrigating before seeding in dry spring. Therefore it plays an important role in overcoming the serious damage to agriculture by drought and fresh water shortage and in ensuring sustainable yield increases. Exploiting and using shallow groundwater, by the use of slightly saline water for irrigation, enables to lower the groundwater depth and to use the shallow groundwater aquifer as a underground water reservoir for regulating rainfall, surface water, groundwater and soil water and to transform the variable nature rainfall into a more reliable water resource for the comprehensive cnntrol of drought, waterlogging, soil salinity and saline water. That is the basic principle for fully regulating and utilizing the local water resources to resolving the contradiction between the nonuniform distribution of precipitation and water requirement of crops in growth period. There are 6 billion m^3 of exploitable resources of saline groundwater in North China Plain (Wenren Xuexing 1992), which could be further exploited and utilized for irrigation together with the soil and water potential raise the yields of food and cotton production to a new level and improving the eco-environment.

REFERENCES

1. Fang Sheng and Sun Xuefeng   Effective Approaches for Relaxing Crisis of Groundwater Proceedings of the Workshop on Water Management: India's Groundwater Challenge   Ahmedabad, 1993

2. Fang Sheng and Hans. W. Wolter   The "Four Water" Concept in China GRID   Issue 2, 1993

3. Hebei Academy of Investigation and Design for Hydrogeology and Engineering Geology Ministry of Geology and Mineral Resources   Hebei General Station of Hydrology: Report on Groundwater Resources Assessment for Hebei Province, 1990

4. Chen Xiuling, Guo Yonchen and Song Wen   Management of Irrigation with Saline Water   Proceedings of International Symposium on Evaluation of Irrigation Systems and Water Management   Wuhan, 1988

5. Chen Xiuling, Guo Yongchen and Song Wen   Water-salt Dynamics of Soil and Crop Yield by Irrigating with Saline Water   Proceeding of International Symposium on Reclamation of Salt-affected Soils   Jinan, 1985

6. Maas E.V. and G.J. Hoffman   Crops Salt Tolerance-Evalution of Current Data Proceedings of International Conference on Salinity   Texas Technology University   Lubbock, 1976

7. Hoffman G.J., J.D. Rhoades, J. Letey and Fang Sheng   Salinity Management Chapter 18 of "Management of Farm Irrigation Systems"   Published by ASAE, 1990

8. Wenren Xuexing   Present Status and Existing Problems of Utilization of Groundwater Resources in North China   Proceedings of Research on China Water Resources   China Publishing House of Science and Technology, 1992

Proc. 30th Int'l Geol. Congr., Vol.22, pp. 80-88
Fei Jin and N.C. Krothe (Eds.)
© VSP 1997

Hydrogeochemical Characteristics and Genetic Analysis of Dongying Basin, China

YANG TIANXIAO
Beijing Research Institute of Uranium Geology, Beijing 100029, P.R.China
WANG MIN, CHEN LIANG
China University of Geosciences, Beijing 100083, P.R.China

Abstract

Dongying Basin is a very important oil-bearing basin, the main oil-bearing stratum is Tertiary, especially Shahe Group(Es). There are a large volume of hydrochemical data of groundwater of Shahe Group. From Es2 to Es4, the average TDS of groundwater increases gradually (35.6 g/l for Es2; 48.8 g/l for Es3; 52.4g/l for Es4), while the order of the highest TDS of ground water is Es2 (322.3 g/l), Es3 (303.3 g/l) and Es4 (223.4 g/l), being contrary to the paleo-salinity (8.2 g/l for Es2, 24.3 g/l for Es3, 32.0 g/l for Es4). It illustrates that vertical groundwater movement probably exists. The probability cumulative curve of TDS shows three different origins of the groundwater: low TDS water, intermediate TDS water and high TDS water. The three types water show regular variations in plane: the low TDS water distributes at the edge of the basin, the high TDS water distributes in the interior of the basin and around the faults, the intermediate TDS water exists in the middle of the Basin. Combining with the rNa/rCl coefficient, we can find that the low TDS water has high rNa/rCl coefficient (rNa/rCL>0.85) , while the high TDS water has low rNa/rCl coefficient (rNa/rCl<0.85). The history of deposit and evolution of the basin show that Dongying basin has undergone sedimentary hydrogeological action, infiltrated hydrogeological action and buried-confined hydrogeological action. Also combining with the structure of the basin, we can conclude that the low TDS water is from infiltrated precipitation, the high TDS water comes from the deeper aqueous stratum which plays a great important role in petroleum accumulation, while the mainly intermediate TDS water are sedimentary water. The groundwater of Dongying basin has three different genesis, i.e., infiltrated precipitation water, sedimentary water and the high TDS water from the deeper aqueous strata.

Keywords: hydrogeochemical characteristics, genetic analysis, Dongying basin, oilfield water

INTRODUCTION

Dongying basin is a pull-apart offshore continental basin formed in Miocene epoch, locating within the southeast of Jiyang depression which forms the Shengli Oilfield in China (Fig.1), with an area of 5700 km². There are more than 1000 faults among the basin and most of them are normal faults, the movement of the faults is intensive . The strata of Tertiary system (E) is nearly 10,000 meters, being the main sedimentary strata, also the main objective strata of oil-gas exploration (Fig.2). From the bottom to the top, the strata of E can be divided into five groups: Kongdian Group (Ek), Shahe Group (Es), Dongying Group (Ed), Guangtao Group (Ng) and Minghuazhen Group (Nm). The stratum of Shahe Group (Es) is the main oil-bearing stratum, making up 87.6% of the total petroleum reserves of Dongying basin. According to the sedimentary character, from the bottom to the top, Es can be divided into four sections: Es4, Es3, Es2 and Es1. The main lithology of Tertiary system strata are sandstone and mudstone[11].

HYDROGEOCHEMICAL CHARACTERISTICS

Shahe Group (Es) is of great significance to petroleum geology, and the level of exploration is high, the hydrochemical data are abundant, while the material of Ek and Ed is insufficient, so we mainly represent hydrogeochemical characteristics of Es in this paper.

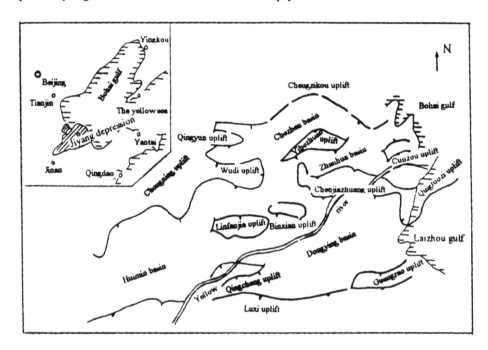

Figure 1. Geographical position map of Dongying basin [11]

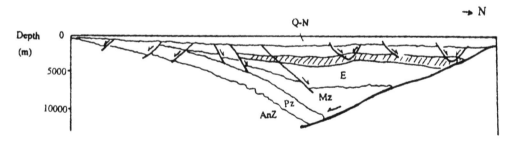

Figure2. Cross section from the south to the north of Dongying basin [10]

Basic Hydrochemical Characteristics

More than one thousand borehole wells were drilled in Dongying basin. Table 1 is the hydrochemical data of groundwater sampling from Tertiary. From the top stratum to the bottom stratum, the hydrochemical characteristics shows regular variations:

1) The TDS of Ng is the lowest. From Es1 to Es2, Es3, Es4, TDS increases gradually, it illustrates that from Ng to Es1, Es2, Es3, Es4, the deeper the strata, the better the confined condition of strata, the more advantageous to keep petroleum.

Table 1. Tertiary hydrochemical characteristics of Dongying basin (unit: g/l)

Strata		Ng	Ed	Es1	Es2	Es3	Es4	Ek
Numbers of sample		50	21	75	195	325	280	4
paleo - salinity				12.6	8.2	24.3	32.0	
TDS	min	0.7	2.3	1.7	1.9	0.6	1.4	6.4
	max	67.8	98.9	76.8	322.3	303.3	223.4	40.8
	ave	10.0	29.8	22.7	35.6	48.8	52.4	25.5
$Na^+ (+K^+)$	min	0.184	0.84	0.46	0.678	0.08	0.13	2.26
	max	20.78	30.27	28.85	113.89	94.59	66.42	14.18
	ave	3.12	9.87	7.40	12.53	16.33	16.72	85.77
Ca^{2+}	min	0.01	0.02	0.01	0.01	0.01	0.02	0.17
	max	3.85	6.10	10.76	24.50	24.65	24.84	1.74
	ave	0.45	12.52	0.80	14.40	21.81	23.41	1.10
Mg^{2+}	min	0.002	0.003	0	0	0	0.003	0
	max	3.01	1.22	1.85	1.56	4.17	3.35	0.25
	ave	0.25	0.23	0.25	0.26	0.31	0.39	0.14
Cl^-	min	0.05	1.16	0.46	0.59	0.02	0.25	3.69
	max	38.13	60.64	41.13	196.91	186.96	137.62	24.98
	ave	5.70	17.66	12.82	21.76	29.33	30.25	15.36
SO_4^{2-}	min	0	0	0	0	0	0	0
	max	5.23	1.08	0.58	1.30	10.11	5.38	0.84
	ave	0.25	0.11	0.055	0.06	0.12	0.21	0.21
HCO_3^-	min	0	0	0	0.06	0.03	0	0
	max	0.95	5.27	6.14	8.98	9.97	52.40	0
	ave	0.3	0.56	0.73	0.81	0.93	1.00	0
CO_3^{2-}	min	0	0	0	0	0	0	0.03
	max	0.89	0.69	0.63	0.41	0.89	1.57	0.27
	ave	0.06	0.10	0.03	0.02	0.03	0.05	0.15

2) From Es2 to Es3, Es4, the average TDS of groundwater increases gradually, while the maximum TDS of groundwater is quite different: the highest is Es2 (322.3g/l), then Es3 (303.3 g/l) and Es4 (223.4g/l), which are contrary to the paleo-salinity (8.2 g/l for Es2, 24.3 g/l for Es3 and 32.0 g/l for Es4). So we can conclude that vertical groundwater movement from the deeper stratum to the upper stratum probably exist and faults are the pathway. According to the petroleum geological character of Dongying basin, Es4 is the oil-generating strata, Es3 is an important oil-generating stratum as well as oil reservoir, Es2 is the oil reservoir, its oil comes from Es3 and Es4. As we know, petroleum migration and accumulation is caused by groundwater movement, it also demonstrates that the vertical groundwater movement exists and faults are the pathway, which plays an important role in petroleum accumulation.

3) According to Shukalev, the water type is Cl-Na. According to Cylin, the dominant water type is $CaCl_2$, $NaHCO_3$ type water exists only in partial area. The content of Cl^- is the highest in the groundwater, the content of HCO_3^- is higher than that of SO_4^{2-} , the content of SO_4^{2-} is almost 0.

4) From Ed to Es1, Es2, Es3, Es4, the content of HCO_3^- increases gradually, it illustrates that from Es1 to Es2, Es3, Es4, the strata are more and more advantageous to keep petroleum.

Characteristics of TDS

TDS of groundwater is the ultimate result of evolution of groundwater chemistry over a long time, it can reflect the regional characteristics of hydrogeochemistry.

Statistical study is made to groundwater chemical data of each layer. Figure 3 is the TDS histograms and probability cumulative curves of TDS of Shahe Group. From the probability cumulative curves of TDS, we can find that the TDS shows normal distribution with three different slopes, which illustrates that the groundwater chemistry has three different origins. The groundwater can be divided into three groups by TDS. The first group is lowly mineralized water (line I , with TDS< 20g/l for Es2, Es3 and Es4, TDS<15 g/l for Es1), the second group is moderately mineralized water (line II , with TDS=40-80g/l for Es2, Es3 and Es4, TDS= 20-45 g/l for Es1), the third group is highly mineralized water (line III, with TDS>120 g/l for Es2, Es3 and Es4, it is not obvious for Es1). Between the first group and the second group, there is an intermediate zone (line I$^'$, with TDS =20-40 g/l for Es2, Es3 and Es4, TDS=15-20 g/l for Es1), which is probably mixed water. Same as the second group and the third group (line II$^'$, with TDS= 80-120 g/l).

Figure 4 is the TDS contour map in plane(Es3 as the example). From the map , we can see that three types water show regular variations in plane. The lowly mineralized water distributes at the edge of Dongying basin, Es1 and Es2 have the largest distribution area, then Es3, Es4 has the smallest distribution area. The moderately mineralized water exists in the middle of the basin and the distribution area of Es4 is the largest. The highly mineralized water distributes in the interior of the basin and around the faults, but the ratio of highly mineralized water is not high (only about 5 %), for Es4, it distributes around Xianhezhuang and Wangjiagang, for Es3, it distributes between the east of Xianhezhuang and the south of Shengtuo, for Es2 it mainly distributes to the south of Shengtuo. In general, the intermediate zone where two different origins water mix is advantageous to petroleum accumulation.

Characteristics of rNa/rCl Coefficient

rNa/rCl coefficient is called genetic coefficient, the hydeogeochemical parameter that reflects the enrichment level of Na^+ . The rNa/rCl coefficient of standard sea water is 0.85.

During the geological event, if Na^+ of the marine sedimentary water exchanges with Ca^{2+} of the stratum, then the content of Na^+ of groundwater will decrease, rNa/rCl coefficient will be less than 0.85. If the marine sedimentary water receives infiltration of meteoric water, rNa/rCl coefficient will be larger than 0.85. Figure 5 shows the contour map of rNa/rCl coefficient (Es3 as the example).

GENETIC ANALYSIS

The genesis of groundwater can be preliminarily determined by rNa/rCl coefficient. When the marine sedimentary water is leached by meteoric water, groundwater with rNa/rCl cefficient larger than 0.85 is formed . Groundwater with rNa/rCl cofficient less than 0.85 has the marine sedimentary feature. For Dongying basin, the groundwater with rNa/rCl coefficient larger than 0.85 distributes at the entrance of the basin and from Es4 to Es3, Es2 and Es1, the distribution area increases.

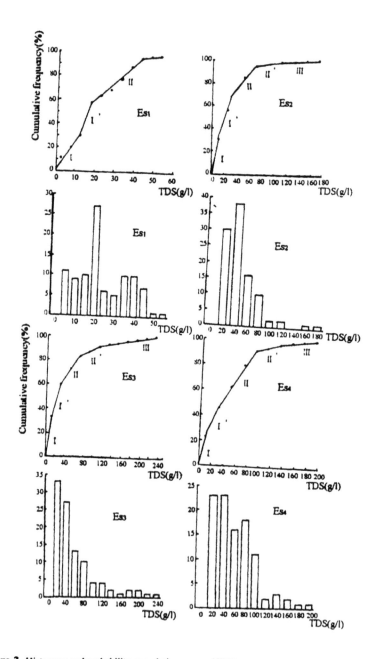

Figure 3. Histogram and probability cumulative curve of TDS

The distribution character of rNa/rCl coefficient is the reflection of hydrogeological condition and hydrogeological action of Dongying basin. During the geological event, Shahe Group of Dongying basin has undergone three hydrogeological actions: sedimentary hydrogeological action, infiltrated hydrogeological action and buried-confined hydrogeological action. Figure 6 shows the hydrogeoloical evolution pattern of each stratum of Dongying basin.

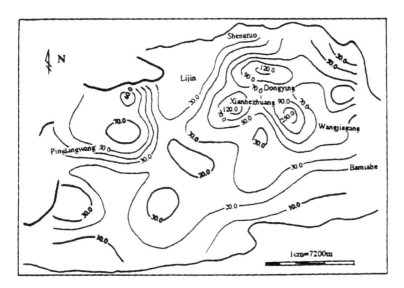

Figure 4. TDS contour map of Es3

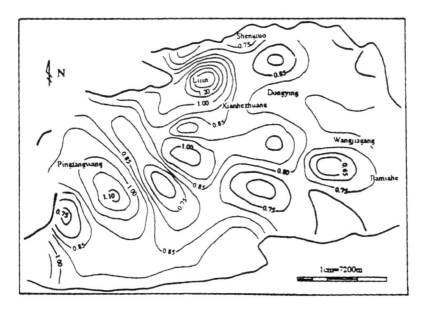

Figure 5. rNa/rCl coefficient contour map of Es3

Sedimentary hydrogeological action. During this period, sediment and sedimentary water formed. The groundwater chemical composition depended on lithologic character of the sediment, the water salinity and ecological features when sedimentation took place. During this period, Es4 had undergone several marine invasions, the sediment and the sedimentary water had the feature of marine facies, rNa/rCl coefficient was about 0.85.

Infiltrated hydrogeological action. During this period, the strata partially or full exposed to Earth's surface to be denuded of, leading to infiltrating of meteoric water and change of hydrogeochemical condition, rNa/rCl coefficient became larger than 0.85.

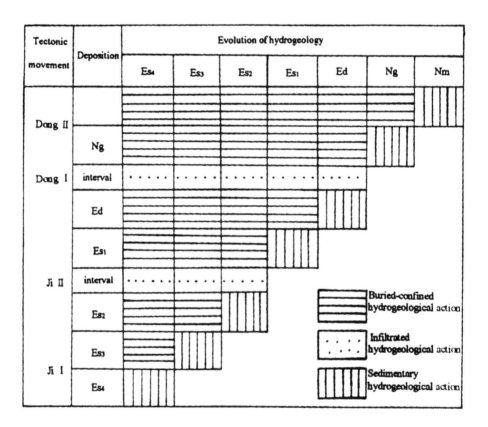

Figure 6. Hydrogeological evolution pattern of each stratum of Dongying basin; symbols: Dong II : Dongying movement stage II ; Dong I : Dongying movement stage I ; Ji II : Jiyang movement stage II ; Ji I : Jiyang movement stage I [5]

Buried-confined hydrogeological action. During this period, the aquifer stratum was covered by new sediment, various metamorphic process and water-rock interaction happened to the groundwater, rNa/rCl coefficient got smaller.

During the evolution process of Dongying basin, tectonic movements were the immediate factor that changed the regional hydrogeological condition of the basin. Since Eocene epoch, Dongying basin had undergone three tectonic movements, and each one had different effect on the basin.

1) *Jiyang movement stage I* at the end of Eocene epoch led to depositional break between Es4 and Es3, resulting in pseudo-concordance between these two strata.

2) *Jiyang movement stage II* at the end of Es2 resulted in the depositonal break between Es2 and Es1.

3) *Dongying movement* at the end of Oligocene epoch involved the whole Jiyang depression and even Bohai gulf area, lasting for 10 million years.

Jiyang movement stage I didn't have strong effect on Dongying basin, only the edge of the basin (e.g. the south edge of the basin), Pingfangwang and the east of Shanjiashi received infiltration of meteoric water. At the end of Es4, earth crust went up, the basin accepted fluvial deposit, forming typical deltaic deposit sequence of Es3. Es3 is not completely covered by sediment of Es2, the stratum of the edge of basin between Es3 and Es2 is discordant. After Paleogene system, the edge stratum of Es3 of the edge of the basin was still exposed to Earth's surface, receiving infiltration of meteoric water.

Jiyang movement stage II resulted in the obvious interval denudation. Chengjiazhuang uplift, Binxian uplift and the south edge of the basin were middle -high mountain, the meteoric water was infiltrated through the edge of paleo-land.

Dongying movement resulted in Ed and Es1 to receive infiltration on a large scale through Chengjiazhuang uplift, Binxian uplift and Qincheng uplift, ect.

Dongying movement produced a large number of faults, linking up geostatic pressure, causing higher-pressured salt water or brine of Es4 to move vertically through faults to lower-pressured Es3, Es2, Es1, even Ed, finally resulted in the mixture between deeper higher TDS groundwater and upper lower TDS groundwater, making the groundwater of upper strata salt. Both the average TDS and the maximum TDS of Es1 is greatly different from the other strata, which illustrates that the deeper highly mineralized water didn't have or weekly mixed with Es1.

CONCLUSIONS

The probability cumulative curves of TDS of Es show normal distribution and three different slopes, which means that the oilfield water has three different origins. Combining with the distribution of rNa/rCl coefficient and hydrogeological evolution of the basin, we can conclude the genesis of the groundwater chemistry of each strata:

1) infiltrated precipitation water from the edge of the basin (lowly mineralized water with TDS<20 g/l);

2) sedimentary water (intermediate mineralized water with TDS=40-80 g/l which is dominant);

3) high TDS water from the deeper aqueous stratum through faults (highly mineralized water with TDS>120 g/l).

Between the interface of two types water, there formed mixed water (with TDS=20-40 g/l and TDS=80-120 g/l, water with TDS=20-40 g/l is mixed by infiltrated precipitation water and sedimentary water, water with TDS=80-120 g/l is mixed by deeper high TDS water and sedimentary water).

Paleo-lake-basin is the main space for petroleum generating, so the oilfield water chemical composition is closely related to the formation and development of paleo-lake-basin, especially the paleo-salinity. During Tertiary deposit, water quality of Es4 is salt water, that of Es3 and Es1 are brackish water and Es2 is fresh water (Tab.1). Different initial conditions of sedimentary water resulted in different chemical characteristics of oilfield water, so , TDS of Es4 is higher than that of Es3, Es2 and Es1. Because of connection of extension faults caused by Dongying movement, the oilfield water was mixed, causing the groundwater of Es3, Es2 and Es1 to become salter.

In conclusion, from Eocene epoch, three tectonic movements undergone by Dongying basin have produced a great impact on the forming of groundwater, evolution and genesis of groundwater chemistry, especially Dongying movement has played a decisive role in groundwater chemical genesis of Tertiary system of Dongying basin.

REFERENCES

1. M.K. Dahlberg. Application of Hydrodynamics to Oil Exploration. *7th Word Petroleum Cong., Proc.,* Mexica City, v.18 (1967).
2. Georg matthess. The Properties of Groundwater (1981).
3 A.G. Kelynce. Oilfield Water Geochemistry. *Petroleum Industry Press,* Beijing (1984).
4. Shen Zhaoli and Wang Yingpu. Several Problems in Oilfield Hydrogeochemistry Study. *Geology Press,* Beijing (1985).
5. Yang Xucheng. Study of Paleohydrogeological Condition of Dongying Basin. *Joural of East China University of Petroleum,* vol.3 (1985).
6. J. Toth. Models of Subsurrface Hydrogeology of Sedimentary Basins: Application to Exploration and Expoitation (1986).
7. Shen Zhaoli. Basic Hydrogeochemistry. *Geology Press,* Beijing (1986).
8. Hydrogeology and Engineering Geology Institute. Paleohydrogeology and Hydrogeochemistry-Jizhong Depression as the Example. *Science Press,* Beijing (1987).
9. John D. HEM. Water-Supply Paper 2254. Study and Interpretation of the Chemical Characteristics of Natural Water, *USGS* (1989).
10. Ren Fasheng. Oilfield Water Chemical Characteristics of Jiyang Depression and Its Relationship with Petroleum Accumulation. *Institute of Geology, Bureau of Shengli Oilfield* (1989).
11. Wang Benghai, Qian Kai et al.. Geological Study and Exploration Practice of Shengli Oilfield Area. *China University of Petroleum Press,* Shandong (1992).

Proc. 30th Int'l Geol. Congr., Vol.22, pp. 89-95
Fei Jin and N.C. Krothe (Eds.)
© VSP 1997

Palaeohydrogeology of the Tuhar Basin (North-West China): Two Stages of Groundwater Development and Their Impacts on Distribution of Petroleum

WANG JIANRONG, ZHANG DAJING, WAN LI, TIAN KAIMING
China University of Geosciences,Beijing,100083

Abstract

Tuhar basin is an intermoutain basin. The fluid flow can be divided into two stages according to the genesis and driving force of water. It is found that the fluid flow is composed of convergent flow of compaction water and meteoric water in the first stage. Petroleum is deposited in the convergent zone originally. In the second stage, it is mainly meteoric water flow by gravity after compaction water is limited. The flow direction is from north to south. Remigration of petroleum is also from north to south. The petroleum deposited in the northern part is redistributed, oil density is higher than $0.80g/cm^3$,TDS is between 4 and 18 g/l, Water type is mainly $NaHCO_3$. But petroleum deposited in the southern part is enlarged, oil density is between $0.71 \sim 0.75g/cm^3$,TDS is higer than 16g/l, water type is $CaCl_2$

Keywords:Palaeohydrogeology, Two stages of groundwater development, Petroleum migration, Finite element simulation, Tuhar Basin

INTRODUCTION

Fluid flow is sedimentary basin may results in a wide variety of physical and chemical processes including the genesis of certain mineral deposits, migration and entrapment of petroleum, and so on . Owing probably to the great variety in the possible geologic situations, different basins may have different fluid flow models and a basin may have different fluid flow model in differmt development stage. According to the assumed nature of the principal driving force of water, fluid flow models can be divided into four kinds (J.Toth,1985), namely : 1) compaction, 2) gravity, 3) compaction and gravity, and 4) thermal expansion and osmosis.

After we read the geological and hydrogeological feature of Tuhar basin. We choice the convergent flow of basinward gravity flow and marginward compaction flow (Roberts III, W.H. 1966, 1982,and Toth and Corbet 1983) as the first stage of ground-water flow models. And the gravity-drive flow (Toth 1962,1963,1978,1980) as the second stage of groundwaer flow models. Petroleum deposited oringinally in the first stage will remigrated in the second stage. Some petroleum pool is damaged and some pool is enlarged. And also, these fluid flow models have influence on hydrodynamic and hydrochemical characteristics.

GEOLOGICAL AND HYDROGEOLOGICAL CHARACTERISTICS

Tuhar basin is an intermoutain basin The general topography feature is that the northern part is higher than the southern part, and the easthern part higher than the western part. And the highest point is in the nortern basin. It's mainly a mesozoic basin. Jurassic strata are the major reservoirs. Some oilfield name can be found in Fig.1.

The hydrostatic pressure gradient is 0.98 ~ 1.2 in Kukuya — Chulin oilfield (north basin) and 0.9 ~ 0.98 in Chudong — Wenjisan oilfield (south basin). The oil density is greater than $0.80g/cm^3$ in Kukuya — Chulin oilfield and it'sbetween 0.72 ~ 0.75 g/cm^3 in Chudong — wnjisan oilfield.Table1 represents hydrochemical feature of J_2s strata in different oilfield. Table2 states the total dissolved solids (TDS) and water type in the above oilfied.

Table1. Hdrochemical elements of J_2s strata

Oilfield	CL	HCO$_3$	SO$_4$	TDS	RSO$_4$/RCL
Kukuya-Chulin	1000--9000	<1000	300--500	4000--18000	<0.60
Chudong-wenjisan	5000--17000	1000--11000	500--1000	16000--40000	0.2--1.4

It states there is a great hydrodynamic and hydrochemical difference between Kukuya — Chulin oilfield and Chudong — Wenjisan oilfield. It is formed by groundwater movement. Because there is more meteoric water exchange in the northern basin, it is not a good hydrogeological condition for petroleum accumulation. The petroleum deposited originally is remigrated. But in the southern basin, it is a better place for petroleum accumulation.

Table 2. TDS and water type of groundwater

Oil field	Well	Depth(m)	TDS(mg/l)	Water type
Kukuya -- Chulin	Ku9	1922.10	10980.00	NaHCO$_3$
		1166.90	4119.00	NaHCO$_3$
	Ku7	896.00	117721.00	CaCL$_2$
	Ku13	2563.60	114927.00	NaHCO$_3$
		1901.00	8398.00	NaHCO$_3$
	Ku6	2471.10	4882.00	NaHCO$_3$
		1939.20	5542.00	NaHCO$_3$
	Lin13	2984.20	2778.30	NaHCO$_3$
	Lin4	2788.70	6763.50	NaHCO$_3$
Chudong -- Wenjisan	Wenxi3	2282.50	33775.20	CaCL$_2$
	Wenxi2	2450.40	27527.70	CaCL$_2$
	Wen8	2413.00	48502.00	CaCL$_2$
	Wen9	2571.90	20095.40	CaCL$_2$
	Wen21	2427.60	11906.00	CaCL$_2$
	Wen23	2683.10	23175.20	CaCL$_2$

TWO DEVELOPMENT STAGES OF FLUID FLOW

The different hydrodynamic and hydrochemical characteristics are created by many factors such as fluid flow models,lithofacies, structures and other factors. But we think fluid flow model is the most important factor.

The first stage :   Combined flow of compaction water and meteoric water.

According to the topography and geology of Tuhar basin. We choice the convergent flow of basinward gravity drive and marginward compaction drive as the first stage of fluid flow when compaction water is voluminous. Fig.1 is the plan characteristics of groundwater flow. The recharge area of compaction water is in the sedimentary center. The flow direction is marginward,i.e. from sedimentary center to margin area. Contrarily ,The recharge area of meteoric water is in basin margin. The flow direction is basinward,i.e. from basin margin to basin center. Because Tuhar basin s not a ideal basin, there are sub-depression centers in basin depression area,so it's rather complicated for discharge characteristics. There are four kinds of discharge sites, (1)The first kind is in Kukuya — Wenjisan and in Shenjinko area. They are both sited in north — west direction. It's mainly compaction water discharge of lateral subdepression. (2)The second kind is in the northern margin area.,it is nearly E — W direction. Because the sub-depression center is near the north margin. The discharge volume of compaction water is small, so the corresponding petroleum accmulation volume is also small. (3) The third kind is in the southern margin area, it is also nearly E — W direction , contrarily,the discharge volume of compaction water is great. And correspondingly, the petroleum accumulation volume is also great. (4) The fourth kind is in the easthern and western margin area, it is formulated in local hydrodynamic condition.The compaction water volume and petroleum accumulation volume are both small.

The second stage :   Gravity-drived flow of meteoric water

As metioned earlier, Tubar basin is a mesozoic basin. Jurassic strata are the major reservoirs, compaction water has already limited. It's mainly gravity-drived meteoric water flow after compaction water is limited. The flow direction is generally from north to south because of topography influence. The regional discharge is in the south. But there is some local discharge in fractured area. The flow direction is from deep to shallow locally.

Fluid flow can be strongly affected by fracture permeability (K.W.Larson,et 1991) . In areas where fracture permeability is important, decisions about hydrocarbon exploration made on the basis of models that ignore frature permeability may be erroneous.

Geochemical study shows that most fractures provided paths of petroleum migration in Tuhar basin. It is evident that groundwater can flow cross-formation through fratures. In the A — A' cross-section (Fig.2) simulation of groundwater movement, we consider fracture as a high permeable zone. The permeability of paleozoic group is very small. It can be considered as basement,that is , it's the boundary of finite-element simulation. Fig.3 shows finite element discretization. There are 365 nodes and 306 elements.Mathematical model is as follows :

$$
\begin{cases}
\mu \dfrac{\partial H}{\partial t} = \dfrac{\partial}{\partial x}\left(K_x \dfrac{\partial H}{\partial x}\right) + \dfrac{\partial}{\partial y}\left(K_y \dfrac{\partial H}{\partial y}\right) + Q, t > 0 \\[2mm]
H\big|_{t=0} = H_0, (x,y) \subset \Omega \\[2mm]
H\big|_{\Gamma 1} = H_1, t > 0 \\[2mm]
Kn\dfrac{\partial H}{\partial t} + rH\big|_{\Gamma 2} = \beta, t > 0
\end{cases}
$$

$$\Gamma 1 \cup \Gamma 2 = \Omega, \Gamma 1 \cap \Gamma 2 = \Phi$$

Here μ — porosity

             H — hydraulic head

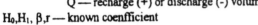

Kx,Ky ,Kn— permeability in x,y and n direction

Q — recharge (+) or discharge (-) volume

H₀,H₁, β,r — known coenfficient

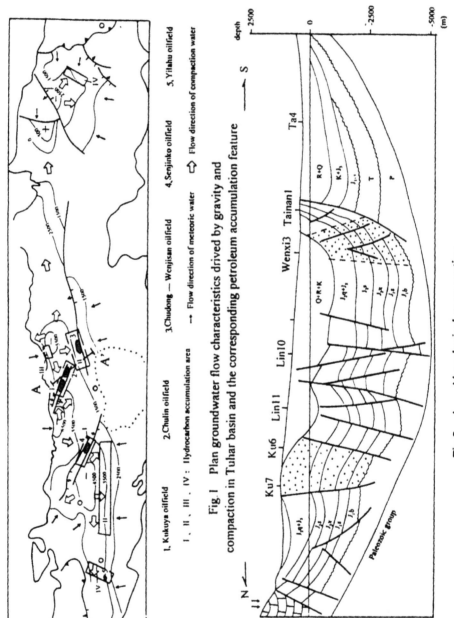

1, Kukuya oilfield   2.Chulin oilfield   3.Chudong — Wenjisan oilfield   4.Senjinko oilfield   5.Yitahu oilfield

I , II , III , IV : Hydrocarbon accumulation area

→ Flow direction of meteoric water   ⇧ Flow direction of compaction water

Fig. 1   Plan groundwater flow characteristics drived by gravity and compaction in Tuhar basin and the corresponding petroleum accumulation feature

Fig. 2   A — A' geological cross-section

The results in Table 3 are the calculated and measured hydraulic head. Besides well Lin 3 and Lin 10, the relative difference between measured hydraulic head and calculated hydraulic head is

Table 3. Hydraulic head of calculated and measured

Well Name	Depth (m)	Measured hydraulic head(m)	Calculated hydraulic head(m)	difference betweem M and C (C-M)/M*100(%)
Lin3	2350.0	855.83	1092.25	27.62
Lin4	2807.5	1050.03	1099.95	4.75
	2120.0	1069.92	1098.79	2.70
Lin10	2930.0	706.79	1099.25	55.45
Lin11	3568.6	1039.96	1099.42	5.72
	3007.2	1027.68	1098.77	6.92
	2686.6	1159.41	1097.27	-5.36
Ku7	2122.8	1181.71	1099.98	-6.92
	1634.5	1241.47	1098.52	-11.51
Ku6	1939.6	1109.26	1098.66	-0.96
Ku10	1397.0	1077.16	1093.31	1.50

generally less than ±10%.So,the finite element simulation results can be accepted. Pressure in well Lin3 and Lin 10 may be formed in sealed system,or it is not the real pressure.

The contour map of calculated hydraulic head is showing in Fig4. Most low pressure area is consistant with petroleum pool. But we still don't find industrial pertroleum in well Ta4 and the sourthern low pressure areas.

This period of water flow has influence on hydrocarbon migration. Petroleum deposited originally in northern basin remigrated to south. Thus,petroleum in northern basin is damaged,but it is enlarged in sourthern basin. Petroleum accumulation characteristics under the hypothesis that in sedimentary basins , the accumulation of hydrocarbons is largely the result of aqeous transport rather than any kind of self-propulsion. According to the paleohydrology of Tuhar basin, hydrocarbon deposited in four kinds of areas (Fig.1) :

I. It is in Kukuya-Wenjisan and in Shenjinko area. The scope is in North-West direction. The North-West part is small, oil density is greater than $0.8g/cm^3$,hydrostatic pressure gradient is between 0.98 and 1.2. TDS is 4 ~ 18g/l. Warter type is $NaHCO_3$. In some sealed system, water type is $CaCl_2$. Kukuya — Chuling oilfield possess partly this feature. But the South — East part is great, oil density is between 0.72 and 0.75g/cm3. Hydrostatic pressure gradient is between 0.92 and 0.98. TDS > 16g/l, water type is $CaCL_2$. Chudong — wenjisan oil field possess this feature.

II. It is in the northern margin area in E — W direction. Because the discharge volume of compaction water is small, the petroleum size is correspondingly small. In the second stage of fluid flow, petroleum remigrated to south part. Therefore this kind of hydrocarbon accumulation not only the size is small, but also the site is near the depression center. The oil density is higer. Pressure gradient is greater than 1.0. TDS <18g/l, water type is $NaHCO_3$.

III. It is in the sourthern area in E — W direction. The discharge volume of compaction water is great, and it is also in the discharge area of the second fluid flow stage. So the petroleum deposit size is not only big but also enlarged.It is a very prosperious area, petroleum and groundwater charactristics is similar to Chudng — wenjisan oilfield.

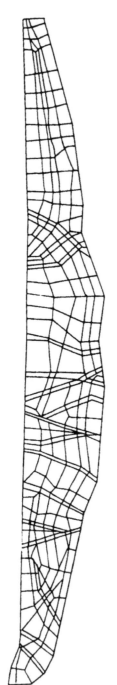

Fig.3   Finite element discretization

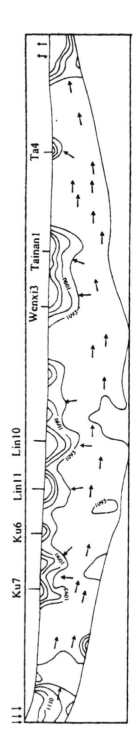

Fig.4   Potentiometric surface simulated by finite element method

IV. It is in the Eastern and Western margin areas. Petroleum deposited under local hydrodynamic condition. The accumulation volume is small, oil density is higer,TDS < 18g/l, water type is NaHCO₃,Yilahu oilfield is belong to this kind.

CONCLUSIONS

It is concluded that there are two major fluid flow stages in Tuhar basin, one is the combined flow of compaction water and meteoric water. Another is gravity drived meteoric water flow. The hydrodynamic and hydrochemical characteristics is highly influenced by groundwater movement. In the southern basin , the pressure gradient is beween 0.92 and 0.98, TDS >16g/l, water type is $CaCL_2$, oil density is between 0.72 and 0.75g/cm³. But in the northern part.There is more meteoric water exchange, the pressure gradient is 0.98 ~ 1.2. TDS <18g/l, Water type is $NaHCO_3$, oil density is greater than 0.80g/cm³. Corresponding to the movement of groundwater, hydrocarbon deposited in four kinds of areas (Fig.1). Each kind of petroluem pool has its own characteristics. The deposit volume, oil density, ressure gradient, TDS and water type are all different. Among those, the southern basin in the most prosperious area of exploration. Two dimension finite element simulation shows that fractures play an important role in the exchange of water and vertical migration of petroleum.

Acknowledgements

We gratefully acknowledge prof. Hu Jianyi, deputy president of Institute Petroleum Exprolation and Development, CNPC. And Senior engineer Zhang Guonai of Institute of Petroleum Geology,MGMR. And prof. Zhou wenzhi for their support of this work. Thanks also to Mr. Li wei for helpful discussion.

REFERENCES

1. Bethke,C.M. 1985. A numerical model of compaction — Driven Groundwater Flow and Heat Transfer and Its Application to the Palaeohydrology of Intracratonic Sedimentary Basin. Journal of Geophysical Research. V.90,No.137.
2. Larson,K.W. Waples,D.W.etc. 1993. Prediting Tectonic Fractures and Fluid flow Through Fractures in Basin Modelling,Basin Modelling:Advances and Applications, Norweigian Petroleum Society. Special Publication No.3.
3. Roberts,W.H.III.1985. Deep Water Discharge : Key to Hydrocarbon and Deposits. Hydrogeology of sedimentary Basin. Alberta.
4. Toth,J.1962. A Theory of Groundwater Motion in Small Drainge Basins in Central Alberta, Canada, Jour.Geoph.Research,V.67,No.11.
5. Toth, J.1980. Cross — Formational Gravity-flow of Groundwater : The Generalized Theory of Petroleum Migration. Problems of Petroleum Migration, W.H.Roberts III and R.J.cordell, Editors. Amer. Assoc. Pet. Geol. Studies in Geol., No.10.
6. Toth,J. 1985. Models of Subsurface Hydrology of Sedimentary Basins, Proceedings Third Canadaian/American Conference on Hydrogeology : Hydrogeology of Sedimentary Basins, Alberta.

Proc. 30th Int'l Geol. Congr., Vol.22, pp. 96-104
Fei Jin and N.C. Krothe (Eds.)
© VSP 1997

Realizing Comprehensive Control of Multiple Objectives for Irrigation Districts by Conjunctive Management of Surface and Groundwater

FANG SHENG   CHEN XIULING
Hebei Institute of Hydrotechnics, 28 Lingyuan Street, Shijiazhuang, P. R. of China

Abstract

There is short of water resources in the northern part of China, carry out conjunctive management of surface and groundwater is the basic approach for rational development and utilization of water resources and comprehensive control of drought, waterlogging, salinity and saline water. For conjunctive management of surface and groundwater, the key lies in regulating the groundwater depth dynamics at a proper level, take the interstratified space of shallow groundwater as the underground reservoir for " four water " [1](rain water, soil water, groundwater and surface water) regulating, to transform the natural rainfall into available water resources in maximum extent, to make the surface runoff fully retained and utilized, to strengthen the salt leaching and water quality freshening by summer rain, then achieved comprehensive control of multiple objectives for irrigation districts. In the well irrigation districts where without fixed canal irrigation water source, retain the excessive rain water in rainy season and diverting river flow after rainy season for recharging groundwater as the water source supplementary. In the river flow irrigation districts where with the condition to exploit groundwater,   should carry out both using wells and canals, optimized regulation of surface and groundwater , to achieve the maximum benefits of water resources.

Keywords: Conjunctive Management, Surface Water, Groundwater, Irrigation Districts, Northern Part of China.

INTRODUCTION

In the arid, semi-arid and semi-humid regions in the northern part of China under the influence of monsoon climate, the distribution of natural rainfall is nonuniform in time and space within one year and between years, and can't meet the proper water demand of crops in their growing period. Without irrigation there will be no agriculture in arid region. In semi-arid, semi-humid regions there will be no stable crop yield without irrigation to make up rainfall deficiency in dry season and the crops always suffered from waterlogging in low-lying land due to concentrated rainfall in rainy season. Affected by drought and waterlogging large area of saline-alkali land and saline water are seen in the region with slow runoff and converged water-salt. Thus. drought. waterlogging, soil salinity and saline groundwater affected one another and caused disasters alternatively, forming the main restraint factors for sustainable development of agriculture. To release the serious affection of drought and water shortage on agriculture, the state has developed irrigation on a large scale.   However, the development of such irrigation project using river flow as Yellow River Diversion Project caused a widespread groundwater table rise and secondary salinization appeared in large area. Since the beginning of 1970's in the northern part of China the groundwater has been expolited and utilized on a large scale to develop well irrigation. As a result, the groundwater table has dropped and the secondary salinization has been controlled in North China Plain. However, as the exploitation amount of groundwater exceeded the replenishment amount, many cone depression of groundwater table formed and the ecological environment was worsened in well irrigation districts. Practice shows that in the northern part where short of water resources, the surface water and groundwater should be taken into consideration as a whole and put them under unified management, well irrigation and well drainage combined with canal irrigation, to optimized regulation of water resources which means making up water deficiency with surplus water. Thus the comprehensive control of drought. waterlogging, salinity and saline water can be realized.

INDEX FOR REGULATING GROUNDWATER DEPTH

The key to realize the comprehensive control of multiple objectives in irrigation districts, lies in regulating the groundwater depth dynamics at a proper level. The groundwater depth is a dynamic process varying constantly, it is the concentrated reflection of the transformed process of growth and decline among atmosphere rainfall. surface water, soil water and groundwater. The groundwater depth should be regulated on proper dynamics for comprehensive control of drought, waterlogging, salinity and saline groundwater in semihumid region, and meet the requirement for drought resistance and salinity control in semi-arid and arid regions. Proper dynamics of groundwater depth in semi-humid region [2] can be represented by the North China Plain in the north of Yellow River. The annual precipitation is 500-700mm in which 70% is concentrates in July and August, the proper dynamics of groundwater depths in different seasons within one year are as follows: 1. In dry season (Oct. to May). the depth of groundwater should be kept on the level for salinity control (2-3 m), so as to reduce the evaporation of phreatic water as much as possible; 2. Before rainy season (June), it should be regulated on the depth for waterlogging prevention and rain water storage (4-6 m), so as to increase the recharge of rainfall infiltration as much as possible and reduce the surface runoff. Thus the waterlogging can be prevented, and the functions of salt leaching and fresfening of groundwater quality by summer rainfall can be strengthened: 3. In rainy season (July to September), it should be kept on depth for wet resistance (0.5-1.0 m), i.e. during certain times the excessive rain water accumulated on surface should be drained off, and the groundwater table raised by rainfall infiltration should dropped below the wet resistant depth.[1] Requirements for regulating groundwater depth dynamics in arid regions can be taken the Hetao irrigation districts as an example. Where annual precipitation is only 190-350mm, using water diverted from Yellow River in Ningxia and Inter Mongolia, the requirements for groundwater table regulation is the process of groundwater depth dynamics with basically stable salt content of soil which is not harmful to crops in the whole year (Table 1), making the root-zone with favorable environment of moisture, fertillity, air and heat.[3]

Table 1. Requirements of regulating groundwater depth dynamics in Hetao irrigation
District in Ningxia and Inter Mongolia

period of irrigation	time	groundwater depth (m)
start of thawing	early March to early April	2.0-2.4
start of spring irrigation	end of April	2.1-2.3
end of 1st watering	1st 10-day of May	1.2-t.5
end of summer irrigation	last 10-tay of July	1.2-1.5
start of winter irrigation	end of Oct.	2.2-2.65
end of winter irrigation	middle 10-day of Nov.	1.3 1.5

FUNCTIONS OF WELL IRRIGATION AND WELL DRAINAGE

1. To against drought by irrigating crops and restraining soil salinization. Drawing
groundwater for irrigation in dry season to overcome the damage by drought timely for
ensuring yield increase. In the mean time it plays an important role in regulation of soil
salt-water dynamics in two aspects: one is irrigation water leached the salt in root zone
of soil which changed the natural process of salt accumulation due to evaporation in dry
season; The other is well irrigation drawdown groundwater table, it enables to restrain
the transportion of salt to surface soil by capillary action.

2. To regulate the groundwater depth. The groundwater table will drop in the process of
pumping water from wells for irrigation functioning as well drainage which enables the
groundwater depth to be on a proper dynamics. The Secondary salinization became
serious after developing canal irrigation before 1963 in Wujiaqiu irrigation district in
Xinjiang in arid region. The groundwater depth rose to 0.5-1.0m from 5-10m. After
1965 the well irrigation was developed in large area the exploited amount of
groundwater account for 65% of the total irrigation water applied, the groundwater
depth gradually dropped to 3-5 m in 1989 (Table 2) the soil salinity reduced to less than
0.2%.[4]

Table 2. Groundwater table dropping by Well irrigation and well drainage Wujiaqiu
irrigation district Xinjiang

	101st regiment	102nd regiment	103rd regiment
wells number	103	129	130
exploitation of groundwater (10^8m)	2.6 (1965-1988)	2.47 (1965-1988)	2.67 (1975-1983)
original groundwater depth (m)	0.8 (1965)	1-2 (1965)	1.5-2.0 (1975)
groundwater depth after well irrigation (m)	7.64 (1985)	3-4 (1988)	3.5-5 (1988)

3. To reduce the evaporation of phreatic water so as to increase the available amount of water resources, and prevent salt accumulation. The available exploitation amount of groundwater in Nanpi Agricultural Development Programme Area was only $434 \times 10^4 m^3/y$ based on the everage groundwater depth in 1984-1986, but it may reach $1259 \times 10^4 m^3/y$ in case of regulating the groundwater depth dynamics on 2.5-5 m, which means that $825 \times 10^4 m^3/y$ water consumed by phreatic evaporation can be transformed into available water resources, i.e. the modulus of exploitable amount will increase $3.5 \times 10^4 m^3/km^2.y$.

4. To increase rainfall infiltration and reduce surface run-off, so that not only prevent waterlogging but also transform precipitation into available water resources. Waterlogging occurred easily in rainy season due to shallower groundwater depth in Nanpi Pilot Area in 1970s. On 23rd July 1977, $p+p_a=156mm$ (individual rain storm + preceding rainfall) and H (groundwater depth) was only 1 m, R (depth of runoff) was 47 mm, and waterlogging occurred. P was 189 mm on 26th August 1987, the groundwater depth was 4.62 m before rainfall, $p+p_a=236.8$ mm, all rainwater infiltratied into subground, no runoff drained out, no waterlogging or subsurface waterlogging occurred. The net increased fresh water for cultivated land, was $5687m^3/ha$[5]. Because the deeper groundwater depth before rainy season and combined with ditch drainage strengthened the functions of salt leaching and groundwater quality freshening by summer rain , the accumulation of salt drained off was 13945 kg/ha in 7 rainy season between 1974-1987, thus promoting the groundwater quality to be freshened. The modulus of exploitation of shallow fresh water increased by 15000 $m^3/km^2.y$.[5]

5. To regulate the exploitation of groundwater resources in multi-year, that is more groundwater can be exploited in dry year, and more replenished in wet year, which means making up the difficiency with the surplus.

RETAIN EXCESSIVE RAIN WATER AND DIVERTING RIVER FLOW TO RECHARGE GROUNDWATER SOURCE

In the northern part of China, there are 298×10^4 wells with a total exploitation amount of groundwater about $641 \times 10^8 m^3$.[6] However, the groundwater resources is not sufficient for meeting the demand of water for industrial and agricultural use. Owing to overdraft in many pure well irrigation districts, the groundwater level dropped contineously and many cone depression of groundwater table formed, even subsidence of ground surface and sea water intrusion were caused. it is necessary to recharge groundwater by retaining excessive rain water and diverting river flow.

The Beijing Institute of hydrotechnics had carried out the experimental research on groundwater regulating and conjunctive use with surface water in the southeast suburb (442.5 km²) Beijing City, since 1978 it has gone on for 18 years. From 1981 in Nangezhuang, Lixian townships, then from 1984 enlarged to all the irrigation districts had turned into using groundwater for irrigation, in the recent 15 years it not only saved water resources more than 1 billion m³ which supplied from surface water reservoir, but also made the most of groundwater reservoir for regulating rainfall, surface runoff and groundwater and comprehensive control of drought, waterlogging and salinity.[7] There was reduced the lost of storm runoff more than 0.2 billion m³, and also reduced a lot of ineffective evaporated consumption, then conserved and enlarged available water resources. From 1980 to 1986, it was a longest dry period in the recent 125 years, its average annual precipitation was only 430 mm. Depend upon using groundwater for irrigation which had met the requirement of water demand for crops. In 1979, owing to there was diverted more river flow for spring irrigation. the groundwater depth was shallower it was only about 2 m before spring irrigation and generally less than 3 m before rainy season. There was a 7 days rainstorm process from 9-15 August, its precipitation was 208.5 mm, there occurred serious waterlogging in farm land, a lot of runoff had lost in Tiantonghe River catchment (279 km²). In 6-12 July, 1994, there occurred another rainstorm process with precipitation 284.3mm, its total amount, intensity and preceding rainfall all larger than that of 1979. Owing to the groundwater depth had regulated at 7-13 m before spring irrigation and 8-15 m before rainy season, the maximum discharge of runoff and the total amount of runoff correspond to 2.4% and 0.1% of 1979 (Table 3), basically non runoff appreared in farm land, and non waterlogging disaster, a lot of runoff transformed into subground for recharging groundwater. The groundwater depth recovered to 6-12 m before spring irrigation in 1995, and rose to 5-11 m before spring irrigation in 1996, it was 2 m higher than that the corresponding period of 1994. The saline-alkali land was occupied 25% and 30%, in the total cultivated land of Lixian and Nangezhuang townships in 1979, after all developed well irrigation, the groundwater table had dropped gradually, almost non phreatic evaporation, increased rainfall infiltration enhanced the leaching of salt, so the saline-alkali land thoroughly reclamated. The grain yield increased 3 folds from 1977 to 1994.

Table 3. Comparation of storm runoff in Tiantonghe River catchment Daxing County (1979, 1984, 1994)

terms	9/8-15/8 1979	6/8-12/8 1984	6/7-12/7 1994	1984/ 1979 (%)	1994/ 1997 (%)
7 days max. rainfall (mm)	208.5	259.1	284.3	124.3	136.4
3 days max. rainfall(mm)	121.5	234.9	178.8	193.3	147.2
1 day max. rainfall(mm)	92.9	141.4	145.9	152.2	157.1
15 days pre. rainfall(mm)	7.7	6.8	61.9	477.9	804.9
max. discharge (m³/s)	148.0	7.35	3.5	5.0	2.4
total runoff (10⁴m³)	3611.2	83.6	5.04	2.3	0.1
average runoff depth (mm)	129.4	3.0	0.18	2.3	0.1
coefficient of runoff	o.62	0.01	0.00004		

In the eastern part of Haihe River plain where is short of surface water, the development of irrigation mainly depends upon exploiting shallow groundwater include slightly saline water. But in the region, there are occassionlly river flow in non irrigation season, if not utilized they could flow to no purpose.. So they should be retained and stored by canals and ditches for recharging groundwater. Longfang city situated in the lower reaches of Haihe river, using rivers, deep canals network, ponds and pump stations to diverted and stored seasonal surface water was $50.24 \times 10^8 m^3$ in 1975-1989, the modulus of water storage was $5.58 \times 10^8 m^3 / km^2$ a. It made the variation range of groundwater table rising to 3.44 m from previous 1.87 m, thus exponding the capability of regulation and storage of phreatic aquifer.[8] In the Nanpi project area mainly irrigated by wells, the groundwater depth generally was on 5-6 m before rainy season which provided the storage capacity for rain water and river flow. In 1990, the precipitation was 511mm in which 360.5 mm was fell in July to September, the river flow came from Zhangweihe River and South Great Canal from July to November, 124 $\times 10^5 m^3$ of water has been diverted through main, branch and field ditches to the Nanpi project area. which recharged the groundwater aquifer, the groundwater depth raising to 2-3 m from 5-6 m.

In order to supplement water source to the eastern part of Haihe River Plain, the Yellow River diversion project has been built from Waishan Shandong province to Linxi Hebei province. It diverts river flow in the non-irrigation season from November to February next year in every winter. Since November 1994 to January 1995 it has begun to diverted Yellow River flow,actual received water amount was 0.4 billion m^3 by Hebei province. the loss of water transfer account for 18%, partial used for winter irrigation. mainly stored in the canals, ditches and ponds and recharging for groundwater.[9]

OPTIMIZED REGULATION OF WATER RESOURCES BY USING BOTH WELLS AND CANALS

In the canal irrigation districts where with the condition to exploit groundwater, the optimized regulation of surface and groundwater can be achieved by using both wells and canals. The people Victory Canal irrigation district was the first large Yellow River diversion irrigation district in the downstream of Yellow River. After 1958 owing to a lot of water diveted for irrigation and without correspondent drainage, till 1961 groundwater depth rose to 1.3-1.5 m from 3-4 m since the starting for irrigation, the secondary salinization developed rapidly. the saline-alkali land from 6,666 ha increased to 18,666 ha. After this the people Victory Canal Management Bureau has controlled the river flow diversion, simultaneously developed well irrigation, carried on well irrigation and well drainage, conjunctive use of surface and groundwater, comprehensive control of drought. waterlogging and salinity. Irrigated by wells in winter and flood season and restricted river flow diversion. The effects of conjunctive use had achieved: 1. well irrigation with suitable water amount timely, realized high yield of crops. 2. well irrigation and well drainage, coordinated with ditch drainage. increased the effects for waterlogging prevention and salinity control. The groundwater table could rose 0.7-0.9 m after once canal irrigation, but it enables to lowered about 1 m after once well irrigation. The groundwater table was controlled below critical depth in great part, the secondary salinization has basically reclamated. 3. Recharging groundwater source, the two third of groundwater exploitation was recharged from canal seepage. The irrigation area increased to 66,666 ha from 40,000 ha when starting for irrigation.[10] People Victory Canal Management Bureau and Xinxiang Institute of Hydrotechnics proposed the regression equation of relationship between groundwater exploited and Yellow River flow diverted for controlling the balance of groundwater level in irrigation district is as following:

$\Delta H = 0.55 - 1.31P$

where ΔH-groundwater table fluctuation

P-ratio of the amount of groundwater to be exploited and total irrigation water

Total irrigation water = groundwater exploited + water diverted from river
+precipitation

When P = 42%, the groundwater table will keep balance.

In Jinghui canal irrigation district Shaanxi province the groundwater depth was 15-30 m in the past, after diverting the Jinghe River flow for irrigation in 1932, the groundwater depth rose by 10-24 m by 1954, the groundwater depth in half area of the irrigation district was less than 3 m, and that in 30% area was over critical depth, and the secondary salinization occurred. At the beginning of 1960's, well irrigation has developed, the area using both well and canal irrigation reached 73,333 ha in 1970's. The annual exploitation amount of groundwater was $1.1 \times 10^8 m^3$ per year, the rising of groundwater level was restrained. When combined with water and salt drainage by ditches the secondary salinization was under control, which is conducive to agricultural production. The adoption of the conjunctive use of surface and groundwater depend on the condition of water supply: 1. Winter irrigation. As the riverflow is sufficient, canal irrigation can be used for salt leaching and fresh water recharge. 2. Spring irrigation. As the groundwater table rose to the peak and salt accumulated is raging, both well and canal irrigation can be used alternatively to prevent salt accumulation. 3. Summer irrigation. As the river flow is deficient in summer and the sand content of river flow exceeds the limit, canal irrigation should be stopped. Well irrigation can be mainly used to lower the groundwater table and vocate a best of groundwater reservoir capacity for waterlogging prevention. Owing to the conjunctive use of surface and groundwater, the comprehensive control of multiple objectives was achieved and the agricultural production increased. The unit grain yield increased to 7,920 kg/ha in 1989 from 1,750 kg/ha in 1949, up three folds. The arable land in the irrigation district accounted for 2.4% of the total in the province, the total grain production accounted for 5.4%, and the total amount of commidity grain accounted for 11% of the total in the province.[11]

OPTIMUM PLAN FOR CONJUNCTIVE US OF SURFACE AND GROUND-WATER

The Hebei Institute of Hydrotechnics has carried on the simulating of calculation of conjunctive use of surface and groundwater by means of the data of 35 years (1955-1989) in the Zhaoquan Pilot Area of Shijing Canal Irrigation District Hebei Province. The mathematical model was composed by an irrigation water demand model and a water amount regulating model in the phreatic aquifer. irrigation with canals or wells only could hardly meet the irrigation water demands of crops, and the comprehensive control of multiple objectives could not realized. The optimum plan of conjunctive use of surface and groundwater (Table 4) can meet the water demands of crops and the comprehensive control of drought, waterlogging and salinity, which can:

1. Enhance the ensurance probability of irrigation water. The number of irrigation times can increase from present 2.2-1.5 to 6-4 times.

2. The groundwater table can be regulated on the proper level with an upper limit of 4 m, and a lower limit of 6 m, within the ideal regulating range of groundwater reservoir, neither soil salinization nor worsen hydrogeological condition for exploitation.

3. Increase the effective use of rainfall. The phreatic evaporation can be reduced by 1,927 m^3/ha per year in nomal year, irrigation water amount can be increased by 1,505- 1,900 m^3/ha per year.

4. Save the amount of surface water diverting. The number of irrigation times under the conjunctive use plan was 4.63 by groundwater and 1.51 by surface water, the ratio is 3:1. The ratio of groundwater exploitation and surface water diverting is 5:3.

5. Relieve the disasters of waterlogging and soil salinity. The groundwater depth was regulated on 0.1-3.29 m for canal irrigation, and 3.22-6.99 m for conjunctive use.

Table 4. Optimum plan for conjunctive use of surface and groundwater in Zhaoquan pilot area

frequency P(%)	5	25	50	75	95	sum 35a.	average
irrigation times							
surface water	0	0	2	2	2	54	1.54
groundwater	4	4	5	5	6	162	4.63
irrigation water use ($10^5 m^3$)	2.74	2.61	4.51	5.20	7.60	173.9	4.97
surface water diverted	0	0	1.39	2.34	3.96	67.5	1.93
groundwater consumption							
exploitation	2.74	2.61	3.29	3.14	4.11	112.6	3.22
phreatic evaporation	1.84	0	0.06	0.10	0.06	11.2	0.32
supplement							
rainfall infiltration	3.59	0.80	1.15	5.17	0	36.3	1.04
canal seepage	0	0	0.42	0.71	1.19	20.4	0.58
irrigation water return	0	0	0	0.15	0.41	58.4	1.70
lateral recharge	0.97	1.71	1.80	1.78	2.00	61.8	1.77
groundwater depth (m)							
maxmum	5.31	6.65	6.99	6.79	7.12	7.12	
minmum	0.76	3.51	3.33	3.23	3.28	0.76	
average	3.37	4.53	4.43	4.46	4.93	4.46	

CONCLUSIONS

In the northern part of China, the Conjunctive management of surface and groundwater is the strategic measure for comprehensive control of multiple objectives-drought resistance, waterlogging prevention, salinity control and saline groundwater freshening and for the rational development and utilization of water resources in irrigation districts. The conjunctive management should be taken the exploitation and utilization of groundwater and developing well irrigation and well drainage as the basis, and diverting surface water as the water source supplementary. Taken the interstratified space of phreatic aquifer as the groundwater reservoir for regulating rain water, surface water, soil water and groundwater, and the groundwater depth on the proper dynamics as the regulation index, to transform the natural rainfall into available water resources to the maximum, to make the surface runoff fully retained and utilized. The surface water and groundwater as a whole, should be put under unified planning, conjuctive management, optimized regulation and rational utilization to buring the benefit of water resources into full play.

REFERENCES

1. Fang Sheng and Hans. W. Wolter   The "four water" concept in China   GRID   Issue 2, 1993

2. Fang Sheng, Chen Xiuling   Inquiry on Indexes for Regulating Water and Salt Dynamics of Soil in Haihe River Plain   Proceedings of The International Symposium on Dynamics of Salt-affected Soils   Nanjing, 1989

3. Tan xioyuan   Inquiry on Regulating Demand of Groundwater Table in Hetao Plain in Ningxia and Inter Mongolia   Select Proceedings of Academic Symposium of Water Conservancy Society Shanxi Province, 1981

4. Foundmental Construct Department of 6th Division of Xinjiang Production and Construction Corps Reclamation of Saline-alkali Land by Vertical Well Drainage in Wujiaqui Irrigation District   Select Proceedings of The Conference for Experience Exchange on Hydrotechnic Measures for Reclamation of Waterlogged and Saline-alkali Land with Moderate and Low Yield in Northern Part Region, 1990

5. Fang Sheng, Chen Xiuling et al.   Study on Comprehensive Control of Saline-alkali Land with Low Yield   Proceedings of 8th Afro-Asian Regional Conference of ICID   Bankok, 1991

6. Chen Meifen   Present Status and Prospect of Development and Utilization of Groundwater in Our Country, Development, Utilization and Management of Groundwater (No. 2)   Publishing House of University of Electronic Science and Technology, 1995

7. Beijing Institute of Hydrotechnics   Study on Conjunctive Use of Surface and Groundwater in Typical Region of Plain in Beijing City, 1994

8. Cui Bingzhong, Wu Zhenying et al.   Diverting River Flow for Recharging is The Effective Approaches for Groundwater Supplement   *Haihe River Water Resources No. 2, 1992*

9. Zhang Xueling   Existing Problems and Countermeasures of Yellow River Diversion Project in Management and Operation   *Hebei Water Resources and Hydropower Engineering No. 1, 1996*

10. Yuan Guongyao. Shan Bingzhong, Cai Lingen   Study on Irrigation by Conjunctive Use of Surface and Groundwater in People Victory Canal Irrigation District   *Proceedings of Scientific Research of Academy of Water Resources and Hydropower No.10, China Water and Power Press, 1982*

11. Zhang Yading   Soil Salinity Control and Its Benefit in Irrigation District   Select Proceedings of Conference on The Experience Exchange for Hydrotechnic Measures for Reclamation of Waterlogged and Saline-alkali Land with Moderate and Low Yield in Northern Part Region, 1990

12. Gao Dianju et al. The Research on The Pattern of The Conjunctive Use of Groundwater and Surface Water   Proceedings of 8th Afro-Asian Regional Conference of ICID   Bankok, 1991

Proc. 30th Int'l Geol. Congr., Vol.22, pp. 105-112
Fei Jin and N.C. Krothe (Eds.)

Groundwater Development for Agriculture in Northern China

JI CHUANMAO

Senior Consulting Center, MGMR , Beijing , 100812 , P.R..China

WANG ZHAOXIN

China Exploration Institute of Hydrogeology and Engineering Geology , MGMR , Beijing ,100081, P.R .China

Abstract

By the end of 1980s' in northern China , total groundwater extraction accounts for 690 × 10⁸ m³/a and agricultural groundwater use is about 70% of total groundwater use. Space distribution of groundwater use is quite uneven. North China plain has the greatest groundwater extraction intensity. Time variation of groundwater use is due to variation of precipitation in dry,normal and wet years. Four genetic types of groundwater regime for irrigated farmland have been identified. It is the reason for us to take main recharge source as criteria for regime type identification that main recharge source is the guarantee for sustainable development of groundwater for agriculture. Positive and negative effects of groundwater development are introduced. Recommendations for further rational groundwater development of various parts of northern China are given as conclusions.

*Keywords: Space distribution , time variation , genetic type , groundwater regime ,
positive and negative effect*

INTRODUCTION

By the end of 1980s' in China , total groundwater extraction accounts for $870 \times 10^8 m^3/a$, in which $720 \times 10^8 m^3/a$ is pumped from water wells . In northern China , total groundwater extraction accounts for $690 \times 10^8 m^3/a$ (80% of that figure for whole China), in which $640 \times 10^8 m^3/a$ is pumped from water wells (89% of that figure for whole China) . These figures indicate the importance of groundwater development in northern China.

In northern China,agricultural groundwater use accounts for 70% of total groundwater use. In Huang-Hai-Huai plain (North China plain) groundwater development intensity occupies the leading place in whole China. Average modulus of exploitable groundwater   resources is about $15 \times 10^4 m^3/(a \cdot km^2)$

SPACE DISTRIBUTION AND TIME VARIATION OF GROUNDWATER UTILIZATION

The distribution of groundwater extraction for farmland irrigation in northern China is as following (Table 1).

Total amount of extracted groundwater for irrigation in northern China has increased   by about

30% from 1977 to 1993.

Table 1. Percentage of extracted groundwater

Region	North China	NE China	NW China
1977	78 %	5 %	17 %
1993	67 %	17 %	16 %

It can be seen from table 1 that : (1) North China has the greatest groundwater extraction intensity; (2) For the last 16 years period , NE China has the greatest growth rate ; (3) Groundwater extraction in NW China is also increasing ,but the growth rate is less than the average growth rate for whole northern China.

Time variation of groundwater utilization is caused by great variation of precipitation during dry , normal and wet years (Table 2).

Table 2. Characteristic values of precipitation in various parts of northern China

Name of station	Xining	Yinchuan	Tongguan	Daxing	Jinan
Average P (mm)	354	195	562	547	664
P_{max} / P_{min}	2.53	7.72	2.58	4.01	3.08
Observation period	1956-1990	1956-1990	1956-1990	1959-1993	1983-1995

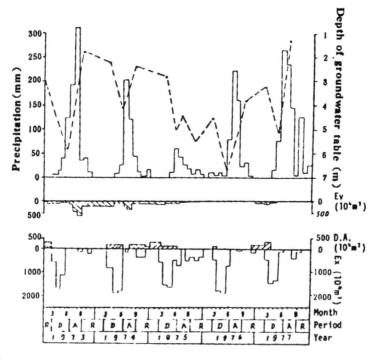

Figure 1. Correlation between the water table fluctuations and main balance elements.
P—precipitation (mm); v—evaporation of groundwater (10^4 m³); D.A.—diverted amount (10^4 m³); Ex—extracted amount (10^4 m³); A—ascending; R—regulating; D—Descending.

Detail study was undertaken in Rongcheng county, Hebei with total area 300km² (Fig. 1). Rongcheng county underwent an extremely dry year in 1975 , with total precipitation of 283mm. The average modulus of extraction within five years (1973-1977) is 18 × 10⁴ m³/(a · km²). The average depth of the water table at the end of the wet season in 1977 was 1.87m, while that in 1973 was 1.78m. Obviously these two values are close to each other and prove that regulation in the use of groundwater over several years is possible.

Northern plain of Henan province has total area 2 × 10⁴ km². Observation data during 1976-1987 show that excellent correlation exists between precipitation and extracted groundwater for farmland irrigation (Fig. 2).

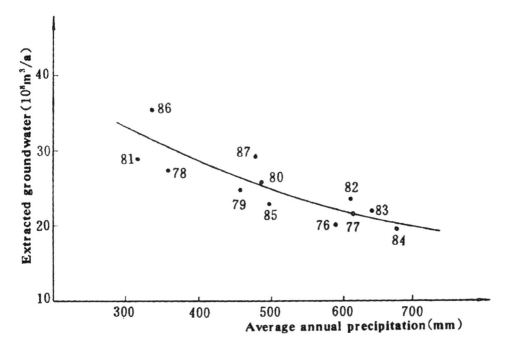

Figure 2. Correlation between precipitation and extracted groundwater for farmland irrigation (Northern plain, Henan province, 1976-1987).

While annual precipitation is rising from 350mm to 700mm, the modulus of extracted groundwater is reducing from 15 × 10⁴ m³/ (a · km²) to 10 × 10⁴ m³/ (a · km²).

GENETIC TYPES OF GROUNDWATER REGIME FOR IRRIGATED FARMLAND

According to main recharge source, four genetic types of groundwater regime have been identified. Each type has certain geographic distribution in specific hydrogeological situation (Table 3).

It is the reason for us to take main recharge source as criteria for regime type identification that main recharge source is the guarantee for sustainable exploitation of groundwater for agriculture.

(1) Dominating vertical water exchange type
In alluvial or alluvial-lacustrine plains in semi-humid zone, groundwater is mainly recharged by

rainfall infiltration and lost through evapotranspiration. This type is characterized by active water exchange among meteoric water , water of aeration zone and groundwater in vertical direction. Huang-Huai alluvial plain in semi-humid zone may serve as example. Quantitative relationship of groundwater balance of Huang-Huai plain within Henan province is as following : Groundwater recharge (100%) consists of rainfall infiltration (81%) and other sources (19%); Groundwater discharge (100%) consists of withdrawal (58%) and evaporation from water table (42%).

Table 3. Types of groundwater regime for irrigated farmland

Name of type	Main recharge source	Example
(1) Dominating vertical water exchange type	Precipitation	North China plain
(2) Dominating horizontal water exchange type	River water from surrounding mountain area	NW peidmont plains
(3) Irrigation-water-feeding type	Irrigation water	Yinchuan plain , Hetao plain
(4) Artificial recharge type	Artificial diverted water	Local area at present time

Within a meteorological cycle , if the total amount of extraction does not exceed that of recharge , the amount of replenishment in a humid year can be used in the next dry or normal year. According to the data on precipitation in Rongcheng county, the frequencies of humid, normal and dry years are 29%, 42% and 29% respectively, that is three humid years, four normal years and three dry years appear alternatively with ten years cycle. Experiences in Rongcheng county of Hebei province, and   Shangqiu county of Henan province, show that to rationally develop shallow groundwater and to control water level seasonal and perennial fluctuation around optimal depth (3.5–4.0) may make sufficient use of water resources in plain areas with greatly varied precipitation.

(2) Dominating horizontal water exchange type
The aquifers receive main recharge from rivers originated from adjacent mountain areas and flowing into the piedmont plain. It is very typical in NW China, where precipitation in plain is rather small. A large number of springs appear downstream and become the important component of the streamflow in the lower reaches of these rivers . Eventually surface water and groundwater are consumed through evaporation. This type is characterized by intensive reciprocal transformation between surface water and groundwater, which from an integrated water resources system (Fig.3).

Groundwater resources in plain areas in Xinjiang, NW China, are estimated as $395 \times 10^8 \, m^3/a$. The components of recharge sources are as follow : (a) River water (flowing out from mountains) infiltration—30.38% ; (b) Irrigation canal infiltration—33.67% ; (c) Farmland infiltration—9.11% ; (d) Lateral underground inflow—17.97% ; (e) Local rainwater infiltration—3.54% ; (f) Piedmont storm infiltration—3.03% ; (g) Reservoir seepage—2.27% . These data prove that river water from surrounding mountain areas (partially through irrigation system) is the main recharge source for groundwater in plain area.

Unified planning of water resources development for upper and lower parts of piedmont plains is a very important principle to avoid ignoring water demand of lower part of the plain, which often leads to serious ecological consequences.

(3) Irrigation-water -feeding type
This specific type exists in Yinchuan plain and Hetao plain in the middle reach of Yellow river. Both plains are basins with thick deposits of Cenozoic age, and have long history of irrigation by using water from Yellow river. Groundwater table rising has caused expansion of secondary soil salinization (Table 4).

During the last two decades, pilot projects have been implemented for promotion of well-irrigation and well-drainage technique.But it has not been popularized due to social-economic reasons.

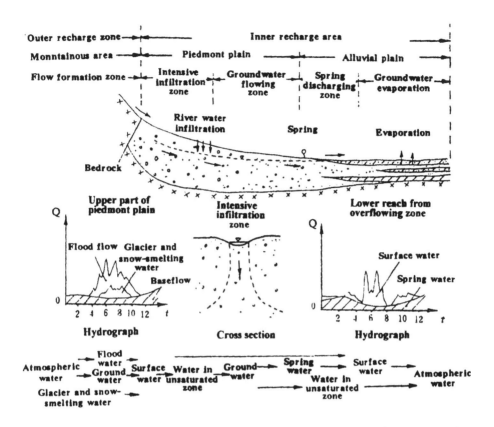

Figure 3. Scheme showing groundwater balance in areas with dominating horizontal flowing transfer between groundwater and surface water

(4) Artificial recharge type

At present time, this type has limited distribution. Huantai county of Shandong province can serve as an example. For the last several years, water-diverting project from Yellow river has already brought benefits to Canzhou district and Hengshui district of Hebei province. In the future, after implementation of water-diverting project from Yangtze river, more territory of North China plain will get considerable supplementary groundwater recharge. This type of groundwater regime will become more important for optimal water resources utilization.

POSITIVE AND NEGATIVE EFFECTS OF GROUNDWATER DEVELOPMENT

Groundwater development has caused both negative and positive effects, the later is often neglected and should be described at first.

Positive effects of groundwater development in rural areas of northern China can be summarized as follows:

(1) Social-economic effects:
—Well-irrigated farmland accounts for 40% of total irrigation farmland in northern China.
—More than 1000 × 10⁴ population of rural areas have got safe drinking water supply sources in severe watershortage areas.

(2) Environmental effects which also bring about huge economic benefits to the people.
--Reduction of salinized soil farmland. Since late 1950s',when groundwater utilization was widely spreading, half of salinized soil has been reclamated in North China plain.
--Enlargement of underground regulatory reservoir, increment of infiltration and reduction of waterlogging. Local surface runoff in 1980 in North China plain is 30% less than that in 1957 due to intensive groundwater development and increment of infiltration.

Negative effects include groundwater table depletion, drying up of springs, seawater intrusion, land subsidence and land collapse, etc. In dry season of 1992 in Hebei province, shallow fresh water, with water table depth more than 10m, accounts for 35% of total plain. In the meantime, deep fresh water, with water table depth more than 30m, accounts for 40% of total plain. Annual rate of water table depletion for deeper aquifer is much more   than that for shallow aquifer, even if the intensity of groundwater extraction of the former is much less than that of the later (Table 5).Furthermore, it should be mentioned that shallow fresh water may get some recovery during extremely wet year .

Table 4. Examples for groundwater regime type with dominating irrigation water recharge

Name of plain	Yinchuan plain	Hetao plain
Area (km^2)	6700	13000
Flow rate of transit Yellow river ($\times 10^8 m^3/a$)	~300	~300
Groundwater recharge :		
(1) Infiltration of irrigation water	87 %	76.5 %
(2) Rainfall infiltration	8.7 %	23.1 %
(3) Others	4.3 %	0.4 %
Groundwater discharge:		
(1) Evaporation	75 %	98.7 %
(2) Others	25 %	1.3 %
(pumpage)	(11.9 %)	

In northern part of Shandong province near Laizhou bay, within studied area 10 $\times$ $10^4 km^2$, during period of 1980_1995, territory with direct seawater intrusion accounts for 250 km^2, territory with underground saline water intrusion accounts for 724 km^2. Linear intrusion rate for underground saline water is as follows: 292 m/a for 1980–1992; 166 m/a for 1993–1995. The cause for underground saline water intrusion is overpumpage,   which reaches about 50 $\times 10^4 m^3/(a \cdot km^2)$ near Shouguang city . Countermeasures are being undertaken to overcome such negative effect.

Table 5. Intensity of groundwater extraction and annual rate of water table depletion (Canzhou, Hebei province, 1986–1990)

Aquifer system	1	2	3	4
Intensity of groundwater extraction ($\times 10^4$ $m^3/a \cdot km^2$)	2.88	0.71	2.51	0.17
Annual rate of water table depletion (m/a)	0.15	0.37	0.86	2.44

RECOMMENDATIONS FOR FURTHER RATIONAL GROUNDWATER DEVELOPMENT

(1) For North China plain:
—Water demand management;
—Reduction of water consumption by using new irrigation techniques (sprinkling irrigation, drip irrigation,etc.);
—Utilization of slightly saline groundwater (up to 4000ppm);
—Artificial recharge and use of transbasin guest water (from Yellow river and from Yangtze river).

(2) For Yinchuan and Hetao plains:
—Promotion of well-irrigation and well-drainage in whole irrigation-drainage system.

(3) For NW arid zone:
—Enhancement of hydrogeological investigation and evaluation by using new techniques;
—Groundwater development strictly in accordance with hydrogeological conclusions in order to avoid overpumpage which may cause serious ecological consequences.

Finally, we should emphasis that technical, economic and legal means should be used conjunctively to promote rational groundwater development.

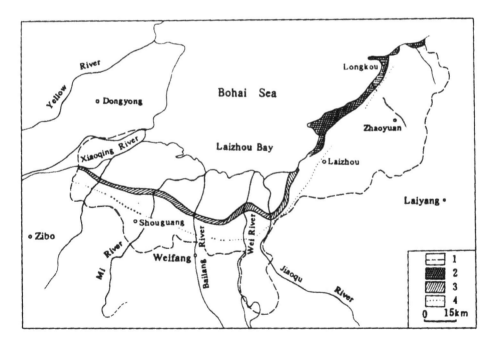

Figure 4. Schematic map showing intrusion areas of seawater and underground saline water near Laizhou bay , Shandong province   1–Boundary of study area ; 2–Seawater intrusion area ; 3–Saline groundwater intrusion area ; 4–Boundary of possible intrusion

REFERENCES

1. Ji Chuanmao. Variation of the groundwater regime under the effects of human activities and it's artificial control. *IAHS Publication 136, Improvement of methods of long term prediction of variations in groundwater resources and regimes due to human activities,* 87–96 (1982).

2. Ji Chuanmao, Hou Jingyan and Wang Zhaoxin. *Groundwater development in the world andguidelines for international cooperation ,* 20–24. Seismological Press. Beijing. (1996).

3. Wang Zhaoxin. Types of formation and regulation patterns of shallow groundwater resources in plain areas (in Chinese). *Geological Review.* 28:5, 453–460. Geological Publishing House. Beijing. (1982).

4. Zhao Desan. Seawater intrusion disaster and countermeasures in coastal zone of Shandong province (in Chinese). *Study on prevention and mitigation of seawater intrusion disaster,* 14–18. Shandong Science and Technology Press . Jinan (1996).

5. Zhou Yuli. Groundwater regime and groundwater utilization in Henan province (in Chinese). *Groundwater utilization and management ,* 6–11. Water and Power Press. Beijing. (1991).

Proc. 30th Int'l Geol. Congr., *Vol.22*, pp. 113-120
Fei Jin and N.C. Krothe (Eds.)
© VSP 1997

Artificially Induced Hydrogeological Effects and Their Impact of Environments on Karst of North and South China[*]

LU YAORU,   DUAN GUANGJIE

Institute of Hydrogeology & Engineering Geology Ministry of geology & Mineral Resources, Zhengding, Hebei, China

Abstract

The basic hydrogeological conditions of karst areas of North and South China are discussed. Two major river development schemes and their potential hydrogeological and environmental impacts are considered: the Shanxi-Shaanxi Gorge (Wanjiazhai Water Project) of Yellow River in North China, and the Three Gorges Project of Yangtze River in South China. The water development constructions may have major hydrogeological witnin the geoecological environmental impacts. The detrimental factors include: karst leakage, earthquake, reservoir margin stability failure, dezertization and pollution etc. The benefits are electrical power generation, flood control and irrigation.

INTRODUCTION

Karst development in both North China and South China is controlled by the climate, the presence of carbonate rocks and the geological structure[1,2,3]. In North China the typical karst regions are the Lulian Shan Mountains, the Taihan Mountains and the Shanxi Plateau. The typical South China karst areas including theYunnan-Guizhou Plateau, the west Hunan, west Hubei and east Sichuan mountain lands plus the Guangxi basin. In the north, the groundwater movement is mainly along fissure flow systems terminating in numerous large karst springs with discharges between 1.5 and 14.5 m^3/s [4,5,6]. These systems have catchments which range from several hundred square kilometres to over three thousand square kilometres. In South China groundwater flow is mainly in large karst passages and underground river systems. In these average water flows range from 1 to several tens m^3/s [2,4,7]; the maximum flow of the underground Du'an River in flood has reached over 300 m^3/s. The basic hydrological parameters for two systems are shown in table 1.

Table 1   Comparison of some karst hydrogeological features between North and South China

Region	Precipitation mm	Percolating coefficient %	Permeability of karstified rock mass m/d	Module of ground run-off L/Km^2.S	Average dissolution rate rate mm/a
North China	400-600	20-50	0.5-5common	2-10	0.02-0.03
South China	1000-2000	30-75	0.5->5	8-35	0.07->0.3

THE YELLOW RIVER, SHANXI-SHAANXI GORGE: KARST HYDROGEOLOGY, ENGINEERING AND ENVIRONMENTAL IMPACT OF THE WANJIA ZHAI

[*] This research project (49172144) is subsidized by the National Natural Science Foundation of China

RESERVOIR

In the northern part of the Shanxi-Shaanxi Gorge of the Yellow River there is a large karst water system feeding the Tienqiao and Luoliuwan group of springs: this system has a catchment of over 3000 km² [2,6,9,10]. The relationships between the karst groundwater and the river water are complex, but three typical situations are expressed diagrammatically in Figure 1.

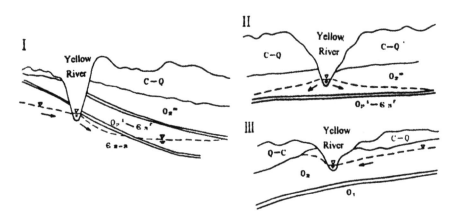

Fig.1 Three water recharge and discharge situations for the Yellow River and karst groundwater in the northern part of the Shanxi-Shaanxi Gorge. I. karst groundwater discharge into Yellow River in one bank and recharge from Yellow River water in another bank; II. Yellow River water recharge groundwater in both banks; III karst groundwater discharge into Yellow River from both banks
C-Q —from the Carboniferous to Quaternary aquifers;
O₂—The Ordovician karstified aquifer

In its natural state the groundwater level in the karst is between tens of metres and about one hundred metres lower than the water in the Yellow River. The natural leakage rates from the Yellow River into the ground are between 0.0006 and 1.13 m³/s.km. One of the most important considerations for the damming of the river and construction of the Wanjiazhai Reservoir is the likely leakage. Table 2 presents leakage information for reservoirs in the karst areas of both North China and South China.

Table 2. Comparison of karst leakage quantities from reservoirs in North China and South China

Reservoir	Dam height m	Real seepage quantity m³/S	Leakage rate per length m³/S.100m	Leakage rate per head m³/D.10m	Location
Taoqupo	61	27.8	8.65	4.55	Shaanxi
Guangting	45	1	0.66	0.22	Hebei
Shuicaozi	40	1.8	0.22	0.45	Yunnan
Zaixiangkou	54.7	20	10.0	3.66	Guizhou

Calculations suggest that the water leakage quantity from the Wanjiazhai Reservoir would be more than 20m³/s. This leakage could impact in many ways, both favourable and unfavourable. Favourable effects might include the increased recharge into the ground which would improve the situation for water abstraction from the aquifer. This extra groundwater could be used for irrigation, industry (including the coal mines and power stations) and as increased yields from geothermal water abstraction boreholes. Unfavourable effects include the loss of generating capacity due to leakage and geological hazards due to the rise in groundwater level. The predicted changes to the karst groundwater levels caused by the construction of the Tienqiao and Wanjiazhai dams are summarised in Table 3. The higher water levels and hydrostatic pressures may also generate landslides and karst collapses (similar to those described below for the Yangtze River). In addition the increased water pressures caused by the construction of the Wanjiazhai Dam may be detrimental to the Tienqiao Dam. Calculations allow the multi-benefit value of the reservoir to be evaluated and the permissable leakage quantity to be determined. The maximum benefit for a reservoir with leakage B_k is:

Table 3. Predit of karst ground water to be influenced by the Wanjiazhai Reservoir

Belt	Yellow River water level in Wanjiazhai m	Yellow River water in Tienqiao m	Karst water table in Tienqiao ml
Natural condition	898	816	817-824
After Tienqiao Reservoir storing	898	835	837-839
After Wanjiazhai reservoir storing	980	835	>840-850

$$B_k = \max F(Q_D) + \max F(K_U \cdot Q_L) \tag{1}$$

B_k--maximum comprehensive benefits of reservoir;

$F(Q_D)$--direct benefit related to water power with the flow quantity of Q_D m^3/s;

Q_L--permissive leakage quantity m^3/s;

K_U--useful coefficient

The permissive leakage quantity with multi-purposeful benefits is :

$$K_U F(Q_L) = K_U[f_1(Q_L), f_2(Q_L) \ldots \ldots f_n(Q_L)] \tag{2}$$

or $\quad K_U F(Q_L) = K_U \Sigma f_i(Q_i) \tag{3}$

Q_i--quantity for using in I term m^3/s;

fi--coefficient in I term.

Using this formula for the Wanjiazhai Reservoir the calculated permissable leakage quantity is about 15 m^3/s. This value is less than the estimated actual leakage quantity of 20m^3/s, thus showing the need to incorporate leakage control measures into the scheme.

THE THREE GORGES OF THE YANGTZE RIVER: KARST HYDROGEOLOGY, ENGINEERING GEOLOGY AND ENVIRONMENTAL IMPACT

The Three Gorges of the Yangtze River lie within the west Hubei-east Sichuan regions. The karst hydrogeological conditions are mainly characterised by numerous karst passages, underground or subsurface rivers[7,8,9,10]. In South China the karst water is usually one of two types. The first is where a surface river disappears underground to form a subsurface river. The second is caused by the collection of water that has percolated into the system from sinkholes, corroded fissures and passages to form an underground river. The hydrological model for a subsurface river is:

$$Q_{0i} = Q_{1,i-1} \pm \sum_{p=1}^{n} q_{p,i-tp} \pm \sum_{f=1}^{m} q_{f,i-tf} + \sum Q_d \tag{4}$$

Q_{0i}--related discharge of the subsurface river in i term m^3/s;

$Q_{1,i-t}$--related flow quantity in entrasnce of the subsurface river in i-t m^3/s;

$q_{p,i-tp}$—related flow quantity of p branch of the surface river in i-t_p time m³/s;

$Q_{f,i-tf}$—related flow quantity of f fissure belt in i-t_f time m³/s;

Σ_{Qd}—related total flow quantity of condensation water into subsurface river form i-t to I time m³/s

The flow model of percolating and underground river is:

$$Q_{0i} = \sum_{s=1}^{k} q_{s,J-ts} + \sum_{p=1}^{n} q_{p,J-tp} + \sum_{f=1}^{m} q_{f,J-tf} + \sum Q_d \qquad (5)$$

$q_{s,i-ts}$—related flow quantity form S depression to collected into the percolating ground river in i-ts time m³/s

The karst water systems contribute to the Yangtze River flow. The Gaoqiao percolating underground river alone adds 139 m³/s with about 16-32 m³/s of this coming from fissures as percolated water. In the dry season the middle branch of the Qingjiang River, in its upper reaches, sinks into the ground and contributes to the karst water flow. The basic hydrological framework of west Hubei-east Sichuan is summarized in Figure 2.

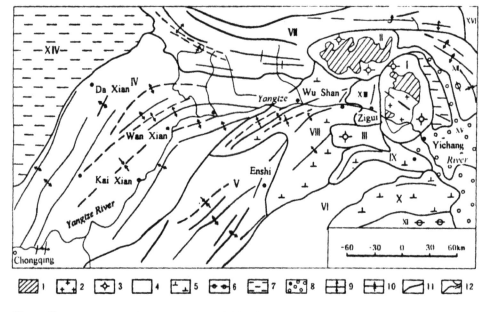

Figure 2. Karst hydrogeological systems in west Hubei and east Sichuan . 1--metamorphic rocks; 2--igneous rocks; 3--dome karst hydrogeological zone; 4--elongated fold karst hydrogeological zone; 5--wide anticlinal karst hydrogeological zone; 6--wide synclinal karst hydrogeological zone; 7--structural basin-buried karst hydrogeological zone; 8--descended plain-buried karst hydrogeological zone; 9--elongated anticline; 10--elongated syncline;11--boundary; 12--river

It is to be expected that the construction of the Yangtze River scheme will have some geoecological and environmental impact on both the upper and lower streams of the river. The main anticipated environmental problems are: earthquakes, slope stability, silting up of the reservoir, karst collapse desertization (caused by population movements and agricultural change) and water pollution.
Eighteen reservoirs in China[14,15,16] have caused earthquakes when they were storing; their statistics are summarised in Table 4.

Three categories of situation for the generation of earthquakes by reservoirs have been identified

(Figure 3); A. loading on faults, B. pneumatolytic pumping, C. air compression in caves.

Table 4   Statistics of earthquakes caused by reservoirs storing in China

Reservoir number been caused earthquakes	Earthquake magnitude	Lithological character in epicenter area	Depth of earthquake focus km	Percent to totalreservoir %
3	2.3-6.1	granite	<2.5-5	16.7
2	3.7-4.8	magnatite	2.5-7.4	11.1
1	2.8	volcanic	0.27	5.5
12	1-4.7	carbonate rock	0.5-6	66.7

The potential earthquake hazard to the Three Gorges scheme can be calculated based on a comparison with the 165 m high Wujiangdu Reservoir Dam in Guizhou. Using the formula below from, it is forecast that the potential earthquake magnitude would be less than Ms = 5.

$$C e = (\sum_{j=1}^{n} R_{ij} \cdot P_{ij} / n + \sum_{i=1}^{n} R_{di} \cdot P_{di} / m) / 2 \qquad (6)$$

Es=Ew · Ce · Kc

C_e—comparative coefficient;

R_{ij}—ratio of j term of indirect factor;

P_{ij}—power corresponding to j term of indirect factor;

R_{di}—ratio of i term of direct factor;

P_{di}—power corresponding to i term of direct factor;

K_c—different coefficient of karst development;

Es—estimated earthquake magnitude;

E_w—coefficient of different earthquake condition.

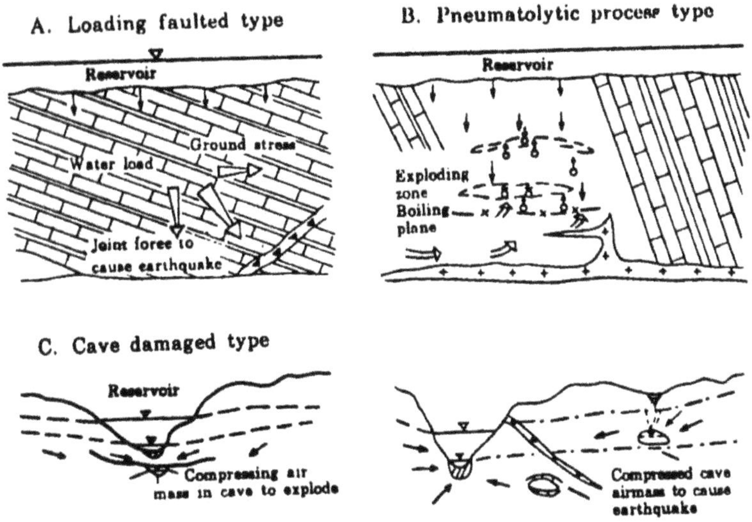

Figure 3.   Analysis of three earthquake types caused in karst areas by reservoir water storage.

Slope stability along the Three Gorges would also be decreased by the construction of the reservoir. Both increased water levels and fluctuations of the water levels, by up to 30m, would contribute to slope failures and landslides. The main instability types are shown in Figure 4.

Table 5.   Separation of water quality zones in reservoir

Zonal position	Surface apper	Middle	Deep	Bottom
Name	Absorpting temperature zone	Changing temperature zone	Smoothing temperature zone	Mixing temperature
Depth m	0-4	4-25	>25-top of lower zone	5 m apper deposit in reservoir bottom

$$Q_T^p = \int\limits_{n=1}^{n}\int\limits_{t=1}^{m}\left[\alpha_{n-1}\cdot\left(D_{n-1}\cdot\nabla^2 q_{n-1,t}^p\right) + \alpha_n\cdot\left(D_n\cdot\nabla^2 q_{n,t}^p\right)\right]dn\,dt \qquad (7)$$

$q_{n-1,t}^p$—the polluted water quantity in n-1 reach at t time;

$q_{n,t}^p$—the polluted quantity in n reach at t time;

D_{n-1}, D_n—comprehensive diffusion coefficient in n-1 and n reaches of the reservoir(mainly caused by density and influenced by temperature and pressure)

α_{n-1}, α_n—self purification coefficient in n-1 and n reaches of the reservoir

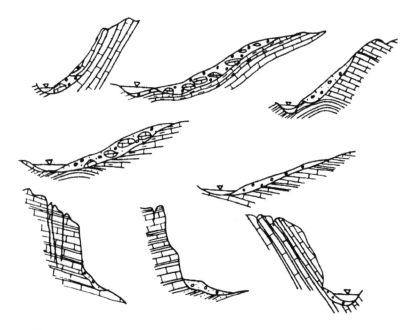

Figure 4. Main landslide types in the upper and middle Yangtze River (classified by main types of failure surfaces). A - Bedding planes dipping steeply towards the river; B - Bedding planes dipping gently towards the river; C - Steeply inclined fissures or joints dipping towards the river; D - Gently inclined fissures or joints dipping towards the river; E - Bedding and fissures/joints dipping gently towards the river; F - Gently inclined bedding dipping towards the river; steeply inclined joints; G - Steeply inclined fissures, gently inclined bedding dipping away from the river ; H - Steeply inclined fissures, steeply inclined bedding dipping towards the river

Using these categories of failure surface some treatment, to prevent landslides, has been carried out for the Three Gorges scheme. Pollution problems in the new reservoirs are also a consideration. Polluted water, in this and other similar karst zones, are commonly contaminated with excrement, urine, organic matter from soil, chemical fertilizers, industrial waste and acid rainfall; many of these can be identified by their nitrogen isotopes.

$$\delta\ ^{15}N=[^{15}N/^{14}N(sample)-^{15}N/^{14}N(standard))]/[(^{15}N/^{14}N)standard] \times 1000\text{‰}$$

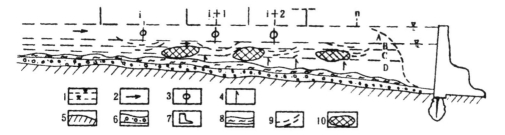

Figure 5. Theoritic analysis of polluted condition in reservoir. 1--reservoir water level; 2--flow direction; 3--polluted water centre mass in different reach of the reservoir; 4--diffused direction of secondary pollution; 5--bed rock; 6--sand and gravels; 7--dam; 8--deposits in reservoir; 9--diffusing polluted water in different reach of the reservoir; 10--opposite dead polluted water body; 11--deeply smoothing temperature zone; 12--bottom zone of mixing temperature

Within standing water bodies, such as lakes and reservoirs, the water is stratified into different layers or zones with different characters; these are summarised in Table 5. The formation of polluted water masses within such a situation, to form a polluted layer, is shown diagrammatically in Figure 5 and can be expressed by the formula below (7). To protect the water quality and to prevent the formation of a polluted water layer within the Three Gorges reservoir a series of waste water treatment schemes and seasonal water management are required.

CONCLUSIONS

1. The karst hydrogeological conditions in South and North China are different. There is strong karstification in South China with complex karst passages and underground rivers. The karst of North China is characterised by corroded fissures and karst springs.

2. In the Shanxi-Shaanxi Gorge of the Yellow River the leakage from the Wanjiazhai Reservoir will be an important geoecological/environmental problem. Control of the leakage,to keep it within the permitted value determined by calculation, is essential. This will yield the maximum benefits and protect the environment.

3. The benefits of the Three Gorges Water Conservancy Project will include hydroelectric power, flood prevention, irrigation and drought prevention. However, associated environmental issues also need to be addressed, they include: the stability of the reservoir banks, earthquake generation, karst collapse and the rate at which the reservoir will silt up. Advanced treatment to mitigate these problems is required.

4. These schemes will have a geoecological impact well into the 21st Century. In order to monitor changes in the geoecological environment it is important to establish the baseline conditions and the thresh hold limits for change. Only then can effective control measures be incorporated into the schemes.

5. The evolution of the geoecological environmental factors may be classified as advantageous, slightly advantageous, slightly detrimental, or harmful. We believe that with appropriate mitigation measures the development of the Yellow and Yangtze rivers can be undertaken to the advantage, or slight advantage of the environment.

Acknowledgement

We express our very thankful feeling to Dr. A. H. Cooper to check-in English of this paper. We also thank so much to Associate Professor Tang Hongcai and Ms Yang Lijuan in this paper.

REFERENCES

1. Lu Yaoru et al. The development of karst in China and some of its hydrogeological and engineering geological conditions. Acta Geologica Sinica. 121-136 (1972).
2. Lu Yaoru.Karst in China, Landscapes . Types . Rules. Geological Publishing House(1986).
3. Lu Yaoru. Karst geomorphological mechanisms and types in China. International Geomorphology 1986 Part II. John Wiley & Sons Ltd. 5:4, 1077-1093(1987).
4. Lu Yaoru. Water resources in karst regions and their comprehensive exploitation and harnessing. International Geomorphology 1986 Part II. John Wiley and Sons Ltd. 4:1-2,1151-1167(1987).
5. Yuan Daoxian et al. Karst of China.Geological Publishing House(1991).
6. Karst Commission, Geological Society of China. Karst and karst water in North China. Press of Guangxi University(1993).
7. Zhu Xuewen. Some karst features in East Sichuan and hydrodynamic conditions of karst evolution. Selected papers from the Second All China Symposium on Karst Sponsored by the Geological Society of China,1978. Guilin,Guangxi,China.Scientific Press.5-13(1982).
8. Lu Yaoru. Effects of hydrogeological development in selective karst regions of China. Hydrogeological karst processes in karst terranes. Preceedings of the International Symposium and field seminar held at Antalya, Turkey. IAHS Publication 207:15-21(1993).
9. Lu Yaoru (ju)and Wang Chaoxin(hsin). The conditions of karst hydro-engineering geology of a dam site in South China. Academy of Geological Sciences, Ministry of Geology, China:Professional papers, Section D Hydrogeology & Engineering Geology. China Industry Press. 1:1-56(1966).
10. He Yubin. Some characteristics of karst water in North China. Carsologica Sinica. 1:2,135-142(1982).
11.Lu Yaoru. Hydrological environments and water resource patterns in karst regions of China. Proceedings of IAH 21st Congress, Karst hydrogeology and karst environment protection. Geological Publishing House. 64-75(1988).
12.Karst Commission, Geoogical Society of China.Human activity and karst environment. Beijing Scientific and Technical Press(1994).
13.Han Xingrui, Lu Rongan. Karst water system,A study on big karst springs in Shanxi. Geological Publishing House(1993).
14. Institute of Geophysic etc. Atlas of the historical earthquakes in China. China Cartographic Publishing House(1983).
15. Xia Qifa. The confluence table of basic parameter of the World reservoir induced seismicity examples. The Chinese Journal of Geological Hazard and Control. 3:4, 95-100(1993).
16. Xia Qifa. The confluence table of basic parameter of the World reservoir induced seismicity examples(contineous). The Chinese Journal of Geoogical Hazard and Control. 1:1, 87-96(1994).

Proc. 30th Int'l Geol. Congr., Vol.22, pp. 121-138
Fei Jin and N.C. Krothe (Eds.)
© VSP 1997

Chinese Karst and Engineering Protection

YAN TONGZHEN,   TANG HUIMING,   LUO WENQIAN

Faculty of Environmental Science and Engineering, China University of Geosciences (Wuhan), Hubei, 430074, P.R. China.

Abstract

Chinese karst is well developed at South west, Central South, South east and the North. Among them Guangxi Province is a typical area with karstic topography. The requisite condition to form karst is the soluble formations themselves. While the sufficient condition must be satisfied to cause the karstic processes developing forever. Chinese always pays attention to the physiographic stages. The traditional theory of double circulations or cycles is introduced and reviewed for karst. Chinese karst development history, earth crust movement and dynamics of ground water circulations are influential one another by flow in the reverse siphons with nonlinear confined aquifer. Recently one Chinese presents assembly flow theory for karstic pipe. It captures ground water from soluble vein pipes and pore spaces. Where there originate in concentrated ground water flow, there are introduced with disciplines of karstology and karst environmentology in China. The comparing karst works are well performed by the foreign geomorphologist and our country experts. Their more attentions are interesting in the mechanism of mixing solution effect in shallow pumping vacua and deeper caves or pipes. For the prospecting of petroleum industry, the conception of pore throat porosity texture should combine statistical distribution character with studying. As for engineering protection in karst regions, three types of slope surface, body and foundation have considered. For the former twos, the karst caves should be abolished with more care where slope stability is damaged. Especially the occurrences of carbonatite formation exist in favorable conditions prone to slide. At these conditions, the easy sliding theory must be cited so as to avoid the huge slide occurring where sliding bed of carbonatite decomposed and formed CO_2 air cushion layer. while all cave pipes or holes damaging slope foundation might be strengthened by engineering protections on the base of safety thickness of cave top arch..

Keywords: Chinese karst, physicgraphic stage, nonlinear confined aquifer, assembly flow theory, formed mechanism, engineering protection, easy sliding theory

INTRODUCTION

Soluble rock masses are requisite condition to form karst. There is 25% of territory areas being belonged by naked and not naked soluble ones, while the naked is only 13%. The sufficient condition must be satisfied to form karst; it is referred to increase concentration of   ions and existed with the indicative ions Na^+, Mg^{2+}, Cl^-, Ca^{2+}, and HCO_3^-. In all coal series strata of China with pyrite, there have to be the decomposition of H_2O and FeS_2. Under these conditions, karstic processes will go on forever in shallow or deep water circulations. For the above mentioned, the dynamic regime flow is considered to form karst basically. In details, this will be done to combine with the conception of mechanism of mixed solution and mixed corrosion effect. The main karst regions locate South west, Central South, South east and North China. Among them, Guangxi Province is a typical region of tropical karst area. While Provinces Sichuan, Hubei, Hunan, and Jiangxi are the regions of naked karst of subtropical zone similar to the Mediterranean Sea type with karstic hill and depressions. As yet the Shanxi, Hebei, and Shandong Provinces are the regions of temperate zone with dominant karstic spring and dry valley being lost rivers underground water flow. The grological epoches of soluble beds include the Sinnian Subera

Table 1. The Characters of Karst development in different regions of China

Stage of Karst	Period	Crust Movement	East Yunnan	Guangxi	Central Guizhou	West Hebei	North China
Ⅲ (S₃)	Quaternary	Himalaya (continued)	Nanpan River Stage, with crust uplift strongly, canyon formed, headward erosion strongly strengthened below the river knickpoint, karst development well within river valley region	Honghui River Stage with west part down-cutting deeply, canyon formed with peaks u-vale, karst depressions and polje hills with east part relatively stable, karst plains, peaks mainly, karst caves of several beds	Wujiang River Stage, with canyon down-cutting deeply, below 300 — 500m karst developed well within river valley region	"Three Canyons" Stage, canyon mt. region cutting deeply, below 300 karst cave beds not dominantly developed, below 500—700m dissolution caves dominant, above 700m with caves, dolines and shafts, great springs widely distributed	Banchan Stage, Qinshui Stage, Huanhai stage, Fenhe Stage with Zhoukoudia cave deposits in carbonate rocks
Ⅱ (S₂)	Neogene	Himalaya (Ⅰ)	Pinnacles Stage with altitude 1100—1600—2200m, thick residuum, cave, polje, doline on surface of karst plain, pinnacles developed at Lunan	Peaks Stage, on the top of peaks about 200 —1000m	Mountain Basin Stage, Pin Chao Substage, with altitude 900 — 1070 — 1170m, mainly karst poljes; Jarzo Substage with altitude 1450 — 1500m, red residuum, wave-like mountain plain composed of karst depressions polje and hill peaks.	Mountain Plain Stage, widely distributed wholly protected; Wangspin Substage, mean altitude 800 — 1000m, karst depressions and poljes, both substages with caves dominant; Zhoujanou Substage, altitude about 1000 — 200m, with hill peaks and karst depressions	Tangxian Stage, altitude 1300 — 1600m, with Bangtao laterite; Taihang Stage (from Qian, 1984), with denudation surface 1600 — 1700m, hills, karst depressions dominant
Ⅰ (S₁)	Eogene	Himalaya (Ⅰ)	Plateau Stage, altitude 2500 — 3000m, with karst depressions and hill peaks	West Guangxi Stage, altitude 300 — 1500m, at top peak elevation high from east to west; with peaks, karst depression at the west part, with single peaks, karst plains at east part	Dalomian Stage, with altitude above 1500m by residuum of peak tops mainly, with doline caverns and underground river below surface about several hundred meters	West Hebei Stage, with karst depressions and hill peaks above altitude 1000 — 1500m, relief, dissolution caves, shafts very dominant	Beitai Stage, altitude about 1500m at the tops of Yeikomhan and Taihangshan remained with top surface, and isolated plateau, with karst depression, vertical cave within Yincben only
O(S₀) (pre-Ceno-zoic)	Cretaceous	Yanshan	Ancient Karst Stage, with residual peak tops and rare relics of karst	Same to the left with 350—400m and higher elevation in East Guangxi and in West Guangxi with 1500m	Same to the left, but with residual hill tops very rare and higher	Similar to the left, but with local residual hill peaks about 2000m, and shallow flat karst depressions and do-lines	?

Note: Simplified table showing the physiographic stages and karst development history of South and North China (modified from Dai,1958;Zhang,1979;Chinese Karst Research [9];1979 and Zou et al.,1994)

to the Paleozoic Era and the Triassic Period of the Platform of South China with thickness of carbonatite sediments about 3000-5000 meters above sea level & for North China, with only 1000-2000 meters. The lithological characters, sedimentary facies and thickness of soluble beds vary strongly in the geocyncline of West China. From viewpoint of environmental resources and engineering geological protections and meteorological backgrounds we can see that the problems are too complicated with different zones showing different tpes of karst landscapes. Hence, there are either common or different problems within the different karst regions waiting for sloved. In these regions the most popular questions are probably karst seepage flow of dammed reservoirs, stability of underground openings and that of foundations and slope stability of cuttings and fillings.

According to the occurrence altitude of carbonate rocks, there are three karst types in China: a.naked karst type where the carbonate rocks outcropped directly on the ground surface, i.e. the main type of karst in China.b. overlay karst type where the carbonate rocks covered with the Quaternary accumulated sediments of the thickness not more than several tens meters. Sometimes, little outcrops of carbonatite may be found around the wide loose sediments or soils. It is mainly distributed at Guangxi Province and Guizhou-Yunnan Plateau with wide karst depressions and plains. c.buried karst type where the carbonatites buried deeply under the overlying unsoluble beds, it distributed at Sichuan basin and North China. The overlying beds may be more than several tens or hundreds even over one thousand meters. Among the three karst types the former twos are tightly related to the slope stability.

COMPARING KARST AND KARST DEVELOPMENT

Karst areas in China and in Europe and elsewhere are compared. *Important Karst Regions of the Northern Hemisphere* [4] in KARST were written by Herak and Stringfield (1972). There are sixteen Chapters served to discuss Histery, Karst of Yugoslavia, Karat of Italy, Karst of France, Karst of Germany, Karst of Austria, Karst of Hungary, Karst of Czechoslovakia, Karst of Poland, Karst of Rumania, Kasrt of the former USSR, Karst of Great Britain, Karst of Jamaica, and Karst of U.S.A.. The last chapter is Conclusion. From the contents, people know that there is no more information of Chinese karst at all. However, Chinese karst is thoroughly investigated and discussed by an other foreign geomorphologist [13] -Sweeting (1995). Chinese methods of karst studying are mostly different from those in the West. Chinese always pays attention to the physiographic stages, karstic development history, crust movement, dynamics of ground water circulations, different circulations with the reversed siphons zone of nonlinear confined aquifer. For the mechanism of mixing solution effect in deeper caves [1] , we are also interesting in the study from Bogli(1969). In karstic pipes and veins or pores, the formed mechanism of shallower pumping vacuums is very dominant and favorable for the research and to form karstology [12] and China karst [17] disciplines by Ren (1983) and Yuan [15] (1993).

Karst Development

The processes and phenomena of karst are the products of development of geological history. This should be to research the regularity of development and distribution of karst with either space or time conception. The circulating replacemen of groundwater is the basic adequate condition of a given region to develop into karst. It is controlled by regional geological structure, geomorphological and topographic features and climate. These factors, in turn, are controlled byrustal elevation and subsidence movement changing with space and time. As soluble rock mass can be only attacked and dissolved when it is uplifted by tectonic movement from sediment vironment to the ground surface, therefore the deposited gap may be served as a feature of ancient karst development. We call ancient karst which had already developed before the Cenozoic Era. The longer the time of deposited gap was, the more the probability to develop into karst. The surface of denudation of ancient karst is a soft weak structural zone easily to cause rock mass failure. From viewpoint of engineering geology it will stress on the karst development of the Cenozoic Era, because of engineering slopes and geostructures are tightly associated with shallow recent karst no less than that of the ancient.

Within the Era of Himalayas Movement, the character of it is dominantly of elevation and subsidence. For each cycle of it, the velocity, altitude and scale are very different which have direct influence on karst development. Based on the regional tectonic research, the stages of karst development may be correlated with the physiographic stages. The outline of tectonic movement for karst development history and the Cenozoic physiographic stages of typical regions in China are shown in Table 1. The karst peneplain and Turm karst (peaks) were developed mainly under ancient climate during the Tertiary Period such as in Yunnan and Guizhou. These may be correlated with those of Alpen in Germany, Sillisia in Poland and Pamir in the former USSR.
From table 1 we can see that different regions and stages had corresponding differences for karst development with more or less different relics of karst phenomena. However, from macro viewpoint there is dominant regularity found. In few words, if regional crust being tectonic stable, the karst development may be reduced gradually from river valley toward drainage divide ridge. Moreover, Some regions did not show this manner such as West Hubei and Central Guizhou where the karst phenomena (depressions, dolines and caves etc.) concentrated at region of drainage divide ridge more stronger than that of river valley. This manner is resulted from crust elevation continuously. At the surface of the Daloushan Stage, such as that of Guadin, Meihuashan divide ridge, the karst development is superposed by the Quaternary dissolution and corrosion processes upon the Tertiary karst of ancient tropic zone. The caves are commonly horizontal and stratified; but their elevations can hardly be compared one by one with river terraces unless they resulted from underground rivers developed within shallow soluble rock mass and remained with their deposits and erosional features alike to those of a river. It is emphasized that stratified caves and their multistratiforms are very complicated due to many variable factors such as geological structures, lithological characters of slouble rocks, ancient geographical and geomorphological conditsand climate changes etc. Multi stratified karst caves in open folding region may be formed. The stratified caves associated with the terraces to(1), $t_1(1)$, $t_2(5)$, t_3 (4 beds of cave) and $t_4(1)$[12]. These mean those the elevation of stratified caves can not be correlated correspondingly with terraces one another. When soluble rock beds interbedded with unsoluble beds i.e. water-bearing bed interbeded with water resisting layers, the karst caves themselves must be stratified and coincided with their bedding planes. The stratified caves may be formed either in subairal saturated zone of openned system. Sketches of Stratified cave are shown in Fig.1.

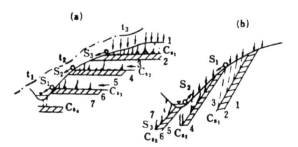

Figure 1. Sketches showing stratified karst caves a)horizontally bedded;b)steeply bedded;1.3;5.7-soluble rocks;2.4.6-unsoluble rocks;C_{s1}, C_{s2}, C_{s3}— stratified caves; S_1, S_2, S_3—springs; t_1, t_2, t_3—river terraces; dotted line-topographic line of oposite bank profile

Some deep karst caves, developed under river bed about several-fold tens to more than one or several hundred meters, have been already discovered by geophysical and mechnical prospecting methods. These deep caves are not yet ancient and without any relationship with terraces. From the mentioned above, the conception of erosion base level and that of corrosion may be established for karst development. For the single and more or less pure slouble rock mass the shallow karst development will be controlled by the erosion base level which is that of local river reach (i.e.the flood level being the limlt). In this case, the elevations of horizontal caves may be approximately to that of the local erosion base level of river terraces. On the orther hand, if the soluble rocks are interbedded with unsoluble beds as shown in Fig.1, then, either shallow (a) or deep (b) karst development must be taken place and controlled by the corrosion base level which is the elevation of the top of unsoluble beds and/or that of the deep limit of soluble rocks of which their joints and fissures are too small to permit the karst water geting its way downward continuously.

Traditional Theory of Double Circulation

By davis [2] (1930), different substages of karst development may be divided. a. When regional elevation takes place causing basic level and under   groundwater to subside, the formerly formed karst cave and under ground river are uplifted above the original position and changed to dry manner. While the karst water passes through structural joint toward deeper level to form new karst channel, and this time interval being called deeper level karst substage. b.Relative stable stage of karst region karst water within staturated zone is fitted with the stable base level and developed its active space by lateral corrosion to enlarge its horizontal channel, this process is called the lateral karst substage.

Table 2. Showing the Substages of Karst development

Substage	Crust Movement	Base Level	Moving Direction of Underground Water	Dynamics of Groundwater
a. Vertical, shallow	Elevation	Descending	Mainly descending vertically	Periodical satrurated or dry
a-b. Vertical-horizontal	Subsidence stable	Uplifting stable	Low water level vertical moving; high water level horizontal moving	Similar as above and/or full of water
b.Horizontal (lateral)	Stable	Stalbe	Mainly horizontal moving	Usually circulating
b-c. Horizontal-vertical	Stable subsiding	Stable uplifting	Region of drainage divide-descending movement, zone of discharge at banksuplifting mainly	Similar as above
c.Vertical, deeper	elevation	Descending	Uplifting or descending or horizontal	Moving slowly moving normally

c. After a stable stage, the earth crust repeats its elevation and the laternal corrosion steps there with empty channel, the karst water sinks down to make its new deep corrosion and this stage is so called the new deep karst substage. The substages of karst development are shown in Table 2. But it is seeing that between a and b, and b and c, there must exist as transition subatages a-b and b-c.

Thus, the five substages make an ideal cycle. When a cycle is finished, there is inherited with a horizontal cave bed in the soluble rock mass. The cycle is so called karst circulation. This karst development conception is traditional and somewhat too idealized. Practically, a cycle not finished may be always superposed with another one which gives rise in complicated phenmena of karst. In the river bank slope the shallow karst circulations may be delayed and associated with river terraces.

Dynamic Characters of Karst Water[8]

Considering of dynamic zones of karst water (macshemovich, 1956, 1978) the idealized profile model of dynamic zones of karst water [12] may be shown in Fig.2.

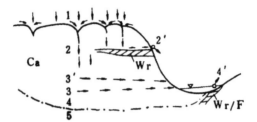

Fig. 2 Showing idealized profile model of dynamic zones of karst water circulations on a bank slope. Ca-carbonatite; Wr/F-water-resisting beds or fault; 1. zone of ground surface of karst opening; 2. zone of vertical seepage circulations; 2'. subzone of hanging water spring; 3. zone of horizontal (lateral) circulations; 3'. subzone of transition of 2 and 3 to replace each other; 4. zone of reversed siphons flow; 4'. zone of lifted spring; 5. zone of deep circulations (modified from Re et al. 1983)

For zones 2 and 3' ,the dynamics of ground water appears to be unstable condition with their dynamic character mainly being gravity gradient flow. However, for zones 4 and 5, their dynamic conditions are stable with mainly being pressure gradient flow. For the naked karst type, zones of 2 and 3' are more thick such as west Guangxi and south boundary of Guizhou; but for the overlay karst type, more thin such as karst plain of Liuzhou. If a given regional crust movement is known to be uplifted with multicycles and a consideration given to combine the conception of Table 2 and Fig.2, there may gain some practical karst sections to explain.

Genesis of karst caves   From mentioned above, the genesis of karst caves is more complicated than that of the idealized ones. We prefer to think of it being controlled by a series of factors such as lithological character, structural patterns, the Pre-Cenozoic and the Cenozoic crust movement, porosity texture types etc. Under the basic condition, these factors take with interaction of corrosion-erosion and corrosion, mixing corrosion in different space of hydrodynamic zones and different time of stages and substages of karst development. This complicated conception for the genesis of karst caves is different from one of those of uniform groundwater level, isolated groundwater flow, and replacement of groundwater circulations in valley.

Mechanism of Mixing Sloution and Mixing Corrosion Effect

It refers that two or more than two kinds of water solution saturated and missed already corrosion capacity meet at a certain point within carbonatite and mixing action is happened there. The mixed solution will get unsaturated condition from the original saturated, and gain a new corrosion capacity [1] to attack soluble rock mass being dissolved. From the result for a lot of caves being investigated by Bogli   (1969), it concluded that genesis of deep caves is resulted from the effect is shown in Fig.3. Suppose two kinds of saturated water W_1 and W_2 are mixed. Their mixed action may be shown as a curve $W_1 W_2$. Unite W_1, W_2 with a straight line and find out its mid point T representing 1:1 in volume of W_1 and W_2. Draw a horizontal line TC to intersect W_1W_2 at point C. The value of TC represents that of loose CO_2 after mixing action. The value of coordinates of point C is (x, y),i.e. (CO_2, $CaCO_3$). It means the value of CO_2 to be balanced and that of $CaCO_3$ content. From chemical equation we know 44mg/l of CO_2 is desired to be wasted to dissolve

100mg/1 of $CaCO_3$, make a ratio line T'A' with 100:44 and from T draw a line TA//T'A', and make AB⊥ TC to gain TB and BC. By BT the gained CO_2 is enewable used to dissolve more $CaCO_3$ with the value of AB after W_1 and W_2 mixed.

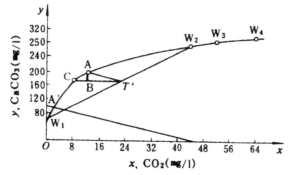

Figure 3.   Diagram showing the effect of mixing solution (slightly modified after Bogli,1969)

For the natural condition of regions of soluble rock mass with joint net and corrosion channels such as doline, shaft sinkhole etc., two or more than two kinds of saturated water solution are easily to meet to mix each other, so the practical effect of mixing corrosin is capable as many examples shown.

The Genesis of Human-Induced Sinkhole in the Shallow Region [14]
It is generally accepted that sinkholes induced by human activities have a close relation with

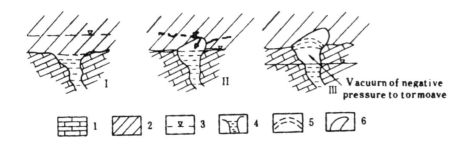

Figure 4. Sketch map showing the mechanism of vacuum suction (modified from Yuan,1987) 1.limestone;2.impermeable cover(semi-soils); 3.water table; 4. waterfilled cave; 5.succion zone;6.collapsed marjin of subsoil

the water table fluctuation, i.e., the fluctuation of the water table in the underlying karstified stata plays an important role in the process of the shallow collapse in the overlying soil. However, the explanations about how groundwater activities induce sinkhole development are controversial, and, of course, they are based on different sets of facts. For instance, some researchers hold that there is a vacuum suction mechanism when the water table in an underlying cavity is transforming from a confined state into a semi-filled state (Fig.4). This could be true when the overlying clay is suficiently stiff and the cavity wall is air-tight. This is equal to the mechanism of pumping in function (J.G.Newton et al.,1983).

ASSEMBLY FLOW THEORY & MAIN PETROLEUM-GAS RESERVOIR BEDS IN KARST REGIONS

Assembly Flow Theory for Karst Pipe (from Zou, et al.,1994) [18]
Karst pipe flow shows itself to have a faster velocity, greater discharge, lesser hydraulic gradient and lower potential of ground water level. It captures ground water from soluble pore spaces and vein pipes. It results in concentrated flow-assemble of ground water. This theory deals with confined water being nonlinear one; and it does not obey Darcy law. Its calculated method for influent seepage of reservoir karst is new one with which permeable coefficient K hardly gained will cite directly the parameters as unit water gradient velocity V and reservoir water head H and karst discharge cross   section W, etc.. So and so the charts gained; and nonlinear problems solved by hyperbola and exponent equation.

In karst regions, both of the flow field and flow net are too complicated due to the chemical-physical processes being added. This theory emphatically concerns the nonlinear morement of karst water in dam reservoirs. Especially it must be to reduce or solve the problems of seepage pressure/discharge/velocity and seepage cropping under dam foundation or around far the dam. The concret investigations ought to study hard with researchers in both field and laboratory. As for the dam foundation, the karstic caves/sinkholes might be to give remedies done as "engineering protections" of below sections.

Main Petroleum-gas Reservoir Beds in Carbonatitos with Pore Ceostructure Research [3]
These problems discussed pore struecture charaster and petroleum gas reservoir beds in carbonatites which distributed with the Permian System, the Triassic System, the Carboniferous System in Sichuan Province and the Sinian System in North China. The pore geostructure research on thin slide section from cast body by scanning electromicroscope study had practically perfect finished. Many parameters of pore ratio, specific surface area and pore distribution gained. Among them, the parameter of the latter is the most of theoretic and practical significance. Key factor is proe-throat size.

Therefore, pores and their combined relationship of throat channel and the ratio of pore vs. throat must be researched in details. The result of it may be used to guide further explaining for karst development. The methods of research of it are based on the geological statistics theory developed by the specialists of petroluem geology. Among them there are methods of measurement in laboratory and mathematic model;for the former including method of pore cas body (body-observing) technique [10] method of cappillary pressure and specific surface area method. Lo et al.(1981) presented a new conception that porosity texture of carbonatite should be described by the method of mixing empirical distribution, i.e. that porosity distribution of carbonatite may be regarded as resulted combination of prosity ditribution in the diagenesis and deuterogenous processes which can not be described by simple normal distribution. For the figure characters of geological mixing empirical distribution, their characteristic information may be represented by processing of rectangle method. The main figure characters of all terms arranged as Table 3.

With the aid of thin section observation under polarized light of cast body method the figure characteristics of porosity texture may be gained. Terms correlated by thin section of pore cast body are ratio of width of porosity   throat, intercoursed No. of pores, combination relationships of basic porosity types, curve characters of capillary pressures predicted and, ratio of porosity vs. Sight-area.

The relationships between porosity combination and permeability property of carbonatites of the Carbonifereous System, South   west and North China are shown in Fig.5.

Talbe 3. Showing porosity distribution in carbonatile with figrue characters

Figure Character	Expression	Meaning
Mean value $\bar{x}$	$\bar{x} = \left[\sum_{iL=1}^{p} x_i f_i / 100 \right]$	Mean position of value-picking for experimental data,i.e.to denote mean value of porosity distribution
Sorting-Coefficient σ	$\sigma = \left[\sum_{i=1}^{x} f_i (x_i - \bar{x})^2 / 100 \right]^{1/2}$	To describe the divergence on figuring axis of experimental data,i.e.to describe the sorting of pore size
Variation Coefficient C	$C = \sigma / \bar{x}$	To describe ratio of porosity sorting and its mean value, and to reflect the goodness or badness of porosity texture
Skewness S_k	$S_k = \sum_{i=1}^{x} f_i (x_i - \bar{x})^2 / 100\sigma^2$	be a measure for distribution asymmetry, to denote porosity distribution shifting from relative mean value, big or little
mxc	Important combination parameter	Showing parameter of summed intercoursed porosity times the goodness or badness of porosity texture
$m./\bar{x}$	Same as above	Showing parameter of summed intercoursed porosity divided by mean value of porosity texture

By the research shown in Fig.5 size varieties of pore throat were discovered to be very tremendous, varying from<0.1 μ to several hundreds of μ. It means that carbonatite beds should be one of the multi medium controlled by fissure pore throat. Physical model of this multi medium may be described with main fissure, big corroded pore, tectonic crack, corroded fissure and porehole, interstice of coarse crystal and of fine crystal [7].

Classification of carbonatites by means of rectangle processing is shown on Fig.6. The figure denotes that geological mixing empirical distribution can be used to describe the complicated porosity texture of carbonatites.

It is noted that the porosity textue is not isolated from that of genetic group, grain size and microcrystalline matrix of soluble rocks, hence the better method need perhaps combine lvith the classes of rock. Based on the classification of soluble rocks, thin section exams must be carried out to research the texture of carbonatites as well as lithologic characters. For dolomite with the molecule form replaces calcite and shrinks its volume into 12-13% to form new pore, so we prefer to emphasize that dolomitzation will promote micropore to develop and enlarge by seepage water, and hence, to strengthen the karst corrosion process. Based on macro-research and micro research, textures of carbonatite may be classified as pelitic (0.002 mm), grain fragment (0.002-0.005 mm), bright crystal (0.05-2 mm), bio-framework and recrystalline replacing texture (>2 mm) etc., For karst development, the texture control of carbonatite may be divided as:

1. control for original pore types and their distribution;

2. control for the corrosion degree of carbonatites. For an example, near Leiwu, shandong, for the limestones with pelitic crystal texture of 2,4,6 members, the karst is not developed well; while for the dolomitic limestone with grain-fragment texture of 1,3,5 members, the karst is highly developed serving as stable water bearing beds. The results of research cause and giude

investigators to pay more attention on the textures no less than chemical or mineral components of carbonatite.

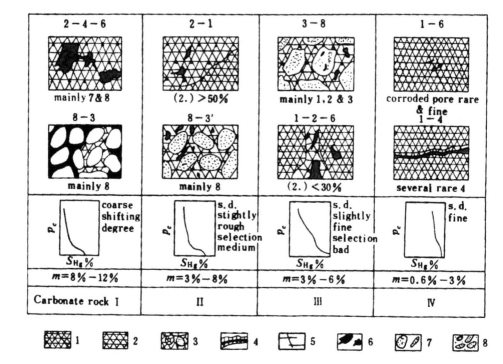

Figure 5. Sketch showing relationships between porosity texture combination and permeability property of carbonatites, c.system,s-w China (after Lo et al.1981). I -Reservoir Permeability rocks good; II -Reservoir Permeability rocks meta-good; III-R. P. rocks medium; IV-R.P.rocks bad. 1-slit of fine crystal,2-slit of coarse crystal, 3-slit of coarse crystal of particles,4-corroded slit, 5-tectonic joint, 6-corroded pore without sorting, 7-corroded pore within particle, 8-pore among particles.

PROTECTIVE MEASURES FOR SLOPE IN KARST REGIONS

The protective measures include principally three types for slope surface, slope body and slope foundation[11].

For slope surface:

All flow and karst flow should be drained from slope engineering surface, especially, when soluble rock of slope is composed of type I and/or II showed in Fig.6. These concern with the drainage of atmosphical precipitatin and its run-off, underground water near the location of slope surface and with rising the capacity to resist the corrosin of all water flow.

For slope body:

All karst caves should be abolished especially for those by which the slope stability is damaged. Thess include filling the dry cave hole with competent material, injecting cement to condense the cave deposit and blasting the irregular holes with fillings etc.

For slope foundation:

All cave channels or other karst holes damaging the slope foundation should be strengthened on the base of evaluating their safty thickness of cave top and safty distance from slope foot.
protective measures for slope

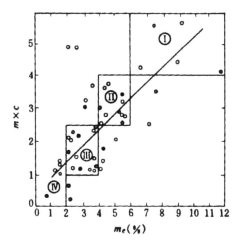

Figure 6. Diagram showing evaluation by classification according to figure characters of rectangle method for soluble rocks of the Carboniferous System, South west China (from Lo et al.1981). I $.m_e$ ·6%, mxc ·4%; II $.m_e$=4~6%,mxc=2.5~4; III $.m_e$=2~4%,mxc=1~2.5; IV $.m_e$ 2%, mxc 1.5; m_e-0.1μ,efective porosity %

For slope surface
Cement coating may be done by injecting method so as to rise the resistence against flow entering in the soluble rock through pore throat; at the same time it is better to concentrate the surface flow with well paved "waterproof" drain ditches arranged on the top border and the foot toe of the slope. Cement coating and ditches may protect the slope free from surface flow to attack and dissolve rock. If there is fault zone these measures or stone revetment set in concret for slope surface should be done.

If there are dry caves discovered on the slope then the stone revetment set with or without concret may be used to block and fill up over there.

For slope body (Fig.7)
In case of there is deeper hole in soluble rock overlying on the unsoluble beds we should block up the hole and put on a waterproof bed around it to stop karst flow development so as to protect for the contact plane of bedding reducing its strength and resulting slide.

The upper course of lateral cave should be block up by means of injection cement through boring holes[6].

Boring hole arragement for injection. When direction of cave channel is known by some prospecting methods, the boring holes are arranged along channel with 10 meters per hole. If position and direction of a lateral cave is not yet known, the boring holes may be arranged zigzag within capable area. The boring hole discovered cave is served as that of cement or clay-sand slush injection. When the size of cave is less than 2m the injection diameter of boring hole may be adopted for 66-108mm. If the slope is important, the "underground wall" resisting flow may be done by cement slush injection under high pressure when the pore-throat of soluble rocks is several times of cement particle (0.09mm).

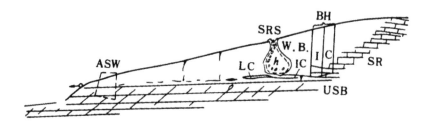

Figure 7. Protective measures for slope body USB-unsoluble beds, SR-soluble rock, W.B.-waterproof bed, srs-stone revetment set, h-hole, LC-lateral cave, BH-boring hole, IC-Injection cemet, ASW-Antisliding wall.

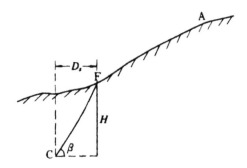

Figure 8. Safty distance from slope foot to karst cave AF-slope, F-foot of slope, c-cave.

In case of an antisliding wall is needed, the design of it is controlled by potential sliding force which may be calculated by force transmitted method.

For slope foundation:
a.Safety distance evaluation from slope foot to karst cave. Suppose there is karst cave near slope foot (Fig.8), once the cave collapses as a doline it will damage the neighbour slope. Hence, an artificial slope cutting must keep a safty distance.

$D_3 = Hcolβ$

Where   H-thickness of apical plate of a cave(m);
          β -diffusive angle of collapse of a cave(degree)
          $β = K(π /4 + φ /2)$
          K-safty coefficient (0.8-0.9);
          φ -internal frictional angle of rock..

If there is soil overlying on the apical plate, then angle β equals to 45° within soil.

b.Safty thickness evaluation of apical plate of a karst cave:

There are three methods to evaluate roughly for it. 1) Evaluation based on force receiving condition of a beam plate.

Consider the apical plate of karst cave as a beam plate receiving general load included beam weight itself and additional loading $Q_1 + Q_2$ (Fig.9). Suppose the width of an additional loading equals to B and of cave to b, then we have section coefficient

$$W = \frac{M}{\sigma} = \frac{BH^2}{6}$$

safty thickness of apical plate of karst cave will be

$$H = \sqrt{\frac{6M}{B[\sigma]}}$$

Where[σ]-bending strength of rock, for limestone, usually equals to $P_d/8$

(Pressing strengt)(kN/m^2);

M-bending moment (kn-m)

When straddle of apical plate contains with joint fracture and with both abutments being stable, then consider it as cantilever beam

$$M = (Q_1 + Q_2) b^2/2$$

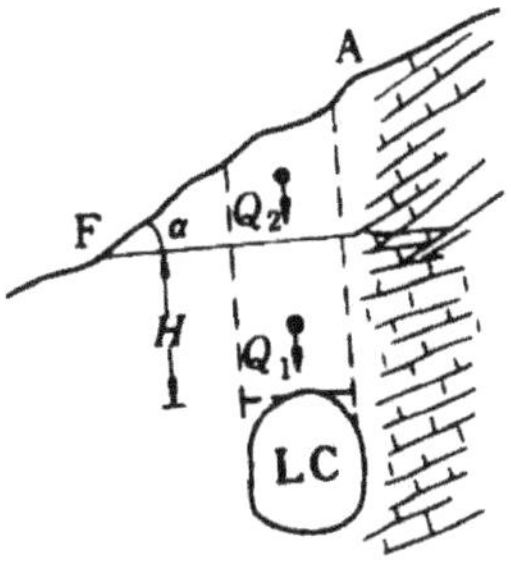

When one of abutments contains with fissure and with the other position of apical plate being stable, then consider it as simple supporting beam and evaluate

$$M = (Q_1 + Q_2) b^2/8$$

If the rock bed is competent to receive force and stable, then consider it as stable beam and evaluate

$$M = (Q_1 + Q_2) b^2/12$$

When the rock bed of apical plate is strong in strength and stable with the thick bedded and almost horizontal, then the safty thickness of apical plate may be evaluated approximately by the expression above to gain H.

2) Evaluation based on shearing strenth conception

Suppose rock mass weight of apical plate of a karst cave equals to Q_1 and with overlying additional weight being Q_2. The thickness H of apical plate need resist shearing action resulted from body weight and additional weight. Under limit equilibrium condition resistence strength T of rock mass of apical plate will be

$$T = Q_1 + Q_2$$

and $T = H \cdot L \cdot \tau$

hence $H = \dfrac{Q^1 + Q^2}{L_\tau}$

Where L-peripheral length of a karst cave in section; τ -resistence strength of rock, for limestone usually equals to $P_p/12$ (permitted pressing strength). This method is adopted for small caves; for H of large caves do not be mainly controlled by shoaring rosistence.

3)Evaluation based on fracturing arch conception (after Plotogyaconov,1908)

Fig. 9. Safty thickness H of apical plate of a karst cave

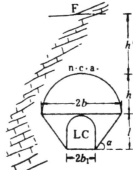

Fig. 10. Showing the height of natural collapse arch (n.c.a) c-lateral cave

Suppose karst cave s rectanglar or with a semiarch (Fig.10). When cave develops the semi-arch must be enlarged upward gradually. Once the arch reaches its highest position it stops to enlarge and cave keeps limit equilibrium condition. The highest height h of the arch of a cave is so called the height of natural collapse.

$$h = \frac{b_1+1 \cdot \tan(\pi/4-\varphi/2)}{f}$$

Where b_1-half of the width of karst cave;

l-height of the cave;

φ-internal friction angle of soluble rock;

$f = P_p/100 = c_m/\gamma H + \tan\varphi_m$;

H-buried depth of the cave top;

c_m, φ_m-mean of weighted mean value of c, φ of soluble rock.

f is termed as competent coefficient or apparent friction coefficient, while the later term is reaCFsonable for it being nondimensional.

If the cave is irregular in shape, the larger sizes of l and bl should be adopted. Rock mass weight above the natural collapse arch is supported by the arch itself. Hence the safty thickness H of the cave equals to k(h+h′) (thickness h′ is needed to support upper loading above the collapse arch; while k is safty coefficient larger than 1.0).

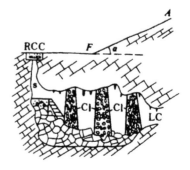

Figure 11. Sketch showing the apical plate of karst cave strengthened by stone concrete columns. cl-columns, s-shaft, AF-slope, lc-lateral cave, RCC-reinforeed concrete cover.

This evaluation method is fitted for fractured rocks buried less than 250-300m. If the safty thickness is not enough to keep slope foundation stable, then the karst cave should be treated by some remedy measures such as Fig.11 shown. In case of the remedy measures not convenience or impossible, we must select the better location for the artificial slope other than the dangerous site.

FOR SLOPE MOVEMENT

Air-cushion Layer Hypothesis

Under discussion for Sherman landslide, Alaska, shreve(1966) suggested firstly lubricative sliding-function concept of air cushion layer. He considered that it is due to the enclosed air cushion layer along sliding bed between moving mass and unmoving one. This causes critical shear stress along sliding plane to be vanished. After the rockslide of Black Chicken Mt. at Mohave desert border of South California researched, he found a hypothesis from above the concept, and considered sliding brecciated zone to be caused by compressive air layer and lubricated with mass moving. This is the very air cushion mechanism. The criterions of this mechanism are high velocity transport, lithological homogeneity in situ, and evidence of air mass emission, and dust clouds at lateral and end border etc.. Others regarded the above criterions being not needed sure even at least one having the existence of air cushion layer, and also comprehensive consideration was the same, there still questions existed. We noticed key point of question lying where there was no physical chemical analysis done yet. And soil rock groups with gypsum and Ca(Mg) CO_3 particles, once the huge slide starts to occur, there must produce the free $CO_2 \uparrow$ and

$H_2O \uparrow$. For the free gas scattered only within rock   soil group of moving shiding beds and hence, the formed air/gas cushion layer is posible.

Granular-fragments Flow Theory
The theory was presented by Hsu(1978). He considered huge slide not to be sliding, but flowing. This viewpoint is equvalent to that of Heim. It is based on the investigated data of the Moon. It is considered that huge slides are similar to those on the Earth; but the former has no water and air, being refused to interpret with air cushion layer. So called fragments flow is similar to huge slide with the gaps of coarse fragments being fullfilled with enormous powdery silt. Powdery silt plays a role as flow and gas masses; and caused normal pressure among granular particles to decline. Then anti   sliding force approaches to lose efficacy. Reserched results denoted that desired initial slope angle of fragments flow to move is less than that of critical angle desired by the whole sliding friction. By theoretical calculation, in the huge fragments flow there existed powdery-silt-dust suspension of different densities; it is very possible. This theory is good for granular-fragments flow on the Moon; but bad for huge slide on the Earth.

EXPERIMENTS AND EASY SLIDING THEORY

The experimental conditions are divided into two aspects-in laboratory and in field. As a whole, they should be combined to analyze. For analysis, some of them may be available, others should be added. It is emphatically focussed on the easy sliding theory (Yan, 1990,1994). Here, we would start at experimental mineralogy and petrology (including groundology), and end in the way to support the air/gas cushion layer theory and scale-dependent rockslide mechanism.

Calcination Experiment with Loess/Fine Soil
Experimental condition is mainly restricted in fields. The aim is known to the process and the results whether they are compared with gliding plane of huge slides. Several trenches were dug on different slopes where calcined experiments might be taken with fire. Combustible materials are coke, natural gas/oil. Temperatures are measured by thermal gauge and thermal electric couple with nichrom and nialumino-mg alloy. The gauge, couple & optical thermometer are used so and so under 0-1000°C or more. Natural water contents in loess/soil are driven off about 200-300°C. The temperatures of calcined materials were reached to 905-1100°C even more. The structures of them are very similar to those of bricks. Analysis of this experiment denoted that initial   mid stage of the process in field may be compared with the practical characters of gliding plane in huge slides; both of them are nearly coincided with each other under 400-600°C. However, the end-mid stage in the experiment process did never be compared with; and could hardly be controlled by workers and investigators. Since calcined bricks did not yet be happened on gliding plane in huge slides. The physical-mechanical parameters in different stages are measured and gained.Some of them are given; disintegration, natural water contents and dynamic sliding possibility could not be correlated; i.e.natural loess/fine soil in huge slide being compensated with water on the gliding plane at lower temperatures.

Calcined Experiment with Gypsum
In laboratory, gypsum is heated about 120°C with the temperature measured by above said instruments. It is lost in a portion of its crystallization water in hydration. Then (from Dana, 1946)

$$CaSO_4 \cdot 2H_2O + heat = CaSO_4 \cdot \frac{1}{2} H_2O + 5H_2O \uparrow \qquad (1)$$

Thus the chemical equation might be reacted once more and again under the condition with plenty gypsum. In field, gypsum rock or fine ground gypsite with the free evapoated water $1.5H_2O \uparrow$ will be replenished continuatively; and it does not scatter along sliding bed. Thus, it serves as a vapour cushion layer to cause huge lide. Here we inroduced chemical potential concept from equation (1) to support water vaporation mechanism.

Calcined Experiment with Limestone/Dolomite

In laboratory and in field, limestone/dolomite, it is representation, separately heated about 898 ℃ /756 ℃ with temperature measured by above said instruments. The decomposition of it, reacts as follows (from Dana, 1946); we inferred from its reaction:

$$CaCO_3 + heat = CaO + CO_2 \uparrow \qquad\qquad\qquad (2)$$

In calcination, equation (2) reacts once and again, it has to be driven off carbon dioxide. In field, lime kiln may be observed as the process in production to obtain CaO and to drive off $CO_2 \uparrow$. Temperatures in the kiln are measured with instruments mainly optical thermometer. In order to reach the full decomposition of limestone/dolomite, the highest temperature surpasses 1000 ℃. Comparing with the sliding plane of huge slides, there is the function introduced to thermodynamics chemical potential in chemical activity being occurred at the deeper zone. This reaction will not continue until calcined experiment or a huge slide event is finished. Due to carbon dioxide released, the gas cushion layer may formed along gliding plane.

Easy sliding theory

It deals with that one in which covering discipline field with slope movement and strike   slipe orogeng is concerned with. It is interestng in rock   soil groups, structural types, argillaceous marls and other interbedding strata prone to slide.

1.Common character of slide and structure

It is emphasized in the gliding plane on which the energy/geostress may be concentrated, adjusted or released. As for slide, our theory is built up for huge one or great slope movement. As for structure, it is, especially, for gravity gliding structure; thrust nappe (reverse gliding), decollement of the sediment lying on the basement complex of the Juras and mountain root above orogenic zones, etc.. The differences of both slide & glide are hardly to be divided. Heterogeneity of material distribution and instability of gravity play an important role in slope or glide movement. Similar basis of both is the soft   "guiding"   surface. So called instability of concept of gravity is designated in material distribution with overcoming potential; but that potential may be released through movement.

2.Reviews for huge slides/glides

Huge slides with high energy and max. horizontal distance are discussed. They represent the typical ones. Among them, there are physical and mechanical data but without any chemical ones. Here, it is the very question. To solve this one means that the formed mechanism of huge lides might be bright in foreground with chemical analysis in the process/chemism. In our subjects they are: chemism   process with chemical equation (1) and (2); chemical energy   thermodynamics potential energy with heat quantity; thermomolecular pressure and thermosiphon efficacy; heat emission and thermal diffusion. Once the process of slide or geostructure to begin, the chemism of eq. (1) or (2) or the couple of the twos must be to react on sliding bed/gliding plane. To start with a moving slide/structure, it is satisfied for requisite condition that in the sliding/gliding plane rear, there the least work principle ought to have satisfied for. For suficient condition, there ought to have been satisfied with sliding condition being waited to be prone to slide. For both of requisite and sufficient conditions are satisfied at same time, a slide/glide will occur certainly and theoretically. Large amounts of $CO_2 \uparrow$ are released to cause gas cushion layer along sliding/gliding plane. so and so the huge slide/glide formed. The rocksoil groups with $Ca(Mg)CO_3$

are the materials of huge slide. Gas cushion layer along sliding bed plays a role to buffer the slid plane with higher temperature in raising manner. So, there is never a slid plane calcined with brick structure like as above mentioned. In practical huge slide, it is kept in with high temperature on the sliding plane due to frictional heat of itself. At the last, other than the reviews we would like to say that many huge slides are occurred in calcareous rock soil groups including carbonatites, marls or soluble rocks.

CONCLUSIONS

1.Chinese karst phnomena are either the environmental resources or the natural disasters which damage the slopes stability.

2.Comporing karst and karst development are concerned with many topics. These have to argue and discuss basic rationale and some theoris, namly, comparing karst principle, double cycles theory, the physiographic stage law related to karst developmemnt and mechanisms of shallw sinkhole and deep pore cave and so on. Comparisons between karst regions in China and in Europe and elsewhere are discussed in details.

3. A simplified table showing the physiographic stages and karst doveloment history in China is fairly well made. It includes several topics shown with Karst stage, Period, Crust movement and the most typical Provinces. From this table the natural regularity of kastology might be summarized.

4. Assembly flow theory results in concentrated karstic water flow. In karst regions, both of the flow field and flow net are complicated due to the chemical-physical processes added. This theory emphatically concerns the nonlinear movement of karst water in reservoirs. It must be to reduce or resolve the problems of seepage pressure and its cropping under dam foundation or around far the dam. In summary, this is a serious environmental protection problem in hydraulic engineering construction.

5. In carbonatite areas for the prospecting of petroleum-gas reservoirs, the pore cast body technqiue with the geological statistics theory is available. The thin section slides are observed under polarized light microscope or by electromicroscope. For the observation, the key technology is focused on the porethroat structure of oil-gas reservoirs.

6. For slope stability in karst regions, the protective engineering remedies are carried out theoretically and practically.

7. It had been introduced as a new concept of chemical energy in the huge slide process with chemism and formed mechanism . Experiment and test in fileld and in laboratory are carried out for gypsum, limstone/dolomit and rock-soil groups contenting carbon componds. Experimental results and chemism concept might be used to support the gas-cushion layer hypothesis and scale-dependent rockslide mechanism. Easy sliding theory might be used to solve either the slide qustions or the ones of glide moving geostructure.

REFERENCES

1.V.A.Bogli,Neue Anschauungen uber die Rolle von Schichtfugen und Kluften i der karst hydrographischen Entwicklung, Geol. Rund. Vol.58,No.2,395-408 (1969).

2.W.M.Davis. Origin of Limestone Carerns. Ggol. Soc. Ame.m Bull. Vol.41, No.3. (1930).

3.E.C.Dahlber. appliod Hydrodanamios in Potrolooum Exploration. Springer-Verlag Hong Kong Ltd.

4.M.Herak and V.T.Stringfield.(eds.). KARST Important Karst Regions of the Northern Hemisphere. Geological and Palaeontological Institute, Faculty of Natural Science. Zagreb, Yugoslavia. Department of the Interior, U.S. Geological Survey, Washington, D.C., U.S.A. Elserier Publishing Company, Amsterdam-London-New York (1972).

5.G.X.Dai. General regularity of karst development in the regions between the Yangzi and the Qing River. Proceedings of Hydrogeology & Engineering Geology (in CHinese) 99   114. Ministry of Geology, Institute of Hydrogeology & Engineering Geology (Edt.). Press of Geology, Beijing (1958).

6.W.R.Dearman, General Report, Session I: Engineering properties of carbonate rocks, Bull. of IAEG. No.24,3-17 Essen, F.R.G. (1981).

7.R.L.Folk, Practical petrographic classification of limestones. Bull. Amer. Assoc. petroleum Geologists,43, 1-38 (1959).

8.Geology-Geography Divistion of Academia USSR et al. Abstract proccedings of karst science reserach USSR (in Russian). Sub-Volume 5, 1-3 Main types of Hydro-dynamic zones of karstwater by G.A.Macshemorich 111-129 (1956).

9.Institute of Geology, Academia Sinica, Chinese Karst Research, (in Chinese) Scientific Publishing House. Beijing (1979).

10.Zhetan Lo et al., Research and progression of porosity texture of main carbonate reservoir beds in China, Petroleum exploration and development, (in Chinese) No.5 40-51 (1981).

11.Main editor: First Designing Institute, Railway Ministry, Road Bed under special conditions, Chapter 16 pp4.157-4.162, People' s railway publishing house, Beijing J.G.Newton et al.(1978).

12.Mei-E Re et al., Outline of Karstology, 170-212,86-99, Business Publishing House, Beijing (1983).

13.M.M.Sweeting Karst in China-Its Geomorphology & Environment. University of Oxfor, U.K. Springer-Verlag Hong Kong Ltd (1995).

14.The Proceedings of 2nd Multidisciplinary Conference on Sinkhole and the Environmental Impacts of Karst/Orlando/9-11 February (1987).

15.T.Z.Yan.Landslide investigation: Rewiews in huge slides and formed meckanism. The 7th Intern. Symp. On Landslides Vol.2 1439-1444 (1996).

16.T.Z.Yan. Karst development and protective measures for slope in China. From Intern Exchange Lectures for Technique University, Clausthal, F.R.G.(1985).

17.D.X.Yuan. China Karst. Press of Geology, Beijing(1993).

18.C.J.Zou (Edt.). Karstic Engineering Geology in Hydranlic & Hydroelectrocity (in Chinese), Press of Hydralics & Hydroelectrocity. Beijing (1994).

Proc. 30th Int'l Geol. Congr., Vol.22, pp. 139-164
Fei Jin and N.C. Krothe (Eds.)
© VSP 1997

ASSESSMENT OF GROUNDWATER CONTAMINATION FROM AN INDUSTRIAL RIVER BY TIME SERIES-, FLOW MODELING- AND PARTICLE TRACKING METHODS

Hongbing Sun
Department of Geology, Temple University, Philadelphia, PA 19122

Manfred Koch
Department of Geohydraulics and Engineering Hydrology, College of Engineering, University of Kassel, Kurt-Wolter Strasse 3, 34109 Kassel

Xinlan Liu
Mecuri and Associates, Inc. Southampton, PA 18966

ABSTRACT

The interaction of an industrial river and the groundwater aquifer in Florida are modeled by time series-, flow modeling- and particle tracking methods. The major purpose of the modeling effort is to delineate the possible contamination corridor in the aquifer as may be caused by the infiltration of polluted water from the industrial river. The interrelations of precipitation, river discharge and groundwater data series are first analyzed by methods of structural time series, in order to quantify the interdependence of the groundwater table and river gauge heights, and thus to statistically examine the hydraulic possibility of aquifer contamination. The interaction of the stream and the groundwater is simulated by the USGS MODFLOW model. For the calibration, simulated water tables are compared with the historical records of monitoring wells in the adjacent aquifer. Lateral 'stagnant' points of the water flow based on the transient simulation are connected to delineate the maximal contamination corridor along both sides of the stream. In addition, particle tracking simulations with varying water sources and forcing conditions are conducted to compute the dynamic movements of water particles out and along the river banks. The modeling results show that lateral migration of contaminants close to the industrial discharge point might be up to several thousands feet in cross-river direction during times when the normally gaining (effluent) stream becomes sectionally a losing (influent) stream. The last situation occurs especially during multiyear-long time spans with less-than-normal precipitation which, for Florida, happens during times between major El Niño occurrences; i.e. during La Niña years.

1.   HYDROLOGICAL BACKGROUND AND GOAL OF THE   STUDY

The small industrial river studied here is the Fenholloway river and is located in Taylor County in northwest Florida (Fig. 1). The stream has an average flow discharge of 132 ft³/s as has been measured at the USGS river gage station 'At Foley' near the US 19 bridge over the last 40 years (USGS, 1992), though with strong seasonal variations. A few hundred feet upstream of this gage station, a paper mill factory has been discharging constantly about 70.5 ft³/s of wastewater into the river since the beginning of its operation in 1954. In fact, the Fenholloway presents a 'present-day anomaly' in the sense that it is the only river left in Florida that is still classified as a 'category-five' river which, by Florida law, is a stream whose only purpose is to serve as an open wastewater channel. The amount discharged by the paper mill is approximately equal to the quantity of water pumped by the factory from the underground by means of 7 to 8 production wells (Fig. 1). Although some additional water entering the hydrological system is produced from the timber decomposition process, it appears to be essentially lost through evaporation. The discharge water from the paper mill factory contains various organic constituents (Watts and Riotte, 1991), that are formed during the cellulose manufacturing process. The organic constituents isolated from the lignin residue of Fenholloway river samples (the 'Fenextract') can be classified into purgeable, extractable and NPTOX (Non-Purgeable Total Organic Halogen) organic compounds.

As most of the rivers in Florida, the Fenholloway river is generally an effluent or gaining stream, which means that the groundwater is recharging the river along its course. However, because of the paper mill's 70.5 ft³/s of contaminated water discharge into the river, a water- mound of about 1-2 feet is created close to the effluent point which may cause a reverse of the natural hydraulic gradient in its vicinity. The severity of this reverse gradient might increase during long dry seasons when the water table in the surrounding aquifer can become very low, while a high stream gage elevation is still maintained by the paper mill's huge effluent. Then the river may become influent in the section close to the discharge point, i.e., it may lose water to the surrounding aquifer and the potential of groundwater contamination by the polluted river water can arise. Also, because the paper mill pumps an average of 70 ft³/s of water from the underground, a wide drawdown depression cone of about 4 miles in diameter is formed around the pumping well gallery (Fig.2). This depression cone lowers the regional groundwater table even further during dry seasons, thus further increasing the lateral inverse hydraulic gradient. It is to be expected that the mounding effect in the vicinity of the paper mill's discharge point is less prevalent during a wet season, when the natural river discharge is already high. We call the above hydraulic situation in the present paper also the 'dry-season' scenario for aquifer contamination from sections of a normally effluent stream that have become influent. Since most of the residents along the Fenholloway river get their drinking water from their own private backyard wells, and given the above mentioned potential for groundwater pollution, it has been proposed by Florida public health authorities to connect the residents within the 'risk-corridor' along the river to public water supply lines. For this reason it is necessary to define the extreme boundaries (or water- lines) of this possibly polluted corridor along the Fenholloway river. This is the major objective of the present study and it will be achieved by

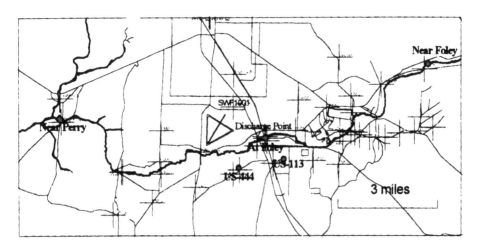

Figure 1. Study region with Fenholloway river and locations of wells (crosses) and streamgages (diamonds). Also depicted is the discharge point of the paper mill's wastewater effluent.

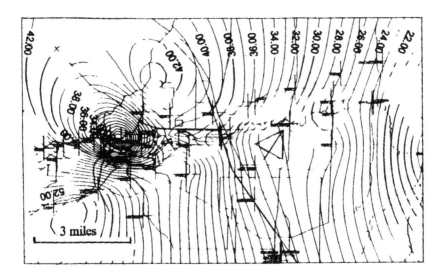

Figure 2. Water table contour map (ft) for 5/11/1994, generated with the SURFER plot package.

means of time series-, flow modeling- and particle tracking methods. Various conceptual models of this system that may lead to hydraulic conditions prone to aquifer contamination from polluted river water are tested by using hydrological and meteorological data from both historical records and obtained from a five-month-long groundwater and stream survey of the study region. Ultimately, 'worst-case' estimates of the possible spatial extensions of a contaminant plume emanating from the river over the short and the long term will be established by these models.

It should be noted, however, that the present study will only consider a hydraulic or advective flow approach to establish conservative water-lines beyond which no groundwater contamination from in-flowing river water is to be expected. Therefore, the classical solute transport mechanisms of hydrodynamic dispersion, adsorption and other chemical and biological decay reactions will not be included in the model. The implications of these limitations on the calculated estimates of the water-lines are discussed in detail in Koch et al. (1995). For example, neglecting adsorption will discard the process of retardation which, depending on the nature of the contaminants, can slow down its migration by factors ranging between 1-3 (Bear, 1979; Maidment, 1993). This means that the water-lines established in the present study by means of an advective travel-path analysis will represent the most extreme (worst-case) bounds. In other words, the proposed corridor of residential wells along the river that is to be connected to the public water supply has an inherent safety factor built in.

2. HYDROLOGICAL TIME-SERIES MODELING AND SYSTEM FEEDBACK STUDY

Time series analysis helps to statistically define the characteristics of the data series involved and helps to quantify interrelations and possible correlations between different data series. The data series investigated here are the stream gages and -flow and the groundwater table elevations and precipitation records.

The gage-height data series analyzed consists of the monthly average records measured by the USGS over the last 20 years at the station ATFOLEY downstream of the paper mill's effluent point. The water table elevation series over that time period is taken from the USGS well #444 which is located about 1000m south of the river and 1500m west of US19. Note that other well-data was also available and could be tested using the same techniques, however this well was chosen because it is located outside the drawdown cone of depression and, therefore, its groundwater table elevations are not significantly affected by possible variations of the pumping rate of the plant. Fig. 3 depicts the time-series plots for both the stream gage elevations and the water table heights. Note in particular the inverse hydraulic gradient between the stream and ambient aquifer during the long drought period of 1988-1992 with anomalously low precipitation. The latter is also shown in Fig. 3 and has been gathered from the Florida Climate Data Center.

Using the Box-Jenkins methodology (Box and Jenkins, 1976; Pankratz, 1991) seasonal autoregressive integrated moving average (ARIMA) univariate models are established for groundwater table elevations, stream gage heights and precipitation levels. For a detailed discussion of the steps involved in the set-up of an ARIMA model, see Sun and Koch (1996) and Sun et al. (1997a).

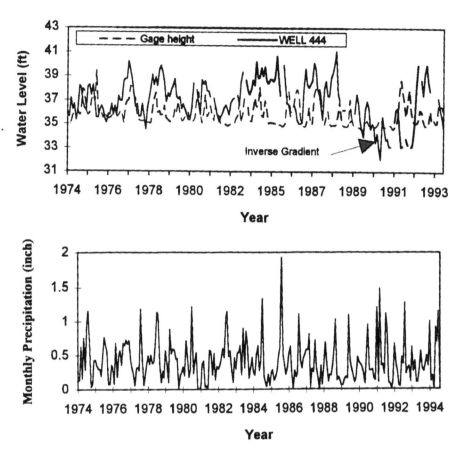

Figure 3. River gage heights and water table elevations at well #444 (top panel) and precipitation in the area (bottom panel) for the last 20 years.

The three autocorrelation plots in Fig. 4 give some insight into the physical phenomena that govern the three processes of precipitation, water table elevations and river gauge heights, respectively. For example, the computed autocorrelations for the precipitation are significant only for very short lag-periods of one to two months which shows that rainfall variations are rather short-term processes. Moreover, the autocorrelations exhibit to some extent the seasonality of the rainfall.

The autocorrelations for the stream gauge heights look somewhat different from what is usually observed for a natural stream (Maidment, 1992), in the sense that correlations are significant over a long period of time. This is the direct consequence of the large permanent industrial discharge into the river. Finally, the autocorrelation for the groundwater table elevations demonstrates illustratively the long-term (low

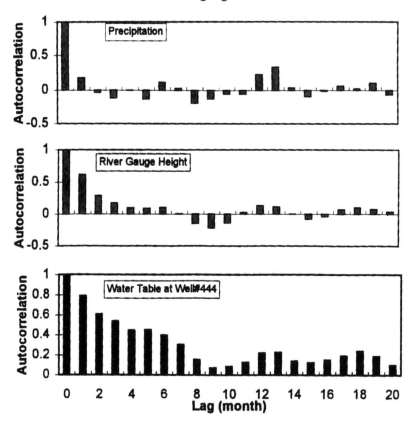

Figure 4. Autocorrelations of precipitation (top panel), river gage height (middle panel) and water table elevations at well #444 (bottom panel).

frequency) processes that govern the recharge of an aquifer. One notices significant autocorrelations over longer lag-periods and the water table changes exhibit less seasonal trend than the precipitation.

As an example we provide in Eq. (1) the ARIMA model for the water table elevation (h_t) of USGS well #444, as computed by means of the SAS statistical package (SAS Inc., 1993):

$$(1+0.2445B^2)\ \nabla h_t\ =\ (1-0.23B^2-0.217B^{10})\ a_t \tag{1}$$

Here B is the backward shift operator with $B\,X_t = X_{t-1}$, ∇ is the differencing operator with

$\nabla=(1-B)$ which is used to make the time series stationary, and a, is the random shock component. The form of the left side of Eq. (1) with the term B^2 illustrates that the water table elevations have a 'memory' of up to two months.

Under a natural condition, the groundwater flow of the effluent river is a one-way process; i.e., water moves from the ambient aquifer into the river and the stream gage elevations depend on the groundwater level. This would imply, when setting up a so-called dynamic regression or linear transfer model (cf. Pankratz, 1991; and Sun and Koch, 1996), that the groundwater table elevation is the input signal and the river gage height the output signal of the model.

On the other hand, with 45 million gallons/day of wastewater discharged into the Fenholloway river, the question arises whether this interaction process in the vicinity of the effluent point is still one-way; i.e., whether the water is still only flowing in one direction from the aquifer towards the stream. If the answer is no, then the question arises about the possibility of reverse lateral migration of river water into the adjacent aquifer. In a time-series analysis, river gage heights (which over rating curves are related to the discharge rates of the river itself) will no longer only be a passive output signal. There is a 'system feedback' from the output to the input signals. The system feedback relates to the interaction (or the interrelationship) of the groundwater table and stream gage levels.

In order to examine this interrelation of the "input and output" data series, both river gage height- and ground water table data series are prewhitened by the above ARIMA model (1). Their residual series are then crosscorrelated. As shown in Fig. 5, significant correlations exist at both positive and negative lags after this prewhitening process. Whereas the significant positive crosscorrelations for positive lag-times imply that the river gage heights are related with the groundwater table

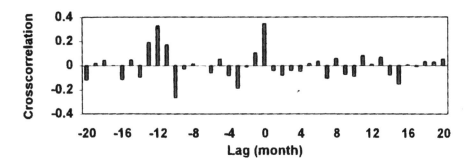

Figure 5. Crosscorrelations between river gage heights at ATFOLEY and water table elevations at Well #444.

elevations, the significant crosscorrelations at negative lags state that feedback from river gage elevations to groundwater table elevations exist as well. Thus, the water table elevations of well #444 and the river gauge heights at gage ATFOLEY at a given time are not independent from each other, leading to the possibility of hydraulic infiltration of polluted river water into the aquifer during certain periods of the analyzed 20 year-long data series.

3. FLOW MODELING OF THE STUDY SITE

To quantify the extent of the previously stated possible lateral migration of stream particles into the surrounding aquifer, the USGS finite difference groundwater flow model MODFLOW(McDonald and Harbaugh, 1984) will be used in the present section.

3.1 Governing equations of the flow system

The groundwater flow equation solved by MODFLOW is the classical groundwater flow equation which, for the purpose of the present 2D (horizontal) application, is written here as (Bear, 1979; McDonald and Harbaugh, 1984):

$$\frac{\partial}{\partial x}\left(k_{xx}\frac{\partial h}{\partial x}\right) + \frac{\partial}{\partial y}\left(k_{yy}\frac{\partial h}{\partial y}\right) - N = S_s\frac{\partial h}{\partial t} \tag{2}$$

where k_{xx} [LT^{-1}] and k_{yy} [LT^{-1}] are the local hydraulic conductivities along the axes x and y (which are taken as isotropic though in the present case; i.e, $k_{xx} = k_{yy}$); h [L] is the hydraulic head; N [T^{-1}] is the volumetric flow per unit volume of water sources and/or sinks; and S_s [L^{-1}] is the specific storage.

Within the study area, the superposition of three major factors will essentially control the piezometric head distribution and, therefore, the overall groundwater flow pattern: (1) The regional groundwater flow (which is mainly driven by the topography of the area and will be generated in the model by means of appropriate boundary condition, as will be discussed further down), (2) the stress of the pumping well field of the paper mill, and (3) the interaction of the Fenholloway river with the ambient aquifer, the understanding of which is, of course, the eventual goal of this part of the study.

Regarding the simulation of the stream-aquifer interaction, the equations which need to be specifically mentioned here are the leakage equations as used in the river package of MODFLOW. The river package allows, on one side, water to flow from the aquifer to the sink reservoir, thereby removing water from the model by seepage into particular reaches of the stream. On the other side, it also permits water to flow out of the stream into the aquifer. The river package uses the streambed conductance (C_{riv}) (L^2T^{-1}) to account for the length (L) and width (W) of the river channel in the cell, the thickness of the river bed sediments (M), and their vertical hydraulic conductivity (K_r) where

$$C_{riv} = \frac{K_r LW}{M} \tag{3}$$

The rate of leakage (Q_{RIV}) (L^3T^{-1}) between the river and the aquifer in the case when the riverbed is fully saturated is calculated from

$$Q_{riv} = C_{riv}(H_{riv} - h) \qquad\qquad h > R_{bot} \tag{4}$$

where H_{riv} is the head in the river (the gage elevation) and h is the head in the ambient aquifer. Depending on whether $H_{riv} - h$ is negative or positive, the river will become a gaining (effluent) or loosing (influent) stream, respectively. However, when the water table in the aquifer falls below the bottom of the streambed (R_{bot}), leakage stabilizes and Q_{riv} is calculated from

$$Q_{riv} = C_{riv}(H_{riv} - R_{bot}) \qquad\qquad h \leq R_{bot} \tag{5}$$

3.2 Calibrated steady state models

3.2.1 Finite difference grid and general set-up of the models

The finite difference grid of the study area is shown in Fig. 6. The area of the model has a size of about 40 x 20 km and is centered in E-W direction close to US 19. The model area covers the paper mill's pumping well field and the ponds, and extends westward close to the gulf coast. The mesh is telescopic with finer grid-sizes in the center of the model domain in the vicinity of the paper mill's discharge point along the river; i.e., the area which is naturally of interest here. The gridding of the mesh results in a resolution of 50 to 100 m in NS direction on both sides of the Fenholloway river and of about 100 m in EW direction in the vicinity of the effluent point.

The primary purpose of the steady-state calibration is to determine optimal values for

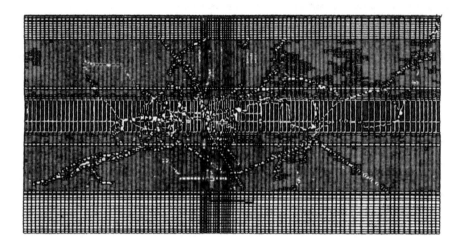

Figure 6.   Finite difference grid used in the MODFLOW models .

the hydraulic conductivity of the aquifer and possibly lateral variations of the latter, as might be indicated by the surficial geological formations in the region. For this reason the model-layer is divided in up to four zones; each of which is a different hydraulic conductivity assigned to, which are refined during the calibration process. The classification of these zones is somewhat arbitrary and is based essentially on the observed pattern of the gradient of the flow and the topography.

As discussed by Anderson and Woessner (1992), selection of appropriate model boundaries and boundary conditions in a numerical model is one of the most crucial, but also difficult tasks. With the location and size of the model grid as shown in Fig. 6, direct natural boundaries of the Fenholloway groundwater flow system are not matched, particularly not for the northern and the southern boundaries of the model. The situation is slightly more favorable for the eastern and western boundaries, although in the east these do not extend fully to the watershed boundary that divides the Fenholloway from the Suwannee river watershed, whereas in the west only the southwest corner of the model reaches the Gulf of Mexico.

3.2.2 Boundary conditions and enforcement of stream gauges

For the steady-state models all four boundaries of the model are set up as constant head (Dirichlet) boundaries, with the head-values extracted approximately from the krigged isoline contour maps of the measured water table elevations of the January-June, 1994, field survey (see Fig. 2). Because of the rather large distance of the boundaries from the central model region of interest, the exact specification of the boundary head values is not too crucial, and the model is mostly constrained by the interior nodal values for the head which are taken from the real observations.

The Fenholloway river is implemented by means of the river and drain packages. The drain package works in much the same way as the river package, discussed earlier, except that leakage from the drain to the aquifer is not allowed and that discharge to the drain is zero whenever heads in the cells adjacent to the drain are equal to or less than the assigned bottom elevation of the drain. Because of this, drains can only be used for the simulation of a permanently gaining (effluent) river, unlike river-nodes, which can mimic both an effluent and an influent stream.

With these considerations, the Fenholloway river is modeled in the region of interest, which extends from about 1000 m east of the paper mill's discharge point over a length of about 7 km to the west, where the river turns toward the north, by river nodes to allow for influent and effluent conditions. Further downstream toward the Gulf of Mexico the river is treated as a set of drains, because it can be assumed that the Fenholloway is an exclusively gaining stream in this section. The river bottom elevations are read from topographical maps, and the river gage heights are calculated through linear interpolation from the values measured at the USGS stream gage station 'AT FOLEY', close to the discharge point, and the station 'NEAR PERRY', about 5 miles further downstream (Fig. 1), and using the average topographic slope of the Fenholloway river.

3.2.3 Model calibrations and hydrological implications

The steady-state model is calibrated based on water table elevation data of the January-June, 1994, field survey. Because of only moderate variations of the precipitation during that time period, no major differences of the piezometric heads are inferred for the different sampling days in this period. Therefore, as the reference data set for the calibration, we have used the elevation data sampled on 5/11/1994 which is rather complete (Fig. 2). In addition to this visual check of the calibration results and optical comparison of the modeled with the measured piezometric contour. map, the residuals between measured and modeled watertable elevations are calculated at the locations of the monitor wells. Initial conductivity ranges were obtained from Bush and Johnston's (1988) discussion of the regional hydrogeology and some aquifer pump test data. Numerous calibration simulations were performed, with values for the hydraulic conductivities k ranging between 50 and 400 ft/day (whereby in one model the k values within the four assigned conductivity zones increase slightly toward the west) and river conductances (taken as constant along the river course) ranging between 20000 ft^2/day and 70000 ft /day. An average aquifer thickness of 1500 ft has been assumed in the calculations.

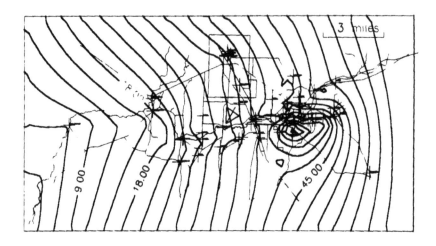

Figure 7. Calibrated steady-state model for elevation data (ft) of 5/11/94.

Results of steady-state four-zone MODFLOW calibration model are depicted in Fig. 7. In addition to providing calibrated conductivities and river conductance values, this steady-state model generates also the initial head condition for the transient simulations to be carried out in the next section.

3.3 'Static' transient model line and the delineation of water lines

3.3.1 Objective and formulation of the conceptional model

The objective of these 'static' transient MODFLOW models is to simulate a pre-specified lowering of the groundwater table due to a recharge deficit and drainage during an extended dry season, as it has been observed at some of the wells in the vicinity of the river during the long drought period of 1988-1991 (Fig. 3). Therefore, the 'static' model attempts to mimic hydraulic conditions that are prone to infiltration of river water into the surrounding aquifer. It should be clear that it is the variation of the natural recharge of the aquifer through rainfall that is the physical trigger mechanism for the groundwater table fluctuations. A deficit of precipitation and increased evaporation during a long dry season is responsible for a groundwater table drop. However, these effects are not included in the present static model, since we are interested here only in the hydraulic implications of a lowering of the water table beneath the sustained stream gage elevation (see below) on the head distribution in the ambient aquifer. Fig. 8 shows a cross-sectional view of the conceptual model

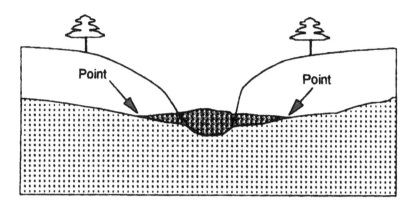

Figure 8. Conceptual cross-sectional model for the delineation of static water lines

which is the basis of the delineation of the static water lines. The lowered groundwater table results in a positive hydraulic gradient from the river bed into the aquifer so that, by virtue of Darcy's law, a river particle will be driven toward the aquifer.

This conceptual model is based on the assumption that the minimum river gauge is sustained during the time span of the simulated drought and during the groundwater table drop. This, of course, is not the case for a natural stream which is in equilibrium with the surrounding aquifer and where the stream-bed might be drying out completely during an extreme drought as is often the case for the section of the Fenholloway river upstream of the paper mill's discharge point. On the other hand, because the industrious plant is constantly recharging the Fenholloway river with approximately 45 million gallons per day, there will be a minimal river gauge height downstream of the discharge point, which should always be sustained. This is supported by the records of the USGS river gauge station "AT FOLEY", which reveal an absolute stage height minimum of 34.56 ft measured during the dry season of 1989-1990 (as compared with a monthly average of about 36 ft over the last 30 years (Fig. 3)). This is the value being used in the MODFLOW river package and it is kept constant throughout the transient static simulation.

3.3.2 Boundary and initial conditions

The boundary conditions used in the transient models are different from those in the steady-state models earlier and are also more intricate to implement into the MODFLOW model. Since the piezometric heads at the boundaries are also changing in time and are not known *a priori*, they cannot be used in the transient simulations.

Instead, flux boundary conditions are employed which are selected to first order in such a way as to sustain approximately the average observed steady-state regional flow. This means in particular that the above combination of flux boundaries was chosen such as to generate an average regional groundwater flow in the southwestern direction, as observed from the piezometric contour lines, deduced from the historical well-data records and from the January-to-June, 1994, field survey (Fig. 2). With this we have for the boundary and initial conditions:

Eastern and northern boundaries: Inflow flux boundaries to simulate horizontal recharge of the aquifer due to the topographic gradient. It is implemented by means of the well-package of the MODFLOW model, whereby 100 injection wells in the north and 149 wells in the east are placed at the nodal points along both boundaries. The amount of the injected water for each well is calculated based on the total incoming recharge at average steady-state conditions and using Darcy's law $q=kA \nabla h$, where q = flow rate, k = conductivity, h = pressure head, A = area, and ∇ =gradient operator. In the present situation ∇h is equal to dh/dx, the regional flow gradient, and is estimated from the observed and previously modeled piezometric head contour lines of the January-to-June, 1994, survey and A is the product of the cell length and the effective saturated aquifer depth b.

Western and southern boundaries: Drain boundaries. The exact specification for the drain depths at the western and southern boundaries is not too important for the purpose of the present static transient simulation of a 'worst-case' scenario of a water-table drop during a long dry season. Therefore, the drain depths have been chosen after various trial-and-error runs in such a way, as to make sure that the groundwater table can drop in the eastern section of the model by about 5 ft and in the area north of the Fenholloway by about 3 feet. Again, the conceptual idea of the transient simulation is to simulate a 'worst-case' scenario of a long dry season during which the groundwater table in the vicinity of the paper mill's discharge point may drop by about 3 or 4 feet.

Initial conditions: The initial conditions for the transient simulations are created from the previous steady-state simulations. As such they would be representative of the piezometric head distribution of 5/11/1994. However, this particular condition is not of importance for the purpose of the present static model which simulates only what could happen hereafter, if the groundwater table were to drop due to a long dry season. Therefore, although the simulations presented start from the 5/11/94 initial head distribution, they simulate a realistic scenario that might have happened in the past, as indicated by the historical records for the water table elevations, or that may occur in the future.

3.3.3 Calibration of the static transient model

The criteria for the calibration of the transient model are: (1) piezometric heads agree satisfactorily with the limited measured historical head data available; (2) the depression cone created by the paper mill's pumping well is adequately represented; (3) the overall mass balance budget of the model is in reasonable compliance with the hydrological constraints of the Fenholloway stream/aquifer system.

The main hydraulic parameters to be adjusted in the calibration process and which directly affect one or more of the above factors are: (1) the hydraulic conductivity k; (2) the storativity factor S; (3) the river and drain conductance k_r; and (4) the effective aquifer thickness b.

The present MODFLOW models show the highest sensitivity to the variation of the hydraulic conductivity k. With increasing k the regional flow might 'wash away' the depression cone that is created by the pumping well field, whereas with decreasing k the latter becomes deeper. The aquifer thickness b needs also to be adjusted in this one-layer unconfined-aquifer flow model which, following the Dupuit-assumption, assumes that the flow in the vertical direction is negligible (Bear, 1979).

In addition to the hydraulic parameters, some numerical parameters that might affect the reliability of the MODFLOW solution are to be adjusted as well. These are the length of the stress period and of the time step increments within one stress period. Although by nature of the numerical implementation (implicit time-integration), the MODFLOW program is unconditionally stable, regardless of the time-step used, time steps too large may cause problems with the convergence during the iterative solution process. Changes of the aquifer thickness b greatly affect the budget analysis. Within the transient calibration, through changes of b, of the enforced boundary conditions, such as the injection wells and the drains (see previous section), and of the storativity S, the amount of drained water can be calibrated.

The mass-balance calculation provided within MODFLOW provides a clue of what are the important in- and outflow components of the aquifer/stream system. The calibration models show that the storage factor has a strong impact on the budget balance. Whereas with relatively small storativities $(S = 0.1)$ the mass budget is fairly balanced, for values of S too large (e.g. $S=0.4$), unbalanced budgets may be produced. On the other hand, larger S facilitate the convergence of the solution process, since large S imply that the unconfined aquifer can yield enough water from the pore-storage without a large drop Δh of the water table.

3.3.4 Model results: Delineation of static water lines

Numerous calibration simulations were executed, during which the hydraulic conductivities, the river conductance and the storativity are varied in the model over a reasonable range of values. Among these model cases we show the results of three particular simulation models, with the most important hydraulic parameters as indicated:

Model case #1: Hydraulic conductivity k=90-120 ft/day, river conductance k_r=20000 ft^2/day, thickness of the aquifer b=1000 ft, storativity S=0.1;

Model case #2: Hydraulic conductivity k=150-180 ft/day, river conductance k_r=20000 ft^2/day, thickness of the aquifer b=1000 ft, storativity S=0.1;

Model case #3: Hydraulic conductivity k=90-120 ft/day, river conductance k_r=20000 ft^2/day, thickness of the aquifer b=1000 feet, storativity S=0.15.

The Figs. 9, 10 and 11 show that, as the dry season goes on, the groundwater table elevation drops slowly because of the water drainage out of the aquifer. This behavior is visualized best by following one particular isoline and watch it moving upstream (in eastern direction) as time increases. However, up to about simulation day 200, the curvature of these isolines in the vicinity of the Fenholloway river is such, that lines orthogonal to the piezometric isolines (which by Darcy's law are the groundwater flow lines) are directed downstream (westward) and toward the river, i.e., the latter is still a gaining stream. From simulation day 240 on, the piezometric isolines close to the river change direction and perpendicular lines will point away from the river, i.e., it becomes a losing stream. Therefore, lateral migration of river contaminants into the aquifer becomes now possible.

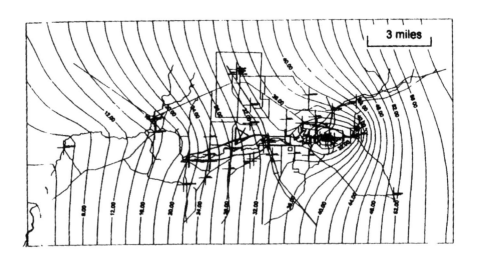

Figure 9. Transient static simulation of water table drop (ft): Model case #1, after 240 days. Solid lines on both sides of the river are the static water lines.

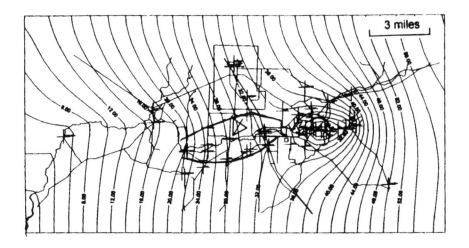

Figure 10. Transient static simulation of water drop (ft): Model case #2, after 448 days.

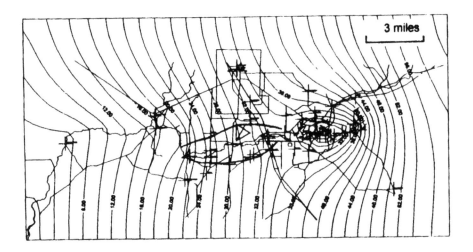

Figure 11. Transient static simulation of water drop (ft): Model case #3, after 448 days.

An envelope for the outgoing flowlines on each side of the river can be constructed by connecting those points on a head isoline where it curves toward the river again. These two envelopes define the hydraulic boundaries which river pollutants would not be able to penetrate under any circumstances if they were to migrate out to these

points. They are called here the *static water-lines* and define the most extreme
corridors of possible Fenholloway river/aquifer pollution. However, they move
further outward, away from the river, as the water table continues to drop with
increasing length of the dry season. After 448 days, when the groundwater table in
the vicinity of the paper mill's discharge point has dropped by approximately 3 feet,
these water lines have extended to about 1.5 to 2 miles in northern and southern
direction from the river and reach about 2.5 miles after 800 days of dry-season
simulation. To illustrate the effects of the different hydraulic conductivities and
storativities, snapshots for model cases #2 and #3 are shown in Figs. 10 and 11,
respectively. After 450 days of dry-season simulation, for model case #2 with a
conductivity k of 150-180 ft/day the inferred water lines are very similar to those of
model case #1, though the depression-cone of the paper mill's pumping well field
becomes shallower (see Figs. 10 and 11). This tendency is in agreement with those
of the previous steady-state MODFLOW models. The 450-days snapshot for model
case #3 (Fig. 11), with a storativity of $S=0.15$, exhibits a slightly narrower water-line
corridor than that obtained for the model case #1 (storativity $S=0.1$). This is due to
the fact that the aquifer can now sustain more drainage out of pores without the need
of a major groundwater table drop.

3.3.5 Discussion and implications of the static flow models

Table 1. Simulated transient water table drops (ft) in several wells to be used for the selection
of the appropriate water lines.

Wells\Days		240	330	448	601(days)
Wells	WaterLevel in 5/11/94	Water Table Drop(ft)			
US#444	36.50	5.20	5.50	6.00	6.30
US#113	37.72	4.72	5.22	6.22	6.70
MW#13	36.18	2.68	3.18	4.18	5.10
MW#09	33.28	0.88	1.08	1.58	2.78
MW#02	32.03	0.90	1.02	1.90	2.03
SWF1001	34.00	2.00	2.20	2.70	3.79

The piezometric head contour plot for 5/11/94 (Fig. 2) represents the initial heads for
the transient model. The mass balance taken from the MODFLOW output pointed
out that the river could be both losing water to and gaining water from the
surrounding aquifer along its course. Figs. 9 to 11 of the results for the transient
model give an idea of what the possible piezometric head configuration might be, if
the water table close to the discharge point dropped by 3 to 4 feet.

Do the real groundwater table elevations ever drop by such an amount below the levels of the 5/11/94 initial water table? The historical records shown certainly indicate this. In fact, as illustrated in Fig. 3, it occurred several times in some of the wells north and south of the paper mill's discharge point since 1960 during the long drought periods. Consequently, Fig. 9 to 11 may represent a realistic historical dry season scenario. Table 1 summarizes the simulated transient groundwater table drops of the Figs. 9 to 11 for several monitor wells with respect to the initial groundwater table. These values indicate that after 450 days of simulation time the water table in the wells listed has decreased more than what has been probably observed in reality in the Fenholloway region over the past 30 years. Therefore, the static water lines shown in Figs. 9 to 11 would represent the outer bounds of a possibly polluted aquifer corridor along the Fenholloway river that are in realistic compliance with the historical observations. The maximum static distance that may be reached by a contaminated river particle might extend to about 1.5 miles north of the river to the Perry airport (large triangle in the maps). From there on, further downstream, the water line begins to slowly converge back toward the Fenholloway river. A similar situation holds for the maximum reach of the static water lines south of the river.

4. DYNAMIC PARTICLE TRACKING OF STREAM/AQUIFER INFILTRATION

4.1 Motivation

The *static water-lines* established in the previous sections by means of the transient MODFLOW simulations are the outer hydraulic flow barriers of horizontally infiltrating stream water; i.e. the latter cannot move beyond these water lines. However, since these barriers were calculated assuming a final groundwater table drop during a certain drought period, they do not account for the actual total movement of a river particle that begins to migrate outwards into the aquifer, as soon as the groundwater table elevation falls beneath the stream gage height and the stream becomes influent. In fact, the same particle might be swept back again towards the river, when during a normal season the latter becomes an effluent stream again. This behavior can only be mimicked by means of a dynamic particle tracking technique that allows the computation of advective travel-paths of conservative tracers emanating from the river under appropriate transient hydrological conditions. For this purpose a semianalytical particle tracking method was developed in Koch et al. (1994) and Koch and Cekirge (1996) and it will be used in the following sections to define, what we now call *dynamic water-lines*.

4.2 Methodology

We only summarize here the most relevant features of the conceptual model underlying the dynamics of the river-to-aquifer particle movement and the final particle tracking technique (see Koch et al. 1994; and Koch and Cekirge, 1996, for details). A 1-D cross-section of this conceptual model is shown in Fig. 8. Assuming a downward hydraulic gradient from the river to the aquifer at a certain time at the beginning of the dry season, a river particle will be advected downgradient into the aquifer. At this stage a *Green's function method* (cf. Tikhonov and Samarskii, 1963) is used for the solution of the one-dimensional (1D) version of the transient groundwater flow equation (2) for the head $h(y,t)$ and applied in the semi-infinite, cross-sectional direction y, that extends from the middle of the stream into the aquifer to the right (see Fig. 8). One of the appealing features of the Green's function method is that it naturally includes (1) time-varying Dirichlet boundary conditions $h(y=0,t) = h_B(t)$ at the left boundary $(y=0)$ of the domain (the river) and which are set equal to the measured stream gage heights; (2) a time- and spatially dependent source/sink term $N(x,t)$ that are related to the effective recharge of the groundwater aquifer (=infiltration due to the difference between precipitation and evaporation); and (3) initial conditions for hydraulic heads $h(y,t=0) = h_i(y)$ that are specified from assumed or measured groundwater table elevations fitted by a functional relationship which is based on Dupuit's theory for the groundwater table change towards the river bed (Bear, 1979).

As the particle moves out of the stream into the aquifer, it will also be dragged downstream by the regional groundwater flow; I .e., the particle moves effectively in a two-dimensional (2D), transient potential field $h(x,y,t)$ that is the superposition of the 1D head-distribution $h_G(y,t)'$ computed above by means of the Green's function, and the regional hydraulic gradient field $h_R(x,y,t)$. The local flow velocity of the particle at a particular point in the 2D regional area can then be computed by applying Darcy's law $v = k \nabla h$, to the total head field $h(x,y,t)$ at that point. The new position (x_1,y_1) of a particle after a small time increment Δt can then be calculated from the previous position (x_0,y_0) by

$$
\begin{aligned}
x_1 &= x_0 + u\Delta t \\
y_1 &= y_0 + v\Delta t
\end{aligned}
\tag{6}
$$

where u and v are the components of the flow velocity v. The final position of a particle is then obtained by integration of Eq. (6) over the total time. This is nothing else than classical particle tracking .

Note that the whole process is fully dynamic, because as the particle propagates into the aquifer at the beginning of a drought, it will sense different piezometric heads which themselves are changing over time, due to the seasonal variations of the rainfall and, ergo, changes in the effective aquifer recharge. In fact, as the drought ceases and the groundwater table rises again because of positive precipitation recharge, the hydraulic gradient will reverse itself and the river particle may partly move back towards the river bed. However, by then it has already irreversibly contaminated a corridor along the river. What we call the *dynamic water-lines* are then the envelopes of the extreme endpoints of all particle tracers on each side of the river. They define the boundaries of this maximal corridor of possible aquifer pollution by the industrial river.

4. 3 Setup of the particle tracking model and selection of model parameters.

The coding, numerical implementation and setup of the Green's function particle tracking model is described in detail in Koch et al. (1994) and Koch and Cekirge (1996). The simulations start at time zero when the river is still effluent. An initial groundwater table $h_i(y)$ is assumed and the boundary condition $h_B(t)$ is set equal to the river gage elevation above the average water table at the bottom of the stream bed. A time-dependent source function $N(x,t) = N(t)$ is applied in the model that mimics the net recharge of the aquifer. The time variation of the source function is one of the most instrumental parameter in the simulation, since it is the form of $N(t)$ that simulates the absolute magnitude and the period of possible cyclic change from a wet to a dry season and vice versa. The form of $N(t)$ used here is

$$N(t) = N_o \sin\left(\frac{2\pi t}{T_o}\right) \tag{7}$$

where N_o is the maximal source/sink rate of the net recharge process and is a parameter to be adjusted in both the calibration and the sensitivity study. T_o is the period of the wet/dry season cycle and, as will be discussed further down, its choice is of ultimate importance for the simulation of 'worst-case' stream/aquifer pollution scenarios.

It should be clear that only cycles with long periods T_o for the wet/dry interchange will lead to a significant lateral penetration of river particles into the surrounding aquifer, as soon as the groundwater table drops beneath the river gage level. If the drought interval of the cycle is too short, the particles barely have time to move into the aquifer, before they are driven back by the reversed gradient that builds up again when the consecutive wet-season interval of the cycle begins.

The visual inspection of the time series of precipitation, river gage heights and groundwater table elevations in Fig. 3 illustrates that these variables have, in addition to the normal seasonal cycle (with a period of about six months), longer cycles with multiyear periods of changes from highs to lows and vice versa. As the more detailed analysis of Koch et al. (1994) and Koch and Cekirge (1996) shows, these long-term variations are related to the El Niño/Southern Oscillation (ENSO), which has been shown to have a profound effect on the global hydrometeorological cycles, and those of the continental US, in particular (Kahya and Darcup, 1993). ENSO results from anomalous cyclic variations of the sea surface temperature as measured in the south-eastern Pacific Ocean. In particular, an El Niño event occurs when the ocean surface temperature is above normal. Historical records show that El Niños have a recurrence rate of about 4 to 6 years. Normal times in between two El Niño events are also called a La Niña.

The effects of El Niño and La Niña events on long-term precipitation on a global scale varies with the region. For example, severe droughts might occur in Australia during an El Niño year, whereas high floods might be prevalent in south America during that time (Ropelewski and Halpert, 1987). For Florida, precipitation above and below the long-term average of up to 40% can be expected in an El Niño and a La Niña year, respectively (Sun, 1996; Sun et al., 1997b). This response of precipitation to the ENSO cycle implies that in a La Niña year, a lower groundwater table can be anticipated within the study region. Since the stream gage elevations for the Fenholloway river remain relatively constant, due to the large industrial discharge (see Fig. 3), a reverse hydraulic gradient from the stream towards the aquifer; i.e., an influent stream, can be created over a relatively long time period. Therefore, a wider corridor of aquifer contamination along the river may be expected in a La Niña year.

4. 4 Application to the delineation of dynamic water lines

Fig. 12 shows a typical result of the application of the dynamic particle tracking to the delineation of dynamic water lines. A recurrence period T_o of 1000 days for the wet/dry season cycle, which mimics approximately the length of the long-term El Niño/La Niña cycle discussed above and, therefore, represents a kind of 'worst-case' scenario, has been used in this calculation. The movements of the river particles have been tracked over a total time of 22 years which comes close to the total time of operation of the paper mill. A hydraulic conductivity of k=50 m/day which, from the previous steady-state MODFLOW calibrations, was found as the most appropriate for the aquifer region under study, is used. The amplitude N_o of the effective recharge function (7) was adjusted during the calibration phase of the particle tracker in such a way that the observed historical, long-period groundwater table fluctuations of Fig. 3 are roughly matched. A detailed description of these and

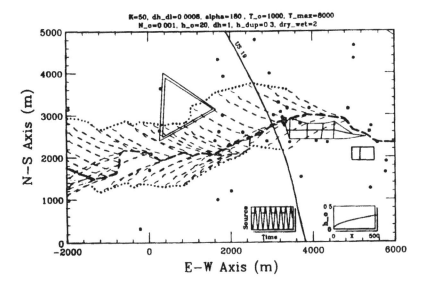

Figure 12. Results of dynamic particle tracking over a total integration time of 22 years, a recurrence period of 1000 days, and using a conductivity of k=50 m/day. The right inset shows the position of the initial free-surface groundwater table. The left inset illustrates the transient cyclic variation of the recharge source function, that mimics the changes from wet to dry years and vice versa.

other considerations that are required for the appropriate calibration and verification of the particle tracking model can be found in Koch et al. (1994) and Koch and Cekirge (1996) where a sensitivity study is also presented.

The envelopes of the extreme endpoints of all particle tracers on each side of the stream delineate the *dynamic water-lines* and 'sandwich' the maximal corridor of possible aquifer pollution caused by the industrial Fenholloway river. Fig. 12 illustrates that this particle tracking simulation (and numerous other ones discussed in Koch and Cekirge, 1996) result in 'pollution corridors' that are close to those obtained from the *static water-lines* that resulted from the transient MODFLOW simulations in the previous section. The 'worst-case' dynamic water lines of Fig. 12 should, therefore, provide water-planning agencies with realistic guidelines for where to draw the 'safest' limits of the proposed public water supply lines for the residents that are living along the Fenholloway river.

5. SUMMARY

Different hydrological modeling methods are applied in this study to analyze a possible contaminated corridor in the aquifer along both sides of the industrial Fenholloway river. Time series analysis and, in particular, ARIMA- and linear transfer models, statistically explain the main hydrological features within the data itself and the interrelation among the data series. For example, the autocorrelations of the river gage heights are significant only for shorter lag-periods than those for the groundwater table elevations which hints of the short time-scale of the surface river flow processes. The crosscorrelations between the river gage heights and the water table elevations show significant positive correlation for lags less than zero, indicating that a feedback exists between the groundwater table and river gage heights. A strong crosscorrelation is also found between the precipitation and the water table elevations. The feedback analysis shows that the river gage and the groundwater table are mutually dependent and, particularly, that the river stage heights are also strongly influencing the groundwater table. This implies, in turn, that the industrial river water may contaminate portions of the aquifer close to the river, especially during long dry seasons when the groundwater table may fall below the river gage elevations and sections of the industrial river may become influent.

Ultimately, extreme estimates of the possible spatial extensions of a contaminant plume emanating from the river over the short and the long term (worst-case scenarios) are established by two explicit modeling approaches. The first one is based on the use of transient MODFLOW simulations and mimics a drop of the groundwater table during a long dry season as has been observed in the area in the past. This approach defines the *static water-lines*. The second modeling option is based on a new model for dynamic stream/aquifer particle tracking, using a Green's function methodology. Using this technique , *dynamic water-lines* are defined by the end-locations of river particles tracked over a long time period for a specified dry- and wet season scenario. It is particularly shown, that the extent of lateral aquifer contamination is mostly affected by the periods of cycles for the wet-and-dry seasons and -years, with the maximum of river-particle migration into the surrounding aquifer for long drought intervals, when the groundwater table drops beneath the river gage level for a significant length of time. It was postulated that the longest cycles of periods of wet and dry climatic variations in Florida are related to the EL Niño/Southern Oscillation (ENSO), which appears to have a periodicity of about 3 to 5 years. During half of the regular ENSO-cycle an El Niño year changes to a La Niña year with, for Florida, a corresponding switch from anomalously high to low precipitation, respectively. Therefore, it is during extreme La Niña years that the potential for stream-aquifer pollution is the highest and the most extreme contaminated corridors, as delineated by the water-lines along the Fenholloway river are established..

ACKNOWLEDGMENTS

The authors are grateful to H. M. Cekirge for his help. We thank Richard Harvey, Paul Lee, John Purcell, Rick Copeland, and Richard Waikowicz for the fruitful discussions and their efforts in gathering most of the field data. Finally we acknowledge the anomalous paper mill factory, the USGS, and the Florida Climate Center for providing us additional data for this study. The initial investigation of this project was financed by Florida Departmental of Environmental Protection. However, this study is solely based on the academic experience and best personal judgement of the authors and does not represent the opinions of any institution or authorities.

REFERENCES

1. Anderson, M.P. and W.W. Woessner, 1992: *Applied Groundwater Modeling, Simulation of Flow and Advective Transport*, Academic Press, San Diego, CA.

2. Bear, J., 1979: *Hydraulics of Groundwater*, McGraw Hill Inc., New York, NY..

3. Box, G.E.P. and G.M. Jenkins, 1976: *Time Series Analysis: Forecasting and Control*, Holden-Day, San Francisco, CA.

4. Bush, P.W. and R.H. Johnston, 1988: Groundwater hydraulics, regional flow, and groundwater development of the Floridan aquifer system in Florida and in parts of Georgia, South Carolina, and Alabama. *USGS Professional Paper 1403-C*.

5. Kahya, E. and J.A. Dracup, U.S. streamflow patterns in relation to the El Niño/Southern Oscillation, *Water Resour. Res.*, *29*, 2491-2503, 1993.

6. Koch, M., H.M. Cekirge, and H. Sun, 1994: Contaminant risk assessment of the Fenholloway river: Physical modeling of the interaction of river and groundwater. *Technical Report to Florida Department of Environmental Protection*, Tallahassee, FL.

7. Koch, M. and H.M. Cekirge, 1996: A transient Green's function analytical flow and particle tracking model to quantify a coupled river-aquifer system: Application to the assessment of possible groundwater contamination from a Floridan industrial river, In: *Advances in Ground Water Pollution Control and Remediation*, M.A. Aral (ed.), NATO ASI Series, Kluwer Academic Publishers, Dordrecht, pp. 127-154, 1996.

8. Maidment D. R., 1993: *Handbook of Hydrology*. McGraw-Hill, Inc.Press, New York, NY..

9. McDonald, M.G. and A.W Harbaugh, 1984: A modular three-dimensional finite difference ground-water flow model, *USGS Open-file Report 83-875*.

10. Pankratz, A., 1991: *Forecasting with Dynamic Regression Models*. Wiley-Interscience Publication, John Wiley & Sons Inc., New York, NY.

11. Ropelewski, C.F. and N.S. Halpert, 1987: Global and regional scale precipitation pattern associated with the El Ni\~no/Southern Oscillation, Mon. Weather Rev., 115, 1606--1626.

12. SAS Institute Inc., 1993: *SAS/ETS User's Guide*, Cary, NC.

13. Sun, H., 1996: Annual Surface Water Resources in Florida in Response to El Niño and La Niña-Southern Oscillation of Sea Surface Temperature. *30th International Geological Congress Abstract*, Beijing, China.

14. Sun, H. and M. Koch, 1996: Time Series Analysis of water quality parameters in an estuary using Box-Jenkins ARIMA models and cross-correlation techniques, In: *Proceedings of the XI International Conference in 'Computational Methods in Water Resources'*, Cancun, Mexico, July, 22--26, 1996

15. Sun, H., M. Koch and W. Jones, 1997a: Application of ARIMA models and Kalman filter techniques to the analysis and forecasting of salinity variations in the Apalachicola Bay, *Journ. Hydraulic Eng.* (In Press).

16. Sun, H., D.J Furbish, and M. Koch, 1997b: Annual Precipitation and River Discharges in Florida in Response to El Niño and La Niña-Southern Oscillation of Sea Surface Temperature, *Journ. of Hydrology* (In Press)

17. Tikhonov, A.N. and A.A. Samarskii, 1963: *Equations of Mathematical Physics*, Pergamon Press, New York, NY.

18. USGS, 1964-1993: *Water Resources Data Florida, Water year 1964-1993*, USGS, Tallahassee, FL.

19. Watts, G.B. and W. Riotte, 1991: Ground Water Investigation Report Number 91-05, Proctor and Gamble Cellulose Perry, Taylor County, *Technical Report to Florida Department of Environmental Protection*, Tallahassee, FL.

Proc. 30th Int'l Geol. Congr., Vol.22, pp. 165-176
Fei Jin and N.C. Krothe (Eds.)
© VSP 1997

Karst and Hydrogeology of Lebanon

H. S. EDGELL
Professor and Consultant, Canberra ACT2603, Australia

Abstract

Karst is very well-developed in Lebanon in thick, exposed, fractured and folded Jurassic, Cretaceous, and Eocene carbonates, as well as in localized, coastal Miocene limestones. This karstification not only results from the predominant calcareous lithology, but is also caused by the high, northerly trending ranges of the country, which cause abundant precipitation, as heavy rain and thick snow, to fall on Mt. Lebanon, Habal Barouk, Habal Niha, and Mt. Hermon, with lesser amounts falling on the Anti-Lebanon, Beqa'a Valley and the coastal hills of the country. Some 80% of precipitation falls from November through February. Five of these are first-magnitude springs (Ain ez Zarqa 11m^3/sec, Ain Anjar (max. 10m^3 /sec), Nabaa Ouazzani (max. 6m^3 /sec), Nabaa Arbaain (max. 3m^3/sec) and Nabaa Barouk (max. 3m^3/sec), plus hundreds of second-and third magnitude springs, and thousands of smaller springs. The large springs are all karstic and contribute to 13 perennial springs in the main Lebanese ranges, and 2 in the Anti-Lebanon. These include major rivers, such as the Nahr el Litani, Nahr el Assi (Orontes) and Nahr el Hasbani (upper Jordan River).

More than two-thirds of the area of Lebanon (i.e. 6900 km^2) is karstified and includes surface karst features, such as poljes, uvalas, dolines, blind valleys, natural bridges, and ponors, as well as smaller features, like karren and hoodoos. Subsurface karst features include many types of solutional shafts and galleries, grottoes, subsurface lakes and rivers and most types of speleothems.

There are at least 15 aquifers in Lebanon, of which 14 are in karstified carbonate atrata. The 1700 m limestone/dolomite core of the ranges and over 2000 m of flanking, or overlying, Cretaceous limestones provide the majority of these aquifers, while significant aquifers are also found in thick Eocene limestones.

High transmissivity values (T=or 1.83 x 10^{-1} m^2) occur in these karstic aquifers, as is shown by the rapid decline in spring flow over the dry summer and autumn months, and their very quick recharge by winter and spring rains and heavy snow on the Lebanese ranges.

INTRODUCTION

Lebanon has so much ground-water, as well as surface springs and rivers, that water can be considered as the most important, and almost the only natural resource of the country.

Precipitation falls in abundance on the main Lebanese ranges (Mt. Lebanon, Jabal Barouk and Jabal Niha) and, also, in considerable amounts on Mt. Hermon (Jabal esh Sheikh), as well as on the Anti-Lebanon mountains bordering Lebanon and Syria. Lesser rainfall occurs in the Beqa'a Valley and on the hills of southwestern Lebanon behind Sidon (Saida) and Tyre (Sour).

Most precipitation in Lebanon is due to the high, northerly trending topography of the country, with 170 km of ranges averaging above 2200 m in height, and lying in the path of westerly winds from the Eastern Mediterranean. A considerable amount of precipitation is in the form of snow, up to 4 m deep on the ranges of Lebanon, and 80% of precipitation falls from Nevertheless, there are 13 perennial streams flowing from the high ranges of Lebanon, which reach a height of 3087 m in

Qornet es Saouda, with a further 2 perennial streams emanating from the Anti-Lebanon and Mt. Hermon, including the headwaters of the Jordan River and the Nahr el Litani, which is the largest of Lebanon's rivers.

Although the total area of Lebanon is 10300 km^2, Mesozoic and some Cenozoic limestones with karstified surface cover an area of 6900 km^2. These limestones exhibit all the typical features seen in the former Yugoslavia. Deep karstification in fractured carbonate strata allows rapid infiltration contributing to an immense reserve of groundwater, which can be seen emerging from a multitude of karst springs, several with first order magnitude discharge according to Meinzer's classification (1923).

The hydrogeology and groundwater resources of Lebanon are mostly related to deep karstification of a sequence of widely exposed carbonate formations ranging from Lower Jurassic to Middle Miocene. Significant quantities of groundwater are also contained in Quaternary alluvium and Neogene terrestrial clastics of the Beqa'a Valley.

The eastern slope of the Anti-Lebanon lies in Syria, so that approximately 67% of the territory of Lebanon is composed of karren limestone surface allowing widespread infiltration and subsurface karstification.

GEOLOGICAL BACKGROUND

The geology of Lebanon basically consists of two great north northeasterly trending horst-anticlines of Mt. Lebanon and the Anti-Lebanon (including Mt. Hermon) separated by the graben-syncline of the Beqa'a Valley. South of the latitude of Beirut, the horst-anticline of Mt. Lebanon changes to the large east dipping flexures of Jabal Barouk and Jabal Niha, separated from lower coastal areas by the northerly trending Toum Fault. On the east side of the main Lebanese ranges, from south of Jabal Niha to the north end of Mt. Lebanon, an immense sinistral fault, known as the Yammouneh Fault, but actually a section of the Dead Sea Transform Fault Zone, separates these westerly ranges from the largely alluvial filled Baqa'a Valley. Further to the east, the Beqa'a Valley is replaced by the large horst-anticline of the Anti-Lebanon, although separated by a saddle near Masna from the high faulted anticline of Mt. Hermon (Jabal est Sheikh).

Almost all the strata which crop out in the Lebanese ranges are carbonates, predominantly Jurassic and Cretaceous shallow water limestones, dolomitic limestones, and dolomites. Large areas of Middle Eocene limestones crop out in the hills southern Lebanon south of Sidon, as well as in the southern Beqa'a. Middle Miocene limestones unconformably overlie Mid Cretaceous and Eocene strata in patches along the Lebanese coast, as seen near Tripoli (Trablous), Nahr el Kelb, and in between Sidon and Tyre. On the western border of the Bequa'a, Miocene terrestrial conglomerates uncoformably rest on easterly dipping Mid Cretaceous limestones. Pliocene marls occur in the embayment between the mouths of the Nahr el Hasbani. The Beqa'a Valley, north northeast of Anjar is largely filled with Quaternary alluvium and Late Neogene lacustrine sediments.

TECTONIC SETTING

Lebanon occupies a unique position astride the Dead Sea Tranform Fault, locally known as the Yammouneh Fault system. The main Lebanese ranges to the west of this fault are part of the African Plate and are generally referred to as the Levant Subplate. However, the Beqa'a Valley and the mountains of the Anti-Lebanon and Jabal esh Sheikh (Mt. Hermon) lie to east of the Dead Sea Tranform Fault and are, thus, part of the Arabian Plate. Considerable uparching and block uplift, together with volcanism and erosion, took place in the Late Jurassic and Early Cretaceous (Berriasian-to Early Aptian). The most active uplift occurred from Middle Miocene through Pliocene due to major left-lateral movement along the Dead Sea Tranform Fault, resulting from opening of the Red Sea Rift and the separation of the Arabian Plate from the African Plate. Although Quenell (1985) has indicated a sinistral movement of 105 km along the Deas Sea Tranform Fault in the northeasterly Rachaya and Serrhaya faults, as well as by closure of the initial aulacogenic rift by folding along the Palmyrene Fold Belt. The originally northerly trend of the fracture that became the Dead Sea Tranform Fault has been cut across in Miocene time by a series of dextral east-west faults so that a distinct northeast trend in the Dead Sea Tranform Fault occurs on the east side of the main Lebanese ranges. This deflected section of the major interplate fault is known as the Yammouneh Fault in Lebanon and has represented a partial barrier during further sinistral movement along the Dead Sea Tranform. As a consequence, there has been only 2 to 4 km sinistral movement along the Yammouneh Fault where much of the movement has been translated into vetical componets (Adams and Barazangi 1984), so that Qornet es Saouda (3087 m) in Mt. Lebanon is the highest point in the Eastern Mediterranean while there is probably a similar amount of downthrow into the Beqa'a Valley graben syncline.

The principal strain axis in the Yammouneh Fault area north northeasterly, almost parallel to the main fault itself (Ron 1987), with the result that two major sets of shear faults are developed at 70 degree (right-lateral) and 320 degree (left-lateral). These two complimentary fault sets and faulting parallel to the Yammouneh extend deep into the carbonate atrata of the Lebanese Ranges and allow for deep and often cavernous karstification.

Basically Mt. Lebanon and the Anti-Lebanon consist of two giant block folded horst anticlines separated by the graben syncline of the Beqa'a Valley. in Lebanon south of the Latitude of Beirut the tectonic character of the Lebanese ranges changes. Mt. Lebanon is separated by the Dahr el Baidar saddle from the high, west dipping, Jabal Barouk flexure, bounded sharply on the east by the Yammouneh Fault. Further south, the narrower Jabal Niha flexure also dips east but is bordered on the west by the northerly Roum Fault and on the east by the NNE trending Yammouneh Fault. The southern part of the Anti-Lebanon consists of a high (2814 m) east faulted anticline of Jabal esh Sheikh (Mt. Hermon) with an exposed core of Jurassic limestones extending down to Liassic. The intervening Beqa'a Valley narrows considerably towards the south and is filled with either Tertiary or Quaternary sediments, and some Pliocene volcanics in the extreme south.

The horst anticline of Mt. Lebanon plunges northward into the Tripoli-Homs Depression, which is also largely filled with Pliocene volcanics and Pliocene sediments towards the coast.

Lebanon thus, consists of seven major structural blocks namely the Mt. Lebanon Horst Anticline, the Anti-Lebanon Horst Anticline, the Jabal esh Sheikh Anticline, the Jabal Barouk-Jabal Niha-faulted flexure, the plateau of southwest Lebanon, the Beqa'a graben syncline and the Tripoli-Homs Depression.

STRATIGRAPHIC SEQUENCE

The deepest exposed strata in Lebanon are Lower Jurassic (Lissic) dolomites seen in the deep valley of Nahr Ibrahim and also in the core of the Jabal esh Sheikh faulted anticline. They are overlain by at least 700 m of gray blue massive interbedded dolomites and limestones, including 350 m of fossilferous limestones and dolomites, known as the Kerouane Limestone, are cliff-forming. They are overlain by the Upper Jurassic Bhannes volcanics about 50 m thick, forming the Falais de Bikfaya. Immediately above this is the Bikfaya Limestone, about 40 m thick, followed by the uppermost Jurassic Salima limestone, some 100 m thick and unnamed volcanics about 50 m thick at the top of the Jurassic sequence. The thickness of carbonates in the Jurassic of Mt. Lebanon is estimated to total 1700 m.

There then follows a profound disconformity with all Berriasian, Valanginian, and Barremenian atrata missing. The Aptian 'basal' Cretaceous sandstone (gres de base) the rest on the partly eroded and paleokarstified Jurassic and are 60 to 70 m thick in Mt. Lebanon and up to 300 m in southern Lebanon. They are succeeded by the Orbitolina Beds (70 to 80 m thick) and by the prominent, cliff-forming, 50 m thick Jezzine Limestone. Marly and fine sandy Knemiceras Beds of Albian age, being 70 m thick and overlain by the prominent Cenomanian Sannine Limestone, some 650 m thick, with very significant karstification. The Ghazir Marls (100 m thick) succeed and are followed upwards by the Hippurites Beds about 200 m thick and then the Chekka Marls of Senenian age up to 800 m thick and seen on the west flank of Mt. Lebanon. Total thickness of Cretaceous atrata in Lebanon is, thus, 2240 m at maximum. A distinct disconformity separates the Upper Cretaceous from the Lower Tertiary (Edgell 1969) with highest Cretaceous and Lowest Paleocene strata being unrepresented.

Paleocene chalk and marl is from 150 to 350 m thick, as exposed to the east of Ba'albek in the Beqa'a Valley, although not present in the Lebanese ranges. Lower Eocene (Ypresian) limestone and marl with flint are found extensively in the southern Lebanese plateau, as well as along the Lebanese coast to the south of Sidon, and in the southern Beqa'a, where they are 55 m thick Middle Eocene nummulitic limestone are up to 850 m thick in southern Beqa'a, and occupy a wide area of Lebanon south of Sidon, where Eocene strata are generally about 200 m thick. A widespread unconformity succeeds the Middle Eocene in Lebanon.

Neogene marine atrata rest unconformably in patches on eroded Cretaceous and Eocene strata in coastal Lebanon. They are best developed around Tripoli from Ras Chekka and Amioun to the basalts of Halba where Miocene limestones and marls are up to 265 m thick. Other outcrops of Middle Miocene algal limestones occur at Nahr el Kelb, east Beirut, and in coastal areas between Sidon and Tyre. Lower Miocene (Burdigalian) strata are known only from small area south of Sidon into Khaizarane. Upper Miocene beds are unknown. Marine Pliocene marls and marly limestones disconformably overlie Middle Miocene limestones to the north and northeast of Jabal Terbol and NE of Tripoli and are from 250 to 400 m thick. In the Beqa'a Valley, and especially along the east flank of Jabal Sannine, Miocene exists as thick continental conglomerates, up to 800 m or 900 m thick near Zahleh, resting with a 20 degree angular unconformity in Lutetian limestones (Dubertret 1955). Coarse, torrential, non-marine, Mio-Pliocene conglomerates fill most of the northern Beqa'a Valley.

Pliocene, marine, sandy marly limestones and marls, with basalt interbeds, up to 407 m thick are encountered in northernmost coastal Lebanon to the west of Halba and have been described from the Sammakieh Water Well (Edgell 1968). Extensive Pliocene basalts occur in northernmost Lebanon.

Quaternary deposits consist of narrow coastal sands, valley alluvium and widespread red soils (terra rosa) are found throughout the central and northernmost Beqa'a Valley.

SURFACE KARST FEATURES

Since 67% of Lebanon consists of exposed, stratified and fractured, carbonate rocks under conditions of relatively high precipitation, surface karst features of all kinds have developed.

The commonest features are karren, especially kluftkarren, but flachkarren, rundkarren and the finer rillenkarren are also widely seen. Karrenfeld best describes the karstified, almost bare, limestone terrain of Mt. Lebanon and Barouk-J. Niha.

Other solution features common on the naked karst of the Lebanese ranges are conical dolines, up to a few hundred meters in diameter, and larger, subcircular poljes with alluvial floors and generally 400 m to 1 km in diameter. The Yammouneh polje is exceptionally large, being 1.25 km wide and 6 km long. Conical dolines are especially common along higher parts of Mt. Lebanon, as between Dahr el Kadib and Jabal Nakiba to the east of Aaqoura. In the Jurassic Kesroaune Limestone are so closely spaced that it is difficult to follow the ridges between them. In this area a natural arch has also been formed, presumably a remnant of a gallery in the limestone strata that existed before further erosion occurred. A blind valley flowing NNE is found a little further to the north and ends in an immense sinkhole, or ponor, some 100 m in diameter and descending vertically through the Kesrouane Limestone for 640 m.

Karstification on the Lebanese Ranges has continued since at least Middle Miocene and the Jurassic limestones also have paleokarst formed during their uplift throughout most of the Early Cretaceous (Berriasian to early Aptian). A result of deep karstification, seen on Mt. Lebanon near Yachouche and north of Aaqoura, is the formation in places of miniature tower karst with isolated hoodoos developed from 5 to 10 m tall.

SUBSURFACE KARST FEATURES

The exposed, fractured, carbonate strata in the Lebanese ranges are some 3940 m thick and up to 4970 m thick if one includes Tertiary limestones, which are now exposed on the flanks of the ranges. Considering a conservative thickness of 3000 m of folded, faulted and strongly jointed Jurassic and Cretaceous carbonates exposed since Middle Miocene to karstification under precipitation of over 1500 mm/y, it is obvious that deep subsurface karst features have developed. Amongst these are ponors (deep swallow holes connected to lateral galleries), cave systems, subsurface rivers and lakes and a great variety of speleothems.

The entire 3000 m of Jurassic and Cretaceous limestones and dolomites exposed in the Lebanese ranges are riddled with shafts and galleries, forming extensive cave systems. Amongst the well-known caves, or grottoes of Lebanon are those of Jeita, Qadicha, Aaqoura, Afqa and the grottoes of Niha and Fakhreddine, the latter two lying on the east side of Jabal Niha.

The Grotto of Jeita, 18 km northeast of Beirut, is the largest and best known cave system in Lebanon and is developed in the Jurassic Kesrouane Limestone. It extends at least 6.2 km northeastward into Mt. Lebanon, as presently mapped, and includes a lower, internal, subterranean lake some 800 m long. All the other cave systems, or grottoes, are smaller, such as the Grotto of Qadicha, just southeast of Bcharre', which is developed in Jurassic limestones. The grottoes of Aaqoura and Afqa are on the upper reaches of Nahr Ibrahim (Adonis Valley) and occur at the base of thick Cenomanian Sannine Limestone escarpments, both being perennial springs. The Grotto of Niha is relatively small and lies on the west flank of Jabal Niha, 4 km northeast Of Jezzine, while the Grotto of Fakhreddine lies 3 km northeast of Jezzine and is developed in Cenomanian limestones near their contact with impermeable sandy argillaceous Albian-Aptian beds.

Some of the cave systems have subterranean lakes, particularly the 800 m long lake in the Grotto of Jeita, and also a smaller river passage about 100 m long in the Grotto of Afqa. The large underground lake in the Jeita cave system is an excellent example of a cave river passage profusely decorated with calcite speleothems. Types of speleothems found in the Grotto of Jeita include delicate strands of stalactites, which curve at the ends like the edges of some huge umbrella. There are also beautiful, terraced and conical stalagmites and areas of smooth flowstone. Erratic speleothems also occur at an angle to the vertical due to crystal growth and are generally small linear types.

EXTENT, RATE AND DEPTH OF KARSTIFICATION

The extent of karstification in Lebanon covers two-thirds of the country due to the widespread outcrops of thick carbonate strata, primarily Jurassic and Cretaceous limestones, and the relatively high precipitation, especially in the mountains of Lebanon.

The rate of karstification is difficult to estimate, even assuming constant environmental conditions, which have certainly not been the case. Judging from Roman limestone sarcophagi in the Tyre area not more than 5 mm have been corroded in the last 1500 years, giving a relatively low rate of 1 mm/300 years. In the Lebanese ranges, where precipitation is much higher, it is considered that karstification and denudation of limestones may have been at least three times this rate, possibly 1 mm/100 years and solutional loss could be as high as 10 mm/100 years. Karst erosion in the Lebanese ranges has been largely autogenic, or autochthonous, on the high ridges of these mountains which are snow-covered for 4 months of the year. On the flanks of the mountains, karst denudation is both autogenic and allogenic and affects mainly Middle Cretaceous limestones, and some Tertiary limestones.

Due to the disconformity between Upper Jurassic and Mid Cretaceous Aptian to Cenomanian limestones, it is clear that the uplifted core of the Lebanese ranges was exposed to karstification for over 40 million years. This long interval of paleokarst may largely explain why karstification is now so well-developed in the Jurassic limestones of Mt. Lebanon. The second phase of karstification took place after the tectonic uplift of Lebanese ranges due to the opening of the Red Sea and movement along the consequent Dead Sea Tranform Fault from Middle Miocene onwards, primarily in the last 11 million years.

Although the Lebanese ranges are high, averaging 2200 m along a length of 170 km, it is clear that deep surface karstification through the thick Jurassic carbonate sequence has taken place. This is shown by the way in which cave systems, such as that of Jeita, exit almost at sea level in Nahr el Kerb. Some large springs even exist below sea level as offshore from Ras Chekka and offshore from Tyre, presumably in response to cave systems that led to a much lower base level (at least -150 m) during the Last Glacial Maximum. The extensive, and up to 650 m thick, Mid Cretaceous, Sannine Limestone usually has its own independent base level determined by underlying impermeable Albian marls. The base of the Sannine Limestone is, thus, the site of a great many of the mountain springs seen in Lebanon, such as Nabaa Afqa and Nabaa Aaqoura.

HYDROGEOLOGY

The two major considerations of the hydrogeology of Lebanon are the surface hydrogeology and the subsurface hydrogeology.

Surface Hydrogeology
Included in this topic are the spring types of Lebanon and their location, discharge and hydrographic fluctuation. The rivers of Lebanon, both perennial and non-perennial are also considered, as well as a discussion of runoff and the estimated recharge of aquifers and their discharge.

There are many hundreds of springs in Lebanon (Fig. 13), mostly karst springs, and found predominately in the ranges. One of the peculiarities of Lebanon is that it possesses numerous large, high-elevation springs. Types of springs known (Guerre 1969) are simple overflow springs, overflow springs at stratigraphic contacts, faults dam springs, large springs associated with a fault-line spring, springs associated with marly-karstic beds, springs at stratigraphic contacts associated with a flexure, artesian springs and contact springs. Apart from these karst springs, there are innumerable low magnitude contact springs in Albian marly sandstones, known as the 'gres de base'.

There are a number of first-magnitude karst springs in Lebanon, according to Meinzer's (1923) Classification of Spring Discharge. This is most striking considering that there are only a few hundred first-magnitude springs in the entire world (Davies and Dewiest 1966) and Lebanon is a very small country. The Ain ez Zarqa spring, on the west side of the northern Beqa'a Valley, has a discharge of 11 m^3/sec (Wolfart 1967) and produces more than a third of cubic kilometer of spring water per year, as well as being the origin of the Orontes River (Nahr al Assi) and emanating from Cenomanian karstified limestones through Neogene cover. Nabaa Arbaain, just west of the Yammouneh Lake   is Sannine Limestone. It has a discharge of over 3 m^3/sec during 6 months of the year, although its discharge diminishes sharply during summer and autumn. The Ain Anjar spring emerges at the contact of Turonian limestone with overlying, impermeable Senonian marls at the foot of the Anti-Lebanon in the central eastern Beqa'a Valley. It has a maximum discharge of 10 m^3/sec, never falling below 2 m^3/sec during summer. Nabaa Barouk fault spring, emerging from the faulted Jurassic Kesrouane Limestone, only becomes a first-magnitude spring at its maximum in February and March and becomes a second-magnitude spring during the rest of the year. In southern Lebanon the contact spring of Nabaa Ouazzani in the valley of Nahr Hasbani (upper Jordan River) discharges up to 6 m^3/sec from the Cretaceous Sannine Limestone but falls to just above 1 m^3/sec during summer month. Flow from first-magnitude springs implies a large area of supply and easy subsurface groundwater communication.

There are a great many second-magnitude springs in Lebanon having discharges from 0.283 to 2.83 m^3/sec. Examples of these are Naba'a el Aassal (1 to 2 m^3/sec), Qoubaiyat (0.5-1 m^3/sec), Nabaa Soukkar (1-2 m^3/sec), Yanabih (0.5-1 m^3/sec), Nabaa Qadicha (0.5-1 m^3/sec), Nabaa Mar Sarkis (0.5-1 m^3/sec), Bkeftine (0.5-1 m^3/sec), Nabaa er Rouaiss (0.5-1 m^3/sec), Nabaa Afqa (1-2 m^3/sec, Nabaa el Hadid (0.5-1 m^3/sec), Nabaa el Liban (0.5-1 m^3/sec), Jeita Grotto (1-2 m^3/sec), Nabaa Sannine (0.5-1 m^3/sec), Chaghour (0.5-1 m^3/sec), Ain ed Delbe (0.5-1 m^3/sec), Ain Zhalta (0.5-1 m^3/sec), Nabaa Jezzine (<500l/sec), Nabaa et Tasse' (0.5-1 m^3/sec), Ras el Ain (0.7-1.5 m^3/sec), Yanabih el Hermel (0.5-1 m^3/sec), Laboue' (1-2 m^3/sec), Nabaa Nahle' (0.5-1 m^3/sec), Nabaa Bardaouni (0.5-1 m^3/sec), Nabaa Ahrous (0.5-1 m^3/sec), Nabaa Moghr-Toffaha (0.3-0.75 m^3/sec), Qab Elias (0.5-1 m^3/sec), Nabaa Chamsin (1-2 m^3/sec), Nabaa Faour (0.5-1 m^3/sec), Nabaa Machgara (0.5-1 m^3/sec), Nabaa Dardara (0.5-1 m^3/sec), Nabaa Joz (0.5-1 m^3/sec), Nabaa Maghara (0.5-1 m^3/sec), Nabaa el Hasbani (0.5-1 m^3/sec), and Nabaa Aammiq (<1.5m^3/sec). the majority of these springs emerge from karstified Jurassic and Cretaceous limestones, as overflow springs near stratigraphic contacts, or as fault-line springs. The Ras al Ain spring and Rachidiye springs near Tyre are unusual, being artesian, supplied spontaneously from subsurface Mid Cretaceous limestones through overlying Senonian marls and limestones, Eocene limestones and Quaternary alluvium. Nevertheless, the Ras al Ain spring has an artesian head of 5 to 6 m. Location of these springs is shown in figure 13. Otherwise, the majority of springs are ordinary springs, mostly of 3rd, 4th or 5th magnitude, or less, emanating as karst springs from the base of the high plateaus of Cenomanian limestones at, or near, its contact with greenish impermeable Albian marls.

There are 13 perennial rivers originating from the ranges of western Lebanon and 2 from the Anti-Lebanon. Most have their sources in large karstic springs, and river flow largely contributes to groundwater supply during winter and spring, mainly as a consequence of the melting of large expanses of snow on the ranges. Eleven of these rivers flow less than 50 km from the heights of the main Lebanese ranges westward to empty into the Mediterranean. The two largest rivers are the Nahr el Litani and the Nahr el Assi (Orontes), which originate in the Beqa'a Valley on the east flank of Mt. Lebanon. Most important is the Nahr el Litani, 132 km long, which flows south southwest and then turns east, cutting a deep gorge at the south end of Jabal Niha and entering the Mediterranean just north of Tyre. Flowing northward, the Nahr el Assi crosses into Syria and flows into the Mediterranean in the Gulf of Alexandretta. It has a discharge of 0.32 km³/year and an overall length of 422 km of which 50 km lies within Lebanon. Another significant river is the Nahr el Hasbani (or upper Jordan River), which rises in Mt. Hermon (Jabal esh Sheikh) and flows southward, supplying water to Israel and Palestine.

SUBSURFACE HYDROGEOLOGY

Considerable groundwater reserves exist in the depths of the mountain massifs of Lebanon and the Anti-Lebanon. This is proven by the regularity of discharge from many large springs, such as Ain ez Zarqa (11 m m³/sec). The amount of flow from first and second-magnitude springs and their broad distribution indicates a large area of supply and easy subterranean communication, while the occurrence of large springs in valley bottoms suggests high permeability and greater depth to the water table. A great part of heavy rain and thick winter snow infiltrates into the fissured carbonates with many solution channels formed by percolating waters.

Multiple aquifers, at least 15 in number, exist throughout Lebanon primarily in karstified limestone strata. There are 4 in Jurassic limestones and dolomites, 7 in Cretaceous limestones, 3 in Eocene limestones, and several minor aquifers in the Quaternary.

Aquifer J 1 occurs in northwest Lebanon, around Douma, in Jurassic carbonates with its main exit at the Dalle' spring at a stratigraphic contact with the overlying Aptian-Albian marls.

Aquifer J II is the main Jurassic aquifer in Lebanon occurring in the massive, 1700 m thick Kesrouane Limestone forming the core of the Lebanese ranges. It is a deeply fractured, fissured and karstified formation extending from Liassic to Upper Kimmeridgian as seen mainly in Nahr Ibrahim and the west slope of the Lebanese ranges. In detail, the Kesrouane limestone consists of basal dolomites, overlain by grey dolomitic limestones with interbedded dolomites and overlain by up to 180 m of the Bhannes Volcanics (basalts and tuffs), especially to the north of the Beirut-Damascus road. The Kesrouane Limestone aquifer is broken up by large transverse faults, causing a partial compartmentalization of the aquifer. Due to the heavy rainfall and thick winter snows over most of this aquifer, in conjunction with its heavy fracturing, infiltration is high. Its main exits are Nabaa Jeita, which emerges as a large subterranean river and also Nabaa Dachouniye'. The aquifers main interval of decrease begins in April or May and ends in October.

Aquifer J III is found in the Middle and Upper Jurassic Mt. Hermon Limestone on the west flank of Mt. Hermon and in the core of the Anti-Lebanon. The karstified limestones of this aquifer are at least 1320 m thick and their groundwater contributes to Wadi Aayoun Jenaim, Wadi el Maaher and their tributaries in Lebanon, as well as Nahr el Aouadj in Syria. The waters of this aquifer decrease from April to September.

Aquifer J V consists of the Kimmeridgian to Portlandian limestones, at least 155 m thick in the Jabal Barouk-Jabal Niha area. Most of this aquifer is uppermost Jurassic and it provides the source for high altitude (1950-1150 m) stratigraphic contact springs, such as Nabaa Barouk, Nabaa Safa, Nabaa Qaa on the west side of this range, and Nabaa Aammiq, Nabaa Qab Elias and Nabaa Machgara on the east flank of the range associated with faulting.

Cretaceous aquifers yield the largest amount of groundwater flow in Lebanon, mostly from karstified Cenomanian-Turonian limestones.

On the northwest slope of Mt. Lebanon, Aquifer C I is found. This aquifer is partly divided into semi-independent reservoirs by transverse faults superimposed on a horst anticlinal structure. It supplies the stratigraphic contact spring of Nabaa Rachaain, near Zghorta and the gravity spring of Nabaa Soukkar, as well as the artesian, submarine springs just offshore from Ras Chekka.

In the high plateaus of central Lebanon, the main Mid Cretaceous aquifer is referred to as Aquifer C II. It has been arbitrarily divided (Guerre 1969) into two subaquifers, namely Aquifer C IIA discharging as contact springs at Albian marls along the west slope of central Lebanon, and Aquifer C IIB which discharges by large, mainly fault-controlled springs along the east side of Mt. Lebanon. The aquifer C IIA exits from the base of the karstified Cenomanian Sannine Limestone as many high springs at altitudes from 1140 m to 1770 m. They include Nabaa el Aassal, Nabaa Roueiss, Nabaa Sanine, Nabaa Jouarzat and Nabaa el Laban. Their hydrographs show greatest discharge between November and May, and then a steady decrease from June to the end of October. The subaquifer C IIB on the Beqa'a side of Mt. Lebanon discharges as large barrier springs, associated with the Yammouneh Fault which flow into the Yammoueh Polje. Two springs are temporary, namely Nabaa Arbaain   and Nabaa Kazzab, while others such as Nabaa Moghr el Toffaha, Nabaa el Aarouss are permanent.

Aquifer C III, in Turonian limestones, underlies the coastal plateaus and plains of southern Lebanon. The outlets of this aquifer are artesian, such as Ras el Ain with a head of 5 to 6 m and Nabaa Rachidiye. Other outlets include the offshore submarine springs near to Tyre.

On the west flank of the Anti-Lebanon and in the eastern Beqa'a Valley, Aquifer C IV occurs in Tuonian karstified limestones capped by impermeable Senonian marls. Discharge from this aquifer is found in Nabaa Laboue', Nabaa Fake'he', and Nabaa Ras el Ain (Baalbek) along a fault. Further south, in the eastern Beqa'a Valley, the same aquifer yields large amounts of groundwater at the stratigraphic contact springs of Ain Anjar and Ain Chamsin, which are amongst the most important springs in Lebanon. Ain Anjar has a discharge of 10 m^3/sec in February and rarely falls just below 2 m/sec in late September.

In the region of Jezzine and Kfar Houne, there is localized aquifer in the Cenomanian Sannine Limestone between the Jurassic core of Jabal Niha and the Roum Fault. This Aquifer C V exits to west in Nabaa Jezzine, which is a boundary spring, and along numerous small springs associated with the Roum Fault.

Aquifer C VI is found in the southernmost Beqa'a and emerges as large water table springs along the valleys of the upper reaches of the Jordan River (Nahr Hasbani), such as Nabaa Ouazzani and Nabaa Hasbani.

A Minor aquifer is located in hills of the western border of the Beqa'a Valley. This is known as Aquifer C VII and originates from Middle Cretaceous karstified limestones after passing through a cover of Neogene terrestrial conglomerates. A small spring near Chtaura emanates from this aquifer with maximum flow in February and then declining through the dry months until the end of October.

Aquifer E I occurs in the central Beqa'a Valley in karstified Middle Eocene (Lutetian) limestones covered by Quaternary clays. It emerges as several small springs (Ain el Fouar, Ras el Ain (Terbol) and Ain el Baida) which are temporary, often being dry in summer.

In the southern Beqa'a, a second Middle Eocene aquifer (Aquifer E II) is found in a broad syncline followed by the southward flowing Nahr el Litani. There are several small water table springs, such as Ain ed Deir, just east of Qaraoun Dam, and others below the dam which contribute to the Nahr el Litani (e.g. Ain el Aaouainat).

South of the latitude of Sidon to the southern border, most of SW Lebanon contains a third Middle Eocene karstified aquifer (Aquifer E III) from which small springs exit, as near Jouaiya and at Nabaa el Hjair (near Qantara). This widespread, low yield aquifer receives an average rainfall volume of 333 Mm^3/year and evapotranspiration of 226Mm^3/year with an effective water volume of 107 Mm^3/year (Mroueh et al. 1996).

The stratigraphically highest aquifers of Lebanon are found in Quaternary sediments in the central Beqa'a Valley, and coastal Lebanon. A spring, which is the source of the Nahr el Litani, emerges from Quaternary sediments 9 km west of Baalbek. In southwestern Lebanon, the total water resources of the Quaternary are estimated by Mroueh et al. (1996) as 13 Mm^3/year and form an aquifer of minor importance along the coast. Perched aquifers occur in Quaternary alluvium on coastal marine terraces as in northern Ras Beirut and provide a temporary shallow aquifer exploited by hand dug wells.

Transmissivity values for the karstified Cenomanian limestones in the Beirut area have been calculated by Ukayli (1971) at from T=$1.83 \times 10^{-1} m^2$/sec to T=$6.1 \times 10^{-2} m^2$/sec from drawdown and discharge in various wells in the Beirut area, where some 400 wells exploit this aquifer. Storativity for this aquifer in the same area is calculated at S=1.7×10^8.

Recharge of the main aquifers in karstified Jurassic and Cretaceous limestones is high and active due to the heavy rainfall, and thick winter snows on the Lebanese ranges. There is no question of 'fossil' groundwater, as in common in most other Arab countries, and active subsurface circulation has been shown by the use of tracers (Hazzaa 1970). Transfer of precipitation over the karstified and fractured recharge surface to the Zone of Saturation is shown by the rapid rise of the water table level during the winter. Calculations (Ukayli 1971) for the Cenomanian C II Aquifer in the coastal Beirut city area give a recharge of 18 mm/year. In the high Lebanese ranges recharge is clearly much greater. For Cretaceous exposures of southwestern Lebanon it is estimated that 271 Mm^3/year are lost by evaporation, and of the remaining 185 Mm^3/year effective water volume some 175 Mm^3/year are infiltrated and 13 Mm^3/year are lost by subsurface runoff (Mroueh et al. 1996). This gives some measure of recharge in an area of south Lebanon where average annual rainfall is relatively low (i.e. 700 mm/year).

Discharge of groundwater from the aquifers of Lebanon is sufficient to maintain 13 perennial rivers, 5 first-magnitude springs and many hundreds of second- and third-magnitude springs. For the Cenomanian carbonate aquifer in Beirut and vicinity, Hazzaa (1970) has estimated that precipitation forms 49% of the water balance, while some 28% is lost by evapotranspiration, 15% goes to infiltration and surface runoff is 6%. Pumpage from over 400 wells in Beirut abstracts 11.3 Mm^3/year (Ukayli 1971). For the Lebanese ranges and their flanks, discharge is clearly higher and is well to remember that the one spring alone, Ain ez Zarqa, discharges 1/3 km/year.

CONCLUSIONS

The carbonate strata of Lebanon, primarily limestones but also including dolomites, are highly karstified. This karstification is seen in the exposed Jurassic core of the Lebanese ranges and the more widespread and thick, flanking Mid Cretaceous limestones, as well as Middle Eocene limestones and to a minor extent in restricted exposures of Middle Miocene limestones.

Due to the high elevation of the Lebanese ranges, which lie almost at a right angles to the prevailing westerly winds, precipitation as rain and thick snow is high and exceeds 1500 mm/year throughout the higher parts of these ranges. Early uplift of the Jurassic core in the Early Cretaceous has caused extensive paleokarst, while the whole ranges have been uplifted and folded, fractured and karstified since Middle Miocene times.

Almost all forms of karst can be found, including widespread karren, dolines, poljes, blind valleys, erosional remnants, such as hoodoos and natural arches, plus extensive solution shafts (ponors) connected to galleries extending at least 6.2 km, often with underground rivers and a wide variety of speleothems.

There are some 14 aquifers in the karstified carbonate strata of Lebanon with 5 first-magnitude springs, hundreds of second-magnitude springs, and literally thousands of third-magnitude of smaller springs. Discharge from these springs gives rise to 13 perennial rivers including major ones, such as the Nahr el Litani, Nahr el Assi (Orontes) and Nahr el Hasbani (upper Jordan River).

Some 67% the area of Lebanon is karstified and the exposed sequence of Jurassic, Cretaceous and Eocene carbonates is at least 4140 m thick, with significant disconformities and interbedded marls. There is no area of the Middle East where karst is so widely developed, while its development to such a depth allows good recharge of aquifers and a well-developed hydrogeological system.

REFERENCES

1. Adams, R.D., and Barazangi, M. 1984. Seismotectonics and seismology in the Arab region: a brief summary and future plans, Bull. Seismol. Soc. Amer., 74,3,1011-1030
2. Angenieux, J. 1951. Une combination de mouvements - verticaux et de mouvements tangentiels dans l'evolution structurale de Liban, Bull. Soc. Geol. France, 6,1,285-301.
3. Besancon, J. 1968. Le polje de Yammoune', Hannon, 3, 1-6, Beyrouth.
4. Beydoun, Z.R. 1977. Petroleum prospects of Lebanon: re-evaluation, Amer. Assoc. Petrol. Geologists, 61,43-64.
5. Bogli, A., 1980. Karst Hydrology and Physical Speleology, 1-284, Springer-Verlag, Berlin.
6. Burger, A., and Dubertrel, L. (eds) 1975. Hydrogeology of karstic terrains, with a multilingual glossary of specific terms, International Union of Geol. Sciences, Ser.8, No.3, Paris.
7. Davis, S.N., and De Wiest, R.J.M. 1966. Hydrogeology, 1-463, John Wiley and Sons, New York.
8. Dubertret, L. 1933. L'hydrologie et apercu sur l'hydrographie de la Syrie et du Liban dans leur relations avec la geologie, Revue Geogr. Phys. Geol. Dynamique, 6,4,347-452.
9. ‒‒‒‒‒‒ 1948. Apercu de geographie physique sur le Liban, l'Anti-Liban et la Damascene, Notes et Mem. Syrie et Liban, 4, 191-226.
10. ‒‒‒‒‒‒ 1943-53. Carte geologique du Liban au 1/50,000 de la Syrie et du Liban, 21 sheets with explanatory notes, Damas et Beyrouth, Ministere des Travaux Publics.
11. ‒‒‒‒‒‒ 1955. Carte geologique du Liban au 200,000 avec notice explicative, Ministere des Travaux Publics, 1-74, 8 pls., 1 map, Imprimerie Catholique, Beyrouth.
12. ‒‒‒‒‒‒ 1963. Liban, Syrie: chaine des grands massifs cotiers et confins a l'Est, 6-144, in Lexique Stratigraphique International, 3, 10c 1, Centre National de la Recherche Scientifique, Paris.
13. ‒‒‒‒‒‒ Edgell, H.S. 1968. Paleontological report on the Samakieh water well, Ministry of Water and Electric Resources, Beirut, 9th Ann. Res. Rept. 1967-68, 22-23.
14. ‒‒‒‒‒‒ 1996. Karst and hydrogeology of Lebanon, 30th International Geological Congress, Beijing, Abstracts, 3,318.
15. Edgell H.S., and Basson, P.W. 1975. Calcareous algae from the Miocene of Lebanon, Micropaleontology, 21,2, 165-184.
16. Fetter, C.W. 1980. Applied Hydrogeology, Merrill Publishing Company, Columbus, 1-487.
17. Ford, D.C., and Williams, P.W. 1989. Karst Geomorphology and Hydrology, 1-601, Unwin Hyman, London.
18. Guerre, A. 1969. Etude hydrogeologique preliminaire des karsts Libanais, Hannon, 4, 64-92, Beirut.
18. Hazzaa, I.B. 1970. Investigations of underground water in Lebanon using radioisotopes, Proc. of Symposium, National Council for Scientific Research, Lebanon, and Middle Eastern Regional Isotope Center for the Arab Countries, Beirut.

19. Jennings, J.N. 1971. Karst, An Introduction to Systematic Geomorphology, 7, 1-251, The M.I.T. Press, Cambridge, Massachusette.

20. Kareh, R. 1967. Les sources sous-marines de Chekka (Liban-Nord) . Exploitation d'une nappe karstique captive a exutoires sous-marins, Hannon, 2,35-39, Beyrouth.

21. Meinzer, O.E. 1923. Outline of groundwater hydrology with definitions, U.S. Geol. Survey Water Supply Paper 494,1-71.

22. Mroueh, M., Jaafar, H., Makhoule, G., and Mansour, K. 1996 Evaluation of the water resources and water balance of Southern Lebanon, 30th International Geological Congress of Beijing, Abstracts, 3, 256.

23. Pfannkuch, H.O, 1969. Elsevier's Dictionary of Hydrogeology- In Three Languages, 168, Elsevier Publishing Company, Amsterdam.

24. Quennell, A.M. 1958, The structure and evolution of the Dead Sea rift, Quart. J. Geol. Soc. 64, 1-24.

25. ──── 1984. The Western Arabian rift system, 775-788 in Dixon, J.E. and Robertson, A.H.F. (esd), The Geological Evolution of the Eastern Mediterranean Geol. Soc. Spec. Publ. No.17.

26. Renouard, G. 1955. Oil Prospects of Lebanon, Bull. Amer. Assoc. Petrol. Geol., 29,2125-2169.

27. Ron, H. 1987. Deformation along the Yammouneh, the restraining bend of the Dead Sea Tranform: Paleomagnetic data and kinematic implications, Tectonics, 6, 653-666.

28. Ukayli, M.A. 1971. Hydrogeology of Beirut and Vinicity. M.S. Thesis, American University of Beirut, 136, Beirut.

29. Wetzel, R. and Dubertret, L. 1951. Carte Geologique au 1:50,000, Feuille de Tripoli avec notice explicative, 1-61, Ministere des Travaux Publics, Beyrouth.

30. Wolfart, R. 1967. Geologie von Syrien und dem Libanon. Beitrage zur regionalen geologie der Erde, 6, 326, Gebruder Borntraeger, Berlin.

Proc. 30th Int'l Geol. Congr., Vol.22, pp. 177-184
Fei Jin and N.C. Krothe (Eds.)
© VSP 1997

The Flushing of Saline Groundwater from a Permo-Triassic Sandstone Aquifer System in NW England

JOHN H. TELLAM and JOHN W. LLOYD
Earth Sciences, University of Birmingham, Birmingham, B15 2TT, UK

Abstract

Using geological, hydrochemical, and geophysical data, a conceptual model is developed in order to explain the occurrence and distribution of saline groundwater in a Permo-Triassic sandstone sequence in northwest England. The saline water is derived from fresh water dissolution of halite, probably sometime prior to the Tertiary. Uplift in the Tertiary lead to erosion which eventually exposed the sandstones, and flushing of the saline water by fresh water commenced. The geometry of the system, and the presence of low permeability faults reduces the flushing efficiency so that saline groundwater exists everywhere in the lower parts of the sandstone body, and even in the upper parts in places. Upconing on a regional scale has also occurred, probably since the end of the Devensian, and flushing is actively occurring.

Keywords: hydrogeology, saline groundwater, Triassic sandstone.

INTRODUCTION

England has sea water intrusion occurring into aquifers in many areas [8]. However, high and very high salinity waters are also present in some major aquifers at depth, often remote from the present coasts. These inland waters pose a problem for water resources development [2, 16]. They also offer an opportunity for studying the long-term development of the aquifers concerned in terms of flow and diagenesis, and provide a means of investigating very long term flushing processes. This article concerns the development of a qualitative conceptual model for the occurrence and past behaviour of the halite-derived saline groundwater present in the Triassic sandstones in an area of NW England (Fig. 1).

GEOLOGY AND REGIONAL DISTRIBUTION OF SALINE GROUNDWATER

Table 1 shows the geological sequence. The study area lies at the northern end of the Cheshire Basin which is underlain by low permeability Carboniferous deposits and contains the complete local succession of the Permian and Triassic in its centre (undifferentiated on Figs. 1, 2). The permeable Permian Collyhurst Sandstone Formation unconformably overlies the Carboniferous, and is in turn overlain by the low permeability Manchester Marl Formation (and its lateral equivalent, the slightly more permeable Bold Formation, around Liverpool). The overlying Triassic sandstones are fluvial red-bed deposits, with occasional thin mudstones, and have a permeability of the order of 1 m/d.   The Basin was inverted sometime in the Tertiary, and subsequent subaerial erosion has removed all younger sediments except for small Jurassic outliers.   The Basin is connected to the Irish Sea Basin in the region of Liverpool (Fig. 1). Much of the discussion will centre around the north of the Basin (Fig. 2). Here the Permo-Triassic sequence dips to the south at around 5^0: and the sub-Quaternary topography is characterised by a channel running along the Mersey Valley and other channels running NW/SE. The till is Devensian in age.

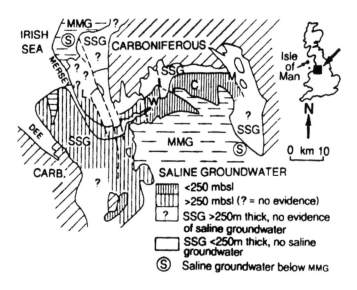

Fig. 1 Saline groundwater distribution in Permo-Triassic sandstones of NW England (L=Liverpool, M=Manchester,   W=Warrington) C=Chat Moss [14,7]

Saline water, of Cl content up to at least 100 g/l, has been found at depth in the sandstones across the area (Fig. 1; [1, 6, 9, 15]).   Beyond the edges of the uneroded Mercia Mudstone Group (MMG), isolated borehole records indicate the saline groundwater to occur below 200-250 m below present-day sea level   (bsl).   However, in the north of the Basin, especially, there is a zone where the saline groundwater occurs at higher elevations and it has been possible to map its occurrence in some detail (Fig. 2 modified after [15]), using amongst other data, drilling core pore water profiles (Fig. 2).   During the Devensian, sea levels at up to 120 m below present sea level would have encouraged regional flushing of the saline water towards the coast which then would have been somewhere north of the Isle of Man (Fig. 1).   This flushing is presumed to have resulted in the 250 mbsl regional level of the saline groundwater: the higher level saline water in the north of the Basin suggests that the known low permeability NW/SE faulting, and in particular the Roaring Meg Fault (Fig. 2)[16, 5, 13], has restricted flushing depths.   Calculation suggests that to retain the saline groundwater over time periods equivalent to the Devensian, the permeability of the Roaring Meg Fault would need to be very low, perhaps of the order of 10^{-12}-10^{-10} m/s.

Table 1. Geological sequence

Period	Formation/Rock Type [Acronym] (thickness) Hydrogeological	General Character
Quaternary	Till, sand, peat (0-80 m)	
Triassic	Mercia Mudstone Group (incl. halite) [MMG] (0->500m)	Aquiclude
	Sherwood Sandstone Group [SSG] (0-600m)	Aquifer
Permian	Bold Fm (Liverpool area)/Manchester Marl Fm [BF/MMF] (0-50m)	Aquitard/ Aquiclude
	Collyhurst Sandstone Formation [CSF] (0->200m)	Aquifer
	Unconformity	
Carboniferous with	Mudstones with sandstones, rare coal	Aquiclude
		local aquifers

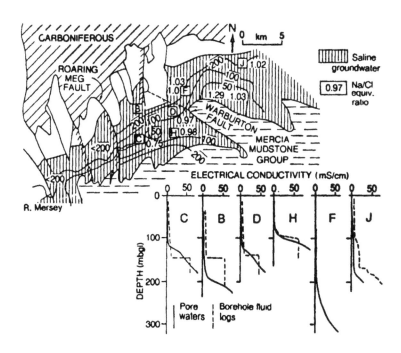

Fig. 2 Distribution of saline groundwater in the Northern end of the Cheshire Basin [14]

SALINE GROUNDWATER ORIGIN

The chemistry of the saline groundwater is consistent with a halite dissolution origin, halite being abundant in the MMG sequence [14]. In particular, Br/Cl and SO_4/Cl ratios are very similar to the ratios for present day waters in the halite sequence (Fig. 3). The $d^{18}O_{H_2O}$ signatures suggest that the H_2O is derived from fresh water recharge under cool climatic conditions (Fig. 3); however, isotope analysis from a sample from the highly saline water present in the CSF at site F (Fig. 2) indicates that the H_2O component of the deepest saline water was recharged under more temperate temperatures. Tellam (1995) [14] suggests that fresh groundwater recharged under temperate climatic conditions dissolved halite at some time between the Triassic and the Quaternary, and that subsequent partial flushing of the upper part of the saline groundwater body occurred during cooler episodes of the Quaternary. Because the lower part of the saline groundwater body appears to have a heavier H and O stable isotope signature, it seems unlikely that the saline water intruded from an ice-dammed lake postulated to have existed in the area in the Devensian: indeed, there appears to be no evidence that the lake was saline, despite the possible direct contact with halite rich areas of the MMG. Taylor et al. (1963) [12] report the base of the Northwich Halite Formation, part of the MMG, to be marked by a 20m zone of brecciated and disturbed strata. They interpret this as resulting from solution "soon" after deposition. Using the stable isotope data available, and assuming that the saline parent brine has a Cl concentration of either 100 g/l (CSF water at F) or 60 g/l (see data in Fig. 4 below), a feshwater $d^{18}O$ of around -12 to -11 can be calculated: using the equation given in [14], the inferred mean atmospheric temperature is around 2-4°C.

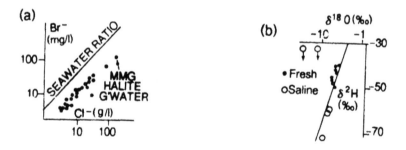

Fig. 3   Saline groundwater chemistry [13]

DISTRIBUTION OF SALINE GROUNDWATER IN THE NORTH OF THE CHESHIRE BASIN

The distribution of saline groundwater in the north of the area has been mapped in some detail using a wide range of data [15]. The fresh saline water interface was found to form a ridge underlying the present-day Mersey Valley/pre-till buried valley (Fig. 2). Two extreme models might explain the ridge: regional-scale upconing below low heads in the centre of the valley, or flushing of saline water by fresh groundwater. Evidence comes from chemical data from the uppermost, diluted part of the saline water body (Fig. 2), from borehole fluid logs (Fig. 2), and from borehole electric logs (Fig. 4 [3]).

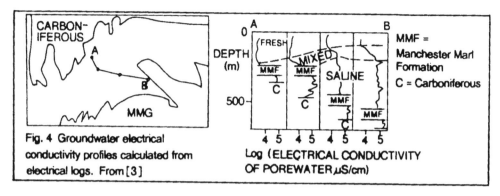

Fig. 4 Groundwater electrical conductivity profiles calculated from electrical logs. From [3]

Fig. 4 Groundwater electrical conductivity profiles calculated from electrical logs. From [3]

Fig. 2 indicates that the saline groundwater in the Warrington area (Fig. 1) has Na/Cl equivalent ratios of <1: Ca and Mg concentrations are unusually high [15], and it is concluded that saline water has migrated into a part of the aquifer previously containing fresh groundwater, and during this process, ion exchange has occurred. In the Chat Moss area (see Fig. 1 for location), Na/Cl ratios are -1, indicating that flushing by fresh groundwater is dominant. In all cases where data exist, the Na/Cl ratios approach 1 at deeper levels. In both areas, the overlying post-glacial fresh groundwater is enriched in Na and depleted in Ca and Mg, strongly suggesting recent flushing of saline groundwater [13]. In Chat Moss, discharge of saline groundwater is known to have occurred at springs in the past [11], supporting the idea of flushing. Chemical data thus indicate that flushing of saline water is dominant in the Chat Moss area. This results in the shallow concentration/depth gradients observed in borehole J. It is suspected that the Warburton Fault (Fig. 2) is of low permeability and thus saline water can only escape the aquifer by being flushed to ground level. In the Warrington area upconing is dominant, but the upper aquifer layers are flushed by fresh groundwater. The upconing results in relatively steep concentration depth gradients, as seen in boreholes B, C, D and H.

Examination of the electrical conductivity logs in Fig. 2 indicates that there is little discernable vertical flow in most of the boreholes in the Warrington area, indicating that upconing has reached an equilibrium state. This conclusion is in agreement with crude Ghyben-Herzberg principle calculations. Taking a fresh water head at 30 m above sea level in the north of the area [13] and the saline water at a depth of 250 mbsl, the inferred saline water density necessary to allow regional upconing to sea level is roughly equivalent to a brine of 100 g/l Cl: this is the concentration in the water present in the CSF at site F.

In the Chat Moss area, head gradients appear to be directed upwards, in keeping with relatively recent discharge of saline water at ground level. Brassington et al. (1992) [3] present data which provide an insight into the variation in concentration at depths below the base of the accessible observation boreholes in the area (Fig. 4). Their results are based on interpretation of electric logs from coal exploration boreholes, and the water concentration estimates are necessarily quite crude. However, assuming the patterns of concentration variation with depth are correct, several important points can be made. Firstly, the dispersion zone observed in all the porewater profiles (Fig. 4) is limited to the top 100 m below the saline/fresh water interface: this zone, characterised by cool fresh water isotopic signatures, represents the deepest flushing level achieved during the low sea level stands in the Quaternary. Below this depth concentrations rise, but at much slower rates. (The oscillations in concentrations suggested by the electric log interpretations, if real, would have diffusion lives less than 100s to a few 1000 years; it is therefore concluded that they are artifacts of the interpretation method). If the rise in concentration with depth has been produced by diffusion, depending on assumptions, a time interval of a few million to a few tens of millions of years is needed - an interval consistent with the time since likely exposure of the sandstone following basin uplift and erosion in the Tertiary. The jump in concentrations across the MMF may be an artifact produced by the interpretation method. However, sensitivity calculations suggest that it may well be a real feature. The MMF may be acting as a semi-permeable membrane: this would invalidate the diffusion calculations outlined above. Alternatively, the concentration jump may be brought about by lateral movement in the saline water body in response to regional upconing, as explained below.

For the Warrington area, assume that initially a saline groundwater body exists in which salinity increases everywhere regularly with depth, and that this body is in diffusive equilibrium with the saline water in the CSF. In response to falling heads in the Mersey Valley, regional upconing occurs, and saline water migrates from the north (and south) to supply the rising ridge of saline water. The northern trailing edge of the saline water is "smeared out", resulting in shallow concentration/depth gradients (eg site F): in the ridge area, the concentration gradient is "sharpened" by upwards movement into a zone of lateral (slow), fesh groundwater movement, thus giving the low Na/Cl ratios in the saline waters, and the high Na/Cl values in the fresh groundwaters. The lateral movement of saline water juxtaposes lower concentration saline water above the MMF and higher concentration saline groundwater below the MMF: the CSF water is prevented from supplying the rising ridge because of the hydraulic barrier of the MMF, and therefore does not move from its original situation. Assuming that the MMF is 50 m thick and that diffusion has not affected concentrations in the sandstones above or below, calculations show that the concentration discontinuity across the MMF occurred not longer ago than around 12000 years. This indicates that the upconing occurred as climate ameliorated and water levels rose. If the upconed ridge of saline water is redistributed, the saline interface in the Warrington area would be at roughly 160-180 m bsl. What is less easy to estimate without more detailed calculations is whether the salinity displacement seen across the MMF is consistent with the amount of upconing. If the salinity profile at site F were entirely due to diffusion, it would have developed over a period of around 200,000 years: this estimate provides an upper bound on the age of the development of the salinity profile, and suggests most active flushing took place in the late Pleistocene, probably since the Ipswichian.

In the Chat Moss area there is no chemical evidence for upconing. The high elevation of the main spring in this area would mean that there would be only a small drop in fresh water head from recharge to discharge areas, and hence little upconing. Flushing appears to dominate. From the evidence of the high Na/Cl ratios in the fresh groundwaters of the region, and from Ra evidence presented elsewhere [7], it is clear that saline groundwater once occupied a large proportion of the aquifer. For the ion exchange still to be recognisable, less than a few pore volumes of fresh groundwater must have flowed through the aquifer (cf. data in [4]). Rough calculations using estimated flow rates indicate a flushing period of 1000-2000 years. This compares with ^{14}C and U-series estimates of water ages [13, 7] of less than a few thousand years. Hence most flushing appears to be post Devensian. It is uncertain why the aquifer was not more flushed in previous interglacials. One factor may be that, provided the Warburton Fault (Fig. 2) has relatively low permeability, saline groundwater can only migrate out of this part of the aquifer by flow out to ground surface whereas in the Warrington area, saline groundwater can flow to the east over the

Roaring Meg Fault in the base of the buried channel which runs along the Mersey Valley, in this region at a depth of around 40 mbsl.

CONCLUSIONS

It is suggested that the deep saline groundwaters at the northern end of the Cheshire Basin have the following history.

Brines formed by freshwater dissolution of halite at some time prior to the Tertiary. The uplift and erosion occurring in the Tertiary exposed the SSG for the first time, and lead also to the formation of a series of low permeability NW/SE faults. Flushing of saline water by fresh groundwater, especially during the low sea levels of the glacial episodes, removed saline waters from above 250 m bsl. This is 130 m below lowest sea level in the Devensian. Most of the flushing appears to have occurred in the Devensian. However, the low permeability faults restricted the flushing in the NE of the Basin. As sea levels rose in the Holocene, a new flow system was established and regional upconing occurred below the Mersey Valley. This resulted in lateral movement in the saline water body. Flushing by fresh groundwater is now continuing across the area.

This conceptual model needs testing quantitatively. It has wider significance in that it is clear that: there has been significant recharge in cool climatic conditions in this area; in the past the active thickness of the aquifer was deeper than at present, due to lower base levels or possibly even due to high sub-ice fresh water heads generated during excavation of buried valleys; and the present active circulation depth is no greater than around 250 m bsl. A basic question yet to be answered is how the saline water orginally got into the aquifer, and why it is not more ion exchanged: have many pore volumes of saline groundwater flowed through the aquifer, or were the clay minerals formed after the saline water was recharged?

REFERENCES

[1] A.D. Allen. *The hydrogeology of the Merseyside area.* Unpub. PhD thesis, University College, London University (1969).

[2] F.C. Brassington and R.J. Ireland. An investigation of complex saline groundwater problems in the Permo-Triassic sandstones of North West England. *Sci. Total Environment,* 21, 261-268 (1981).

[3] F.C. Brassington, P.A. Lucey, and A.J. Peacock. The use of down-hole focussed electric logs to investigate saline groundwaters. *Q.J. Eng. Geol.,* 25, 343-349 (1992).

[4] H.F. Carlyle. *The hydrochemical recognition of ion exchange during seawater intrusion at Widnes, Merseyside, UK.* Unpub. PhD thesis, Earth Sciences, University of Birmingham, 275 pp (1991).

[5] K.W.F. Howard. Beneficial aspects of sea-water intrusion. *Ground Water,* 25, 398-406 (1988).

[6] F.T. Howell. The significance of the ingress of saline waters into the Permo-Triassic aquifers of south Lancashire. *Wat. and Wat. Engineering,* 70, 364-365 (1965).

[7] M. Ivanovich, J.H. Tellam, G. Longworth and J.J. Monaghan. Rock/water interaction time scales involving U and Th isotopes in a Permo-Triassic sandstone. *Radiochim. Acta,* 58/59, 423-432 (1992).

[8] J.W. Lloyd. Saline groundwater associated with fresh groundwater reserves in the United Kingdom. *A Survey of British Hydrogeology,* Spec. Pub., Roy. Soc., London, 73-84.

[9] P.A. Lucey. *The hydrochemistry of groundwater in the West Cheshire aquifer, North West England.* Unpub. MPhil. thesis. University of Lancaster (1987).

[10] C.F.Mitchell. Raised beaches and sea levels. In: F. W. Shotton (ed), *British Quaternary Studies: Recent Advances*, Clarenden Press, Oxford, 169-186 (1977).

[11] G.W.Ormerod. Outline of the principal features of the saltfield of Cheshire and adjoining districts. *Q.J. Geol. Soc. Lon.*, 4, 262-288 (1848).

[12] B.J. Taylor, R.H.Price and F.M. Trotter. *Geology of the country around Stockport and Knutsford*. Mem. Geol. Surv. Gt. Brit.,(1963).

[13] J.H. Tellam. The groundwater chemistry of the Lower Mersey basin Permo-Triassic sandstone aquifer system, UK: 1980 and pre-industrialisation-urbanisation. *J. Hydrol.*, 161,287-325(1994).

[14] J.H. Tellam. Hydrochemistry of the saline groundwaters of the Lower Mersey Basin Permo-Triassic sandstone aquifer system, UK. *J.Hydrol.*, 165,45-84(1995).

[15] J.H. Tellam, J.W. Lloyd and M. Walters. The morphology of a saline groundwater body: its investigation, description and possible explanation. *J. Hydrol.*, 83,1-21(1986).

[16] University of Birmingham. *Lower Mersey Basin saline groundwater study*. Unpub. report to North West Water Authority, 81 pp.(1981).

Proc. 30th Int'l Geol. Congr., Vol.22, pp. 185-203
Fei Jin and N.C. Krothe (Eds.)
© VSP 1997

A Model for Coupled Fluid Flow and Multicomponent Chemical Reactions with Application to Sediment Diagenesis

MING-KUO LEE
Department of Geology, Auburn University, Auburn, AL 36849, U.S.A.

Abstract

This paper describes a numerical model for the calculation of the rates of minerals that precipitate or dissolve in basin strata as groundwaters migrate along temperature and pressure gradients. The calculation incorporates chemical reaction and mass balance equations into a transient flow model that predicts flow velocities, flow directions, and temperature and pressure distribution along flow path. The model integrates predicted groundwater flow patterns with geochemical reaction path modeling; this approach allows us to predict the rates of precipitation and dissolution in geochemical systems open to mass and heat transfer. In steady-state runs, the model calculates the instantaneous rates of precipitation and dissolution; in transient runs, the model can trace the volume of cements that precipitate in basin strata from migrating groundwater as well as the volume of minerals that dissolve. The program then uses the net volume of precipitation and dissolution to modify porosity. Such an approach accounts for the effect of precipitation and dissolution on sediment porosity, which in turn affects permeability and fluid flow. The program is integrated with the stochastic model to evaluate how the spatial variability in permeability affects the pattern of diagenetic reactions. The model is used to predict chemical reactions during several hydrologic processes, including (1) diagenesis of quartz and calcite by flow through a wavy sandstone, (2) cementation of amorphous silica and its feedback effect on thermal convection, and (3) regional diagenesis by migrating brines in the Illinois basin. The sample calculations shed light on the rates and patterns of chemical diagenesis that likely accompany fluid migration in sedimentary basins. When the predicted results can be compared to diagenetic patterns observed in nature, the model places important constraints on the origin of diagenesis.

Keywords: diagenesis, reaction path modeling, groundwater flow, sedimentary basin.

INTRODUCTION

Since the early 1960s, geologists have recognized that many diagenetic reactions take place in open systems [13,14,15,24,49]. A quantitative prediction of the rate of diagenesis requires the treatment of hydrologic transport and geochemical reactions as coupled processes. Helgeson [16] and Helgeson et al. [17,18] were the first to use a computer to trace the irreversible reaction of a geochemical system. Since then, geologists have developed a number of numerical models to study processes in geochemical systems that are open to mass or heat transfer [6,36,39,40,42,51,52]. Reaction path models have found widespread application in the studies of weathering, diagenesis, hydrothermal alteration, and ore deposition. Although geochemical models can be extended with configuration implicitly accounting for hydrologic processes such as fluid mixing and flushing [6], they do not include mass transport by hydrodynamic dispersion, nor do they describe the variation of chemical reactions in space.

In the past decade, numerical models accounting for both hydrologic and geochemical processes, with varying degrees of sophistication, have been developed [21,34,37,44,46]. These models simultaneously solve for equations of mass transport and chemical reaction, thus providing a great improvement in simulating interactions among migrating fluids and minerals. More recently, new modeling techniques have been developed and solved in quite large time and distance scales [27,41,47], this can account for complex geochemical processes in sedimentary basins. Some models are developed to examine how reaction induced porosity and permeability changes affect flow patterns [47]. Although significant progress has been made in formulating numerical methods for the study of reactive flow, very few of these models have been applied to field problems of diagenetic reactions in sedimentary basins.

In this study, a numerical model is developed for the calculation of the rates of minerals that precipitate or dissolve in basin strata as groundwaters migrate along temperature and pressure gradients. The model calls on a reaction path model to calculate how mineral's solubilities evolve with temperature over a polythermal path; this linking accounts for the influence of common-ion effect and ion-strength effect on mineral's solubility in multicomponent geochemical systems. The quantitative estimates of mineral solubility are then incorporated into the predicted flow system. The model is formulated and solved in geologic-time and basin distance scales, and can therefore be applied to study regional diagenesis related to basin-scale fluid migration. The calculation can adjust sediment porosity from the net volume of precipitation and dissolution, therefore accounting for the feedback effects of chemical diagenesis on porosity, which in turn affects permeability and fluid flow. The model integrates the Turning Bands method [29] to generate statistical permeability fields. Different realizations of permeability fields can be incorporated into the reactive flow model to examine the effects of permeability heterogeneity on diagenetic patterns.

I apply the model to predict precipitation and dissolution reactions during several hydrologic processes, including (1) diagenesis of quartz and calcite by flow through a wavy sandstone, (2) diagenesis of amorphous silica and its feedback effect on thermal convection, and (3) diagenetic reaction driven by migrating brines in the deep aquifers of the Illinois basin. In the first example, the calculations give insight into how a basin's geometry and permeability heterogeneity affect diagenetic patterns. The maximum alteration occurs where groundwaters move across temperature isotherms at a sharp angle. The calculation results in which the stochastic permeability model is used better approximate the irregular pattern of diagenetic alteration observed in nature, but differ markedly from the uniform pattern predicted by the homogeneous model. In the second example, the model shows that the interaction between fluid flow and chemical diagenesis can prevent the buoyancy-driven convection from maintaining a steady state over geologic time. The convective cell tends to propagate from zones of precipitation to zones of dissolution. In the last example, the predicted results can be compared to the diagenetic patterns observed in a basin, therefore providing constraints on the origin of the diagenetic cements.

MODEL DEVELOPMENT

The model development is based on two sets of partial differential equations: the chemical equilibrium equations and the mass balance equations. The equations of chemical equilibrium provide a basis for solving the equilibrium solubility of minerals in basin strata through geologic time; the mass balance equations incorporate chemical reactions and various solute transport processes including advection and hydrodynamic dispersion. These equations are incorporated into a basin hydrology model Basin2 [1,5] to trace the volume of minerals that precipitate or dissolve in basin strata as groundwaters migrate along temperature and pressure gradients.

Chemical Equilibrium Equations
The calculation is based on the assumption that minerals maintain local equilibrium with migrating groundwaters and their solubilities depend on temperature T and pressure P. Temperature and pressure change with time in basin strata in response to basin subsidence, sediment compaction, erosional unloading, and groundwater advection. As temperature and pressure change, minerals precipitate or dissolve so that the pore fluid adjusts to a new equilibrium composition. The change of solubility C_k (moles/cm^3 fluid) of mineral k with time t can be described by

$$\frac{dC_k}{dt} = \frac{\partial C_k}{\partial T}\frac{dT}{dt} + \frac{\partial C_k}{\partial P}\frac{dP}{dt} \tag{1}$$

Similarly, the variations of mineral's solubility in space are given by

$$\frac{dC_k}{dx} = \frac{\partial C_k}{\partial T}\frac{dT}{dx} + \frac{\partial C_k}{\partial P}\frac{dP}{dx} \tag{2}$$

$$\frac{dC_k}{dz} = \frac{\partial C_k}{\partial T}\frac{dT}{dz} + \frac{\partial C_k}{\partial P}\frac{dP}{dz} \tag{3}$$

Equations (2) and (3) depict that the concentration gradients (dC_k/dx, dC_k/dz) in the horizontal (x) and the vertical (z) direction are proportional to temperature gradients (dT/dx, dT/dz) and pressure gradients (dP/dx, dP/dz) along the flow path.

Mass Balance Equations
Subsurface processes of mass transfer and chemical reactions must satisfy the principle of conservation of mass. For a two-dimensional problem, the principle equation that describes the transport of non-reactive solutes by dispersion and advection [11] is

$$\frac{dC_k}{dt} = \frac{\partial}{\partial x}\left[D_x\frac{\partial C_k}{\partial x}\right] + \frac{\partial}{\partial z}\left[D_z\frac{\partial C_k}{\partial z}\right] - v'_x\frac{\partial C_k}{\partial x} - v'_z\frac{\partial C_k}{\partial z} \tag{4}$$

where v'_x and v'_z are the lateral and vertical groundwater velocities (cm/yr) in curvilinear coordinates, and D_x and D_z are coefficients of hydrodynamic dispersion (cm^2/yr), which account for mixing due to both molecular diffusion and mechanical dispersion. The model calculates the coefficients from

$$D_x = D^* + \alpha_L v'_x + \alpha_T v'_z \tag{5}$$

and

$$D_z = D^* + \alpha_L v'_z + \alpha_T v'_x \tag{6}$$

here D^* is the diffusion constant (cm^2/sec), α_L and α_T are the dispersivities (cm) in the longitudinal and transverse directions (or along and across the direction of flow). According to equations (5) and (6), the effect of dispersion is proportional to the flow velocity. Diffusion is the only mass transfer process under no flow conditions. Because the diffusion coefficient is relatively small (about 10^{-6} cm^2/sec) compared to dispersivities [11], the effects of dispersion dominate those of diffusion at even modest flow rates.

The advection-dispersion equation (4) can be extended to include the influence of chemical reactions

$$\frac{dC_k}{dt} = \frac{\partial}{\partial x}\left[D_x \frac{\partial C_k}{\partial x}\right] + \frac{\partial}{\partial z}\left[D_z \frac{\partial C_k}{\partial z}\right] - v'_x \frac{\partial C_k}{\partial x} - v'_z \frac{\partial C_k}{\partial z} - \frac{dR_k}{dt} \tag{7}$$

where $\dfrac{dR_k}{dt}$ is the rate at which mineral k precipitates or dissolves (moles/cm^3 fluid yr).

Substituting equations (2) and (3) into (7), by rearranging, the transport equation becomes

$$
\begin{aligned}
\frac{dR_k}{dt} = & -\frac{dC_k}{dt} \\
& + \frac{\partial}{\partial x}\left[D_x\left[\frac{\partial C_k}{\partial T}\frac{\partial T}{\partial x} + \frac{\partial C_k}{\partial P}\frac{\partial P}{\partial x}\right]\right] + \frac{\partial}{\partial z}\left[D_z\left[\frac{\partial C_k}{\partial T}\frac{\partial T}{\partial z} + \frac{\partial C_k}{\partial P}\frac{\partial P}{\partial z}\right]\right] \\
& - v'_x\left[\frac{\partial C_k}{\partial T}\frac{\partial T}{\partial x} + \frac{\partial C_k}{\partial P}\frac{\partial P}{\partial x}\right] - v'_z\left[\frac{\partial C_k}{\partial T}\frac{\partial T}{\partial z} + \frac{\partial C_k}{\partial P}\frac{\partial P}{\partial z}\right]
\end{aligned} \tag{8}
$$

Predicting Cementation Rates

In diagenetic problems, it is more useful to know the change in a mineral's volume than its mass change because the volume change gives information regarding how porosity evolves in reaction. The dimensionless solubility V_k, or the dissolved volume of mineral k per unit volume of groundwater, can be calculated from the mineral molal solubility m_k (moles/kg $\cdot$ H$_2$O) as

$$V_k = \frac{m_k \cdot M_v \cdot \rho\left[1 - \dfrac{TDS}{10^6}\right]}{1000} \tag{9}$$

here M_v is the molar volume of the mineral (cm^3/mole), ρ is the density of the fluid (g/cm^3), and TDS is the total dissolved solutes (mg/kg) in the fluid. Because the greatest uncertainty in predicting cementation rates comes largely from the estimation of flow velocity, it is sufficient to assume the term TDS as zero and ρ to be one in equation (9). Expressing m_k in unit of moles/cm^3 fluid (C_k), the equation (9) can be further simplified as

$$V_k = M_v \cdot C_k \tag{10}$$

Similarly, X_k, the volume of the mineral precipitated or dissolved per unit volume of the formation, is given by

$$X_k = \phi \cdot M_v \cdot R_k \qquad (11)$$

where ϕ is the porosity of rock. Therefore, the following equation relates the cementation rate $\dfrac{dX_k}{dt}$ to R_k

$$\frac{1}{\phi}\frac{dX_k}{dt} = M_v \frac{dR_k}{dt} \qquad (12)$$

Multiplying (8) by M_v, substitute (1), (10), and (12) into (8), the cementation rate of mineral k is

$$
\frac{1}{\phi}\frac{dX_k}{dt} = -\left[\frac{\partial V_k}{\partial T}\frac{dT}{dt} + \frac{\partial V_k}{\partial P}\frac{dP}{dt}\right]
$$
$$
+ \frac{\partial}{\partial x}\left[D_x\left[\frac{\partial V_k}{\partial T}\frac{\partial T}{\partial x} + \frac{\partial V_k}{\partial P}\frac{\partial P}{\partial x}\right]\right] + \frac{\partial}{\partial z}\left[D_z\left[\frac{\partial V_k}{\partial T}\frac{\partial T}{\partial z} + \frac{\partial V_k}{\partial P}\frac{\partial P}{\partial z}\right]\right]
$$
$$
- v_x'\left[\frac{\partial V_k}{\partial T}\frac{\partial T}{\partial x} + \frac{\partial V_k}{\partial P}\frac{\partial P}{\partial x}\right] - v_z'\left[\frac{\partial V_k}{\partial T}\frac{\partial T}{\partial z} + \frac{\partial V_k}{\partial P}\frac{\partial P}{\partial z}\right] \qquad (13)
$$

The first two terms on the right side of equation (13) describe the effects of temperature and pressure changes, as might occur in response to burial and exhumation or changes in heat flow, on precipitation and dissolution. In a steady-state flow regime there are no such effects, because T and P do not change. In this case, reaction is driven only by hydrologic mass transport and chemical diffusion. The third and fourth terms describe cementation by chemical diffusion and mechanic dispersion. The last two terms account for cementation by groundwater advection, which is the dominant transport mechanism over the distance scale considered in sedimentary basins [46] ranging from tens to hundreds of kilometers wide.

Equation (13) can be solved explicitly by a finite difference approach. The model calculates the reaction rate dX_k/dt of a mineral k directly by evaluating the right hand side of (13). The model first employs Basin2 to calculate the flow velocities and T and P distribution in basin strata. To evaluate the derivatives $\partial V_k/\partial T$ and $\partial V_k/\partial P$, however, requires the knowledge of mineral's solubility. The model uses established solubility correlation from literature to calculate mineral solubility. For example, the correlations of Rimstidt and Barnes [43] give the molal solubility of silica minerals at different temperature (Table 1). By these correlations, the solubility of silica mineral increases with increasing temperature. Alternatively, the model can call on a reaction path model React [6,25] to calculate how mineral solubility varies with temperature. The React calculation accounts for chemical interactions such as the common ion effect and other factors that might affect a mineral's solubility. For example, calcite solubility varies with temperature, CO_2 fugacity, and salinity of the fluid. By linking React to the model, we can account for these factors.

The model predicts instantaneous rates of reaction dX_k/dt (volume % per m.y.) in a steady-state run and, in a transient run, cumulative volume change of minerals (volume %). In a transient run, the model can adjust porosity, from the net volume of cementation or dissolution by

$$\phi = \phi_0 - \sum_k \Delta X_k \tag{14}$$

here ϕ and ϕ_0 are the adjusted porosity and pre-cement porosity, respectively. ΔX_k is the change in volume of mineral k by cementation and dissolution. Since linear correlations are commonly observed between the logarithm of permeability and porosity of rocks [5], we can use the modified value of porosity to recalculate horizontal permeability k_x (µm or darcy) after precipitation/dissolution as

$$\log k_x = A\phi + B \tag{15}$$

here A and B are the slope and intercept of the correlation of permeability k_x with porosity ϕ. Thus, the program can account for the effect of cementation on sediment porosity, which in turn, affects permeability and hence the flow field.

Table 1. Solubility of silica minerals as a function of absolute temperature T_K [43]

Minerals	Equilibrium concentrations (molal)
Quartz	$\log m = 1.881 - 2.028 \times 10^{-3} \, T_K - 1560 \, T_K^{-1}$
Chalcedony	$\log m = 2.219 - 2.323 \times 10^{-3} \, T_K - 1547.8 \, T_K^{-1}$
α-Cristobalite	$\log m = -0.0321 - 988.2 \, T_K^{-1}$
Amorphous.silica	$\log m = 0.338 - 7.889 \times 10^{-4} \, T_K - 840.1 \, T_K^{-1}$

DIAGENESIS OF QUARTZ AND CALCITE IN A WAVY SANDSTONE

Groundwater migrating through a basin alters sediments through which it flows because mineral solubility varies with temperature and pressure along the flow path. Quartz, for example, in increasingly soluble with rising temperature. Groundwater saturated with quartz precipitates quartz cement as it migrates along a path of decreasing temperature, whereas it dissolves quartz as it migrates toward higher temperature. In this study, the solubility correlation established by Rimstidt and Barnes [43] is used to calculate quartz solubility at different temperature. Calcite, on the other hand, has no unique solubility correlation because its solubility depends strongly on the fluid's pH and salinity. For the calculation in this section, I assume that calcite is in equilibrium with a fluid having a seawater composition, and the initial fugacity of CO_2 of the fluid is set to reflect the atmosphere pressure of $10^{-3.5}$. The geochemical model React is used to calculate calcite solubility at different temperature. In contrast to quartz, calcite solubility decreases with increasing temperature [23].

In the first example, I examine how a gravity-driven flow drives diagenetic alteration of quartz and calcite in a wavy sandstone. Figure 1 shows a hypothetical basin consisting of a wavy sandstone sandwiched between two aquitards. The sandstone and aquitards have homogeneous permeability of 1 and 10^{-4} darcy, respectively. All calculations in this study assume the diffusion coefficient D^* to be 1.5×10^{-5} cm^2/sec; and the longitudinal dispersivity (α_L) and transverse dispersivity (α_T) are 500 cm and 50 cm, respectively. The calculation shows that fluid migrates from the left to the right in response to a slight uplift to the left. Because temperature increases with depth in the aquifer, groundwater encounters varying temperature as it flows through the wavy sandstone. As a result, the advecting fluid diagenetically alters the wavy sandstone because mineral solubility varies with temperature along the flow path.

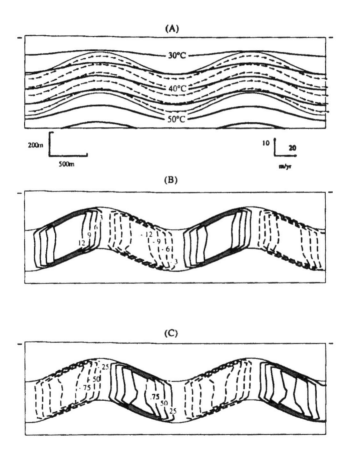

Figure 1. (A) Cross section showing temperature distribution (contours, in °C) and flow pattern in a basin with a wavy sandstone. The model assumes heat transfer by conduction only. Arrows show direction and velocity of groundwater flow (see scales). (B) and (C) show calculated rates of quartz cementation and calcite cementation (contours, in %/m.y.), respectively, along the homogeneous wavy sandstone. Solid contours show zones of precipitation with positive reaction rates; dashed contours indicate zones of dissolution with negative reaction rates. Quartz and calcite show opposite diagenetic patterns.

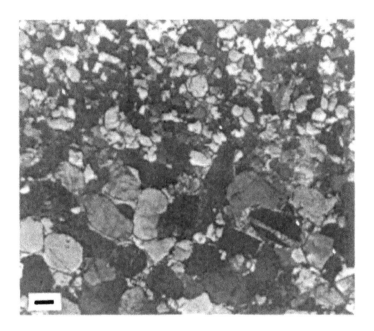

Figure 2. Photomicrograph showing calcite and quartz cementation in the Permian Lyons sandstone, Denver basin (Sample 13398, numbered by Chevron Oil Field Research). Scale bar = 100 μm. The distribution of cements is controlled by permeability heterogeneity. Cementation tends to take place in the more porous portion of the sandstone with larger grain size. Less cements are found in the fine-grain, less porous portion of the sandstone.

The model predicts that quartz, because of its prograde solubility, precipitates along the up-dip limb of the wavy sandstone where the fluid ascends and cools; and it dissolves along the down-dip limb in which the fluid descends toward the troughs. The model predicts an opposite pattern for calcite because of this mineral's retrograde solubility. Quartz precipitates or dissolves in the sandstone at a maximum rates about 0.15% per m.y. The dissolution/precipitation rate for calcite is about 1% per m.y., but may vary with fluid salinity and CO_2 fugacity, as previously discussed. The most extensive alteration occurs along the flanks where fluids move mostly sharply across temperature isotherms. Little dissolution or precipitation takes place near the crests or the troughs, on the other hand, where the fluid migrates almost parallelly to the isotherms.

In real world cases, diagenetic alteration often show scatter, irregular pattern (Fig. 2). Such irregular pattern cannot be quantitatively described by models in which sediment is assumed to have a uniform permeability. A stochastic permeability model is used to evaluate how the spatial variability in permeability affects the distribution of diagenetic reactions. Two-dimensional heterogeneous permeability fields are generated by the Turning Bands method [29]. The permeability data are incorporated into the model to simulate flow and diagenetic patterns in the wavy sandstone. Two stochastic permeability models are presented, both have geometric mean of 1 darcy and variance of log permeability of 2.0. The first model has a horizontal and vertical correlation distance of 850 m and 35 m, respectively (Fig. 3A), and the second model also has a horizontal correlation length of 850 m but a vertical correlation distance of 175 m (Fig. 3B).

The modeling results show that flow through heterogeneous sediments results in irregular diagenetic patterns (Fig. 4), which contrast sharply with the vertical cementation contours predicted by the homogeneous model (Fig. 1). Despite constant mean fluid fluxes in all simulations, the spatial variation of permeability causes local focused flow and creates irregular diagenetic patterns. Areas with high-permeability are more altered than the adjacent low-permeability areas. Diagenetic zones are wider in the vertical direction when sediment has higher vertical correlation length. The results argue that irregular and scatter diagenetic patterns may be caused by groundwater flow through heterogeneous media. If a stochastic modeling approach is carefully employed, heterogeneous permeability map may be constructed to approximate the spatially distributed diagenetic alteration.

(A)

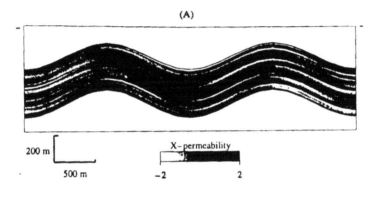

(B)

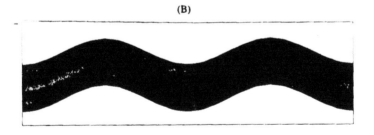

Figure 3. Two-dimensional stochastic permeability fields of 350-m-thick wavy sandstone. Colors map the permeability from 10^{-2} (white) to 10^{+2} (dark gray) darcy. All permeability fields have mean of 1 darcy, variance of log permeability of 2, and horizontal correlation length of 850 m. Model A has a vertical correlation length of 35 m and model B has a vertical correlation length of 175 m.

EFFECTS OF CEMENTATION ON HYDROTHERMAL SYSTEMS

In the second example, the effect of diagenetic reactions on groundwater flow is illustrated. I calculate at transient state how cementation of amorphous silica affects thermal convection. The dynamics of convective flow and related diagenetic patterns have been investigated previously [8,28,35,47,53], but most of these studies did not examine the potential feedback effects of diagenesis on the flow field.   In this example, a numerical experiment is conducted to examine how reactions and fluid flow couple through porosity and permeability changes in rocks.   Figure 5 shows a cross section composed of a homogeneous formation.   I set porosity to a constant value of 30 %, and lateral and vertical permeability, respectively, to 10^{-2} and 10^{-3} darcy.   There is no topographic relief across the basin's surface.   Heat flow is set to be 4 heat flow units (1 HFU = 10^{-6} cal/cm^2 sec) at the left of the cross section and 1 HFU along the right side.   The variation in basal heat flow creates a lateral temperature gradient that drives fluid convection.

(A)

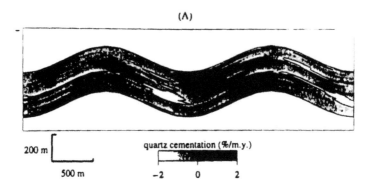

(B)

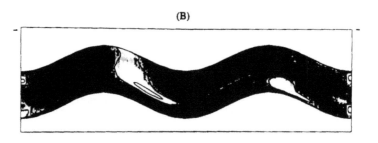

Figure 4.   Predicted cementation pattern of quartz along a heterogeneous wavy sandstone.   The cementation in stochastic permeability fields of Fig. 3A and Fig. 3B are shown in (A) and (B), respectively.   Colors map the precipitation rates from -2 (white) to +2% per m.y. (dark gray).   The irregular diagenesis patterns contrast strongly with the vertical cementation contours predicted by the homogeneous permeability model (Fig. 2).

Figure 5 shows at three times (1, 10, and 20 m.y.) the evolution of the flow field, temperature distribution, and cementation pattern of amorphous silica.   A plume of ascending flow develops along the left side of the section, above the point of the highest heat flow.   The fluid becomes cooler and sinks as it migrates toward the shallower part of the section.   Amorphous silica precipitates mainly along the ascending limb of the flow where fluids begin to cool, and dissolves in the lower part of the section where fluids are descending from cooler regions to hotter regions. Because cementation reduces the available pore space and rock permeability near the left part of the section, the convective cell gradually propagates downward and to the right part of the section where dissolution occurs.   In zones of dissolution, sediment becomes preferentially more porous and permeable. In turn, flow is increasingly focusing toward those zones, and dissolution can be enhanced there. Such a reaction-transport feedback effect can produce channeled flow in the lowest part of the section.

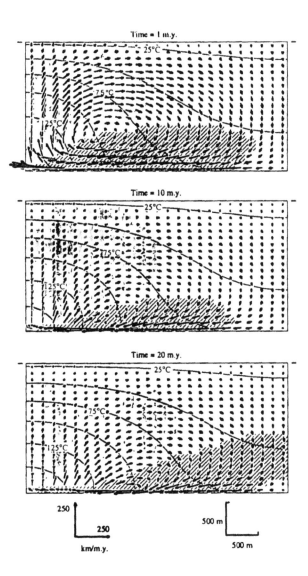

Figure 5. Feedback effects of cementation of amorphous silica on thermal convection. Diagrams show cementation pattern, flow field (arrows), and temperature distribution (contours) at three points in time in a

convective flow system. Stippled pattern maps zones where amorphous silica precipitates; rules pattern shows zones of dissolution.

This example demonstrates that the interactions between fluid flow and chemical reaction can result in changes in flow field, rates of reaction, and locations of precipitation within a flow system. These changes are generally considered as self-organization phenomenon in hydrological systems [34]. The examples argues for the importance of accounting for the feedback effects of chemical reaction in groundwater modeling.

DIAGENESIS BY MIGRATING BRINES IN THE ILLINOIS BASIN

In the last example, I consider the origin of dolomite and K-feldspar cements in the Illinois basin. The origin of these cements is of special interest because they provide important information concerning the basin's paleohydrology and brine migration.

The entire northern flank of the Illinois basin, including the Upper Mississippi Valley (UMV) area which hosts world-class zinc-lead deposits, is extensively dolomitized (Fig. 6). Late dolomite cements are found mainly in strata adjacent to the late Cambrian Mt. Simon and middle

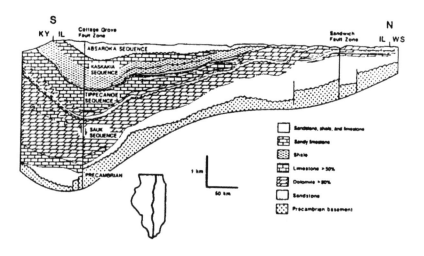

Figure 6. Simplified lithofacies distribution of Illinois basin (D. Kolata, personal communication). Paleozoic carbonates are almost entirely dolomite in the northern portion of the basin, whereas they are mainly limestone to the south.

Ordovician St. Peter sandstone aquifers, where brines are believed to have migrated in the past. Late generation of dolomite cements found in the Ordovician Platteville and Galena Formation can be correlated over long distance [26]. Authigenic K-feldspar is also distributed over the region in the Mt. Simon sandstone [9,10,32] and in the St. Peter sandstone [19,33,38]. The pattern of regional dolimitization and feldsparthization suggests that migrating brines played a role in forming these cements.

K-bearing cements have been the subject of many chronological studies [3] to date the episode of brine migration in the Illinois basin. The radiometric studies have yield three ages, early Devonian, late Devonian, and Permian, which are inferred to represent three episodes of brine migration events associated with Taconic, Acadian, and Alleghanian/Ouachita orogenies. A recent study of dating sphalerite from the UMV ores yielded Permian ages [7]. Paleomagnetic studies of Paleozoic carbonate rocks in the mid-continent also indicated that the reset of paleomagnetic pole positions occurred at Pennsylvanian-Permian time [30,45,49]. The coincidence of Permian radiometric and paleomagnetic ages is consistent with brine migration initiated by Alleghanies/Ouachita orogeny. This migration event was also coeval with the beginning of the rise of the Pascola arch in the southern part of the Illinois basin.

Different hydrologic models have been proposed to explain the Permian episode of fluid migration in the Illinois basin. Garven and Sverjensky [12] suggested that ore deposits in the UMV area might have formed by brines originally driven out of the Appalachian foreland or Ouachita basin during the Permian. Bethke [2] argued that the regional migration of brines was related to uplift of the Pascola arch in the southern Illinois basin. Bethke et al. [4] used Basin2 to model the groundwater flow set up by the tectonic uplift to the south. In the reconstructed basin, the hydraulic gradient created by the surface relief drives groundwater flow northward. Permeable basal Cambrian-Ordovician aquifers focus flow toward the Wisconsin arch on the north, near the UMV ore district. The hydrologic model predicts a flow rate of meters per year. Migrating at such a high velocity, fluids can avoid losing heat by conduction to surface and effectively carry heat from the deep basin to the UMV ore district on the north.

Bethke and Mashak [3] and Lee [23] used React to simulate the chemical reactions that likely accompanied brine migration in a hypothetical aquifer consisting of carbonates and sandstones. They used chemical analyses of modern brines in the Mt. Simon sandstone [31] and the assumption of mineral saturation to constrain the chemistry of the brines. Their model predicted that, as the brine cools from 250 to 20°C, K-feldspar and dolomite precipitate at the expense of quartz, calcite, and clay minerals (Fig. 7). The predicted reaction is

$$2.8 \; SiO_2 + 2 \; CaCO_3 + 0.1 \; CaSO_4 + 0.4 \; Ca_{.165}Mg_3Al_{.33}Si_{3.67}O_{10}(OH)_{10}$$
$$\text{\textit{quartz} \quad \textit{calcite} \quad \textit{anhydrite} \qquad \textit{smectite}}$$

$$+ 0.7 \; KAl_3Si_3O_{10}(OH)_2 + 1.5 \; K^+ + 0.4 \; HCO_3^- + 0.2 \; H_4SiO_4(aq) + 0.5 \; H^+$$
$$\text{\textit{muscovite}}$$
$$\rightarrow 2.2 \; KAlSi_3O_8 + 1.2 \; CaMg(CO_3)_2 + 0.9 \; Ca^{2+} + 0.1 \; SO_4^{2-} + 2 \; H_2O$$
$$\text{\textit{K-feldspar} \qquad \textit{dolomite}}$$

The progressive cooling of brines drives diagenetic reactions due to the retrograde solubility of calcite and formation of K-feldspar at lower temperatures. Quartz fails to precipitate, as might be expected from its prograde solubility, because the overall reaction consumes silica to form K-feldspar. The calculation predicts that a total of about 67 cm^3 of K-feldspar and 26 cm^3 of dolomite will precipitate from each kilogram of brine.

The results from the geochemical simulation are integrated into the flow system predicted by

Bethke et al. [4] to quantify the cementation rates of minerals in basin strata. The calculation is based on the volume changes of minerals at different temperature (dV/dt) predicted by the geochemical simulation (Fig. 7), as well as flow rates and the distribution of temperature in the predicted flow system (Fig. 8). The model allows us to estimate the distribution and degree of diagenetic alteration accompanying brine migration. The predicted cementation rates then can be compared to the volumes of authigenic cements present in the basin, providing information about the duration of brine diagenesis and, potentially, a check on the accuracy of the model.

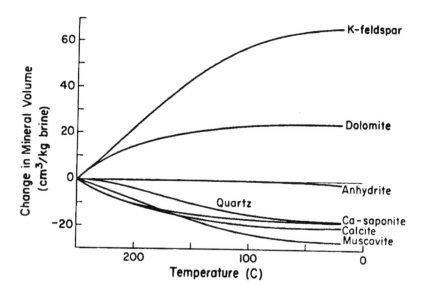

Figure 7. Predicted overall diagenetic reaction accompanying the cooling of brines during migration in the Illinois basin. The calculation is modified after Bethke and Marshak [3] and Lee [23]. The model equilibrates 1 kg of brine at 250°C and cools it to 20°C. Over the course of calculation, the maintains local equilibrium with minerals in a hypothetical aquifer, which consists of quartz, K-feldspar, muscovite, Ca-smectite, calcite, dolomite, and anhydrite.

The coupled model predicts that migrating groundwaters precipitate dolomite and K-feldspar in the deep aquifer and along the basin's northern margin where fluids ascend and cool (Fig. 8). Dolomite and K-feldspar precipitate in the deep aquifers, according to the modeled results, at rate as large as 1 and 3 vol % per m.y., respectively. Late dolomite cements usually comprise less than about 1% of Ordovician carbonate today [26]. K-feldspar overgrowths also occupy up to about 2.5% of the Mt. Simon sandstone in the northern basin today [9]. Therefore, reaction over a time period of about one m.y. or less is needed to account for the volume of each cement in the deep aquifers. The predicted interval of diagenesis agrees generally with a study of zinc distribution in carbonate rocks which suggests that mineralization in the UMV district formed over a short period of about 0.25 million years [22]. The result might also be compared to paleomagnetic studies on Paleozoic carbonates of the Appalachian foreland, which suggest that remagnetization by migrating brines occurred during relatively short interval of geologic time [30,45]. A preliminary modeling of solute transport indicates that a pulse of hot and saline groundwater discharged through the northern margin of the Illinois basin for a short interval of 10^5 to 10^6 years [20].

It is important to point out that the predicted reaction rates reflect groundwater velocities predicted by the flow model, which are poorly constrained. The predicted flow rates could vary over orders of magnitude, depending on the choice of topographic relief or permeability correlations assumed in the hydrologic model, which comprise the greatest uncertainty in calculating flow velocities. For example, authigenic K-feldspar averages from 18 to 35% volume of the granitic basement rock in the UPH-3 deep hole in the northern basin [9]. High cement's volume could be explained by a high flow rate than predicted in the calculations; higher flow rates may result when fluid flow is channeled along fractured basement highs in the northern basin. Nonetheless, flow velocities of the order of magnitude assumed in the model required to transport heat from deep strata to basin's margin [2].

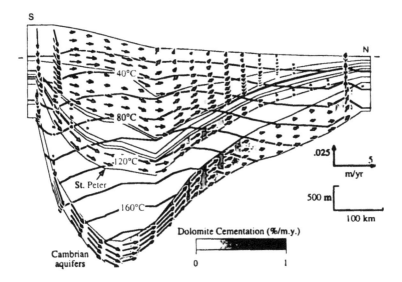

Figure 8. Calculated distribution of dolomite cementation in the Illinois basin during the Permian, assuming the southern basin was uplifted 700 m above the elevation of basin's northern margin. Colors map the precipitation rates from 0 (white) to +1% per m.y. (dark gray). The most extensive dolomitization occurs along the deep aquifers and basin's northern margin where the warm fluid ascends and cools. K-feldspar precipitates in a similar pattern (not shown), at a rate as much as 3 % per m.y. The hydrologic model is modified after Bethke and others (1991). Cross section and hydrologic properties (porosity and permeability) of hydrostratigraphy units used in the simulation are described in [4] and [23].

CONCLUSIONS

In this study, I develop a numerical model to predict the rates and patterns of chemical diagenesis that accompany fluid migration in sedimentary basins. The model demonstrates how a basin's geometry and permeability heterogeneity affect the spatial distribution of diagenetic reactions. The model allows us to examine the reaction-transport feedback of chemical cementation-dissolution on thermal convection. In field applications, the predicted results can be compared to the diagenetic patterns observed in basin strata, therefore providing important constraints on the origin of diagenetic alteration.

The patterns and rates of diagenetic reaction in hydrologic systems are controlled mainly by mineral's solubility, flow velocity, flow direction, and distribution of temperature and pressure along the flow path. The maximum alteration occurs where groundwaters move across temperature isotherms at a sharp angle. Very little diagenetic reaction occurs where groundwaters migrate along the isotherms. The results have important implications for the origin of ore deposits associated with basin-wide flow. Mississippi Valley-type ore deposits, for example, often occur in mild deformed regions or over basement highs [48], where flow is channeled and gradient reactions are enhanced. The calculation results in which the stochastic permeability model is employed more closely represent the irregular pattern of diagenetic alteration observed in nature, but differ markedly from the uniform pattern predicted by the homogeneous model. The result argues for the importance of using geostatistics approach in assessing diagenetic processes in sedimentary basins.

The interaction between fluid flow and cementation of amorphous silica can prevent the buoyancy-driven convection from maintaining a steady state over geologic time. The convective cell tends to propagate from zones of precipitation to zones of dissolution. In zones of dissolution, sediment becomes preferentially more porous and permeable. In turn, flow is increasingly focusing toward those zones, and dissolution can be enhanced there. The result argues for the importance of accounting for chemical diagenesis in modeling groundwater flow.

Dolomitization and feldspathization in the Illinois basin probably occurred as a pulse of hot and saline brine discharged through the northern flank of the basin during a short interval of one m.y. or less. Brines ascended northward from the deep strata at rates of m/yr in response to the uplift of the southern basin. As brines cooled along the flow path, they reacted to form dolomite and K-feldspar at the expense of quartz, calcite, muscovite, and smectite. According to the modeled results, dolomite and K-feldspar precipitates in the deep aquifers at rates as much as 1 and 3 vol % per m.y., respectively. Late dolomite cements occupy up to about 1% of carbonate strata today. Authigenic K-feldspar comprises about 2.5% of the Mt. Simon sandstone in the northern basin. Therefore, reaction over a time period of about one m.y. or less is needed to account for the volume of dolomite and K-feldspar cements in deep aquifers.

Acknowledgments

I thank Dr. Craig Bethke for his thoughtful advice and help in developing the mathematical and numerical models. Special thanks are due to Kurt Larson for many helpful discussions during the study. Dennis Kolata provides an unpublished cross-section map of the Illinois basin. This study was supported by NSF grant EAR 85-52649 and EAR 86-01178 (Craig Bethke, Principle Investigator) and Auburn University Grant-in-Aid 96-022 to Ming-Kuo Lee.

REFERENCES

1. C.M. Bethke. A numerical model of compaction-driven groundwater flow and heat   transfer and its application to the paleohydrology of intracratonic sedimentary basins, *Jour. Geophy. Res.* **90**, 6817-6828

(1985).

2. C.M. Bethke. Hydrologic constraints on the genesis of the Upper Mississippi Valley mineral district from Illinois basin brines, *Econ. Geol.* **91**, 233-249 (1986).

3. C.M. Bethke. and S. Marshak. Brine migrations across North America - The plate tectonics of groundwater, *Ann. Rev. . Planet. Sci.* **18**, 287-315 (1990).

4. C.M. Bethke. J.D. Reed. and D.F. Oltz. Long-range petroleum migration in the Illinois basin, *Amer. Assoc. Petrol. Geol. Bull.* **75**, 1925-1945 (1991).

5. C.M. Bethke. M.-K. Lee. H.A.M. Quinodoz. and W.N. Kreiling. Basin Modeling with Basin2, A Guide to Using Basin2, B2plot, B2video, and B2view: Hydrogeology Program, University of Illinois, Urbana, 225 p. (1993).

6. C.M. Bethke. The Geochemist's Workbench^TM Version 2.0, A Users Guide to Rxn, Act2, Tact, React, and Gtplot. Hydrogeology Program, University of Illinois, Urbana, 213 p. (1994).

7. J.C. Brannon. F.A. Podosek. and R.K. McLimans. A Permian Rb-Sr age for sphalerite from the Upper Mississippi zinc-lead districts, Wisconsin, *Nature.* **356**, 509-511 (1992).

8. S.H. Davis. S. Rosenblat. J.R. Wood. And T.A. Hewett. Convective flow and diagenetic patterns in domed sheets, *Amer. Jour. Sci.* **285**, 207-223 (1985).

9. M.E. Duffin. M.-C. Lee. G. deV. Klein. and R.L. Hay. Potassic diagenesis of Pre-Cambrian granitic basement and Cambrian sandstones in UPH-3 hole, Upper Mississippi Valley, U.S.A, *Jour. Sed. Petro.* **59**, 848-861 (1989).

10. N. S. Fishman. Diagenetic studies of the Mt. Simon sandstone - Implications for paleohydrology (abs.), *U.S.G.S. Open File Rep.* **92-1**, 11-12 (1992).

11. R.A. Freeze and J.A. Cherry. Groundwater, Englewood Cliffs, New Jersey, Prentice-Hall, 604 p (1979).

12. G. Garven. And D.A. Sverjensky. Hydrogeology of regional flow systems associated with the formation of Mississippi Valley-Type ore deposits in the Mid-continent, *Geol. Soc. Amer. Abs. with Progs*, **21**, A9 (1989).

13. R.L. Hay. Stratigraphy and zeolitic diagenesis of the John Day Formation of Oregon, *Uni. California Pub. in Geol. Sci.*, Berkeley, 199-261 (1963).

14. R.L. Hay. Zeolites and zeolitic reactions in sedimentary rocks, *Geol. Soc. Amer. Spec. Papers*, **85**, 130 p (1966).

15. P.P. Jr. Hearn. J.F. Sutter. and H.E. Belkin. Evidence for Late-Paleozoic brine migration in Cambrian carbonate rocks of the central and southern Appalachians: Implications for Mississippi Valley-type sulfide mineralization, *Geochim. Cosmochim. Acta*, **51**, 1323-1334 (1987).

16. H.C. Helgeson. Evaluation of irreversible reactions in geochemical processes involving minerals and aqueous solutions - I. Thermodynamic relations, *Geochim. Cosmochim. Acta*, **32**, 853-877 (1968).

17. H.C. Helgeson. R.M. Garrels. and F.T. Mackenzie. Evaluation of irreversible reactions in geochemical processes involving minerals and aqueous solutions - II. Applications, *Geochim. Cosmochim. Acta*, **33**, . 455-481 (1969).

18. H.C. Helgeson. T.H. Brown. T.H. Nigrini. And T.A. Jones. Calculation of mass transfer in geochemical processes involving aqueous solution, *Geochim. et Cosmochim. Acta*, **34**, 569-592 (1970).

19. D.J. Hoholick. Porosity, grain fabric, water chemistry, cement, and depth of burial of the St. Peter sandstones in the Illinois basin, Unpublished M.S. Thesis, University of Cincinnati, Cincinnati, Ohio, 72 p (1980).

20. K.W. Larson. C.M. Bethke. and M.-K. Lee. Heat and solute transport and the role of basement during brine migration in the Illinois, *EOS Trans.* **73**, 125 (1992).

21. A.C. Lasaga. Chemical kinetics of water-rock interactions, *Jour. Geophy. Res.* **89**, 4009-4025 (1984).

22. N.G. Lavery. and H.L. Barnes. Zinc dispersion in the Wisconsin zinc-lead district, *Econ. Geol.* **66**, 226-242 (1971).

23. M.-K. Lee. Quantitative models of fluid flow, chemical reaction, and stable isotopic fractionation and their applications to sediment diagenesis and hydrothermal alteration, Unpublished Ph.D. Dissertation, University of Illinois, Urbana, Illinois, 179 p (1993).

24. M.-K. Lee. And C.M. Bethke. Groundwater flow, late cementation, and petroleum accumulation the Permian Lyons sandstone, Denver basin, *Amer. Asso. Petrol. Geol. Bull,* **78**, 217-237.

25. M.-K. Lee. and C.M. Bethke. A model of isotopic fractionation in reacting geochemical system, *Amer. Jour. Sci.* In press.

26. W. Li. Mechanisms of dolomitization in the seawater and the origin of Middle Ordovician dolomites in the Illinois basin, Unpublished Ph.D. Dissertation, University of Illinois, Urbana, Illinois, 129 p (1994).

27. P.C. Lichtner. The quasi-stationary state approximation to coupled mass transport and fluid-rock interaction in a porous medium, *Geochim. Cosmochim. Acta,* **52**, 143-165 (1988).

28. A. Ludvigsen, A., E. Palm. and R. McKibben. Convective momentum and mass transport in porous sloping layers, *Jour. Geophy. Res.* **97**, 12315-12325 (1992).

29. A. Mantoglou and J. Wilson. The Turning Bands method for simulating of random fields using line generation by a spectral method, *Water Res. Res.* **18**, 1379-1394 (1982).

30. C.R. McCabe. R. Van, der Voo. D.R. Peacor. C.R. Scotese. and R. Freeman. Diagenetic magnetite carries ancient yet secondary remanence in some Paleozoic sedimentary carbonates, *Geology,* **11**, 221-223 (1983).

31. W.F. Meents. A.H. Bell. O.W. Rees. And W.G. Tilbury. Illinois oil-field brines, Illinois State Geological Survey, *Illinois Petrol.* **66**, 38 p (1952).

32. T.A. Metarko. Porosity, grain fabric, water chemistry, cement, and depth of burial in the Upper Cambrian Mt. Simon and Lamotte sandstones of the Illinois basin, Unpublished M.S. thesis, University of Cincinnati, Cincinnati, Ohio, 88 p (1980).

33. I.E. Odom. T.N. Willand. and R.J. Lassin. Paragenesis of diagenetic minerals in the St. Peter sandstones (Ordovician), Wisconsin and Illinois, *Soc. Econ. Paleont. Mineral. Spec. Pub.* **26**, 425-443 (1979).

34. P. Ortoleva. E. Merino. C. Moore. And J. Chadam. Geochemical self-organization I: Reaction-transport feedback and modeling approach, *Amer. Jour. Sci.* **287**, 979-1007 (1987).

35. E. Palm. Rayleigh convection, mass transfer, and change in porosity in layers of sandstone, *Jour. Geophy. Res.* **95**, 8675-8679 (1990).

36. D.L. Parkhurst. D.C. Thorstenson. and L.N. Plummer. PHREEQE - A computer program for geochemical calculations, *U.S.G.S. Water-Res. Inv. Rep.* **80-96**, 210 p (1980).

37. O.M. Phillips. Flow and Reactions in Permeable Rocks,: New York, Cambridge University Press, 285 p (1991).

38. J.K. Pitman. M.B. Goldhaber. And T. Shaw. Regional diagenetic patterns in the St. Peter sandstone, Illinois basin: Evidence for multiple episodes of fluid movement during the late Paleozoic, *U.S.G.S. Open File Rep.* **92-1**, 45-47 (1992).

39. L.N. Plummer. D.L. Parkhurst. and D.C. Thorstenson. The development of reaction models for ground-water systems, *Geochim. Cosmochim. Acta,* **47**, p. 665-686 (1983).

40. L.N. Plummer. D.L. Parkhurst. G.W. Fleming. and S.A. Dunkle. PHRQPITZ - A computer program incorporating Pitzer's equations for calculation of geochemical reactions in brines, *U.S.G.S. Water-Res. Inv. Rep.* **88-4153**, 310 p (1988).

41. J.P. Raffensperger. and G. Garven. The formation of unconformity-type uranium ore deposits. 2. Coupled hydrochemical modeling, *Amer. Jour. Sci.* **295**, 639-696 (1995).

42. M.H. Reed. Calculation of multicomponent chemical equilibrium and reaction processes in systems involving minerals, gases, and an aqueous phase, *Geochim Cosmochim. Acta,* **46**, 513-528 (1982).

43. J.D. Rimstidt, and H.L. Barnes. The kinetics of silica-water reactions, *Geochim. Cosmochim. Acta,* **44**, 1683-1699 (1980).

44. J. Rubin. Transport of reacting solutes in porous media: relation between mathematical nature of problem formulation and chemical nature of reactions, *Water Res. Res.* **19**, 1231-1252 (1983).

45. C.R. Scotese. R. Van der Voo. and C. McCabe. Paleomagnetism of the Upper Silurian and Lower Devonian carbonates of New York State: Evidence for secondary magnetizations residing in magnetite, *Phys. Earth Planet. Int.* **30**, 385-395 (1982).

46. C.I. Steefel. and A.C. Lasaga. Putting transport into water-rock interaction models, *Geology,* **20**, 680-684

(1992).

47. C.I. Steefel. and A.C. Lasaga. A coupled model for transport of multiple chemical species and kinetic precipitation/dissolution reactions with application to reactive flow in single phase hydrothermal systems, *Amer. Jour. Sci.* **294**, 529-592 (1994).

48. D.A. Sverjensky. Genesis of Mississippi Valley-type lead-zinc deposits, *Ann. Rev. Earth Planet. Sci.* **14**, 177-99 (1986).

49. D.T. Symons. and D.F. Sangster. Paleomagnetic age of the Central Missouri barite deposits and its genetic implications, *Econ. Geol.* **86**, 1-12 (1991).

50. E.N. Wilson. L.A. Hardie. and O.M. Phillips. Dolomitization front geometry, fluid flow patterns and the origin of massive dolomite: the Triassic Lateman buildup, Northern Italy, *Amer. Jour. Sci.* **290**, 741-796 (1990).

51. T.J. Wolery. Calculation of chemical equilibrium between aqueous solution and minerals: The EQ3/6 software package, *Lawrence Livermore Lab. Rep.* UCRL-52658, 41 p (1979).

52. T.J. Wolery. EQ3NR, A computer program for geochemical aqueous speciation-solubility calculations: user's guide and documentation, *Lawrence Livermore Lab. Rep.* UCRL-53414 (1983).

53. J.R. Wood. and T.A. Hewett. Fluid convection and mass transfer in porous sandstones - a theoretical model, *Geochim. Cosmochim. Acta*, **46**, 1707-1713 (1982).

Proc. 30th Int'l Geol. Congr., Vol.22, pp. 204-212
Fei Jin and N.C. Krothe (Eds.)
© VSP 1997

THE PHYSICAL AND ECONOMIC IMPACT OF AQUIFER OVER-EXPLOITATION AT HANGU, CHINA

T.R. SHEARER, B. ADAMS, R. KITCHING & R. CALOW
British Geological Survey, Wallingford, UK
X.D. CUI, D.J. CHEN
Tianjin Bureau of Geology and Mineral Resources, Tianjin, PRC
R. GRIMBLE
Natural Resources Institute, Chatham, UK
Z.M. YU
University of Nankai, Tianjin, PRC

Abstract

During the second half of this century, the deep unconsolidated sedimentary aquifer under the coastal city of Hangu has been heavily exploited by the local industry, agriculture and domestic users. This has caused large piezometric head reductions and has resulted in significant land subsidence. The proximity of the sea makes this particularly undesirable.

In order to better understand the relationship between the groundwater abstraction and the aquifer compaction, a computer model of the aquifer system which simulated the settlement was created. The nature of the local geology which included thick, slow-draining, clayey aquitards meant that the compaction was delayed in relation to the reduction in pressure in the more transmissive, sandy layers. The reproduction of this delay required the development of a special model package. The modelling indicated that a significant reduction in abstraction was required to restrict the amount of future subsidence to an acceptable level.

In parallel with the specific hydrogeological investigation of subsidence, a wider socio-economic study of aquifer over-exploitation used the Hangu situation as a case study. This concluded that water scarcity threatens future regional development, but that more efficient water allocation and efforts to manage groundwater demand might go some way towards putting water use on a more sustainable footing and reduce the need for supply augmentation.

Keywords: aquifer, subsidence, compaction, over-exploitation, modelling, Hangu

INTRODUCTION

The city of Hangu is situated on the coast of the Gulf of Bohai at the eastern end of the North China Plain (Fig. 1). Historically important as a port for the capital city Beijing, it has developed over the last fifty years as a major centre for the Chinese chemical industry. The industrial development and associated population growth has led to a significant increase in demand for water: a situation common to many cities of the region and throughout the world.

Hangu relies for its water supply on a very deep, multi-layered aquifer system consisting of unconsolidated, Quaternary and Upper Neogene deposits. There has been a sharp decline in piezometric head in the confined aquifers which has resulted in compaction of the formation. Over one metre of land subsidence has been observed in the 35 years since records began with maximum rates of fall of 90 mm per annum. This rate is regarded as unacceptable and the local authorities wish to restrict the subsidence to a maximum of 25 mm pa.. They are considering replacing the local abstraction of groundwater with surface water from neighbouring districts: a solution with significant economic implications.

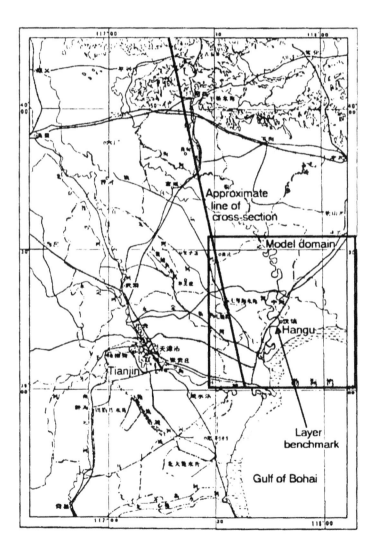

Fig. 1 Tianjin showing model domain and alignment of geological section shown in Fig. 2.

A collaborative study of the physical and economic impact of over-exploitation at Hangu was undertaken by the British Geological Survey, the Tianjin Bureau of Geology and Mineral Resources (TBGMR), the Natural Resources Institute and Nankai University [1]. The purpose of the study was to provide a framework for improved water resource management. The main components of the study were: i) the development of a digital hydrogeological model to simulate land subsidence due to groundwater abstraction, and ii) a socio-economic analysis of the costs and benefits of groundwater use in the area. This paper concentrates on the modelling component of the project. The study was funded by the British Overseas Development Agency and TBGMR.

THE HYDROGEOLOGICAL SETTING

In the vicinity of Hangu the unconsolidated Quaternary and Upper Neogene deposits have a thickness exceeding 1000 m. They consist of a layered sequence of horizontally discontinuous fine sands and clays which are mainly of marine origin. Ten separate aquifer units have been identified; the upper units are shown in Fig. 2. The uppermost, unconfined unit contains saline water and is not used. The lower, confined, units have good quality water (TDS 0.3 - 0.6 gm/l) and are being progressively exploited. In 1989/90, Unit 2 (50 - 180 m bgl) produced 20 MCM and Unit 3 (180 - 270 m bgl), 16 MCM: 40% and 33% of the total groundwater abstraction respectively. There are almost 700 boreholes in the area some of which reach 1000 m.

28 MCM (56% of total) was used within urban Hangu. Of this 43% was used by the Hangu Chemical Plant and 32% was used for domestic purposes. Despite using river water for cooling and recycling a proportion of the consumption, the production of the chemical plant is presently constrained by water shortages. Domestic users are effectively rationed by restricted periods of supply; three intervals totalling eight hours per day. In rural areas, 77% of the water supply was used for irrigation. Southeast of the city, near the coast, is an extensive area used for salt pans which provides raw material for the chemical works. In this area there is little groundwater abstraction.

Annual abstraction peaked at 56 MCM in 1983/84 when consumption was curbed by improved regulation. In 1986/87 abstraction was 33 MCM. However further industrial development has caused a rise more recently.

As the exploited aquifer units are all confined, there is no direct recharge. The intense industrial abstraction has led to the formation of substantial cones of depressions around the wells and these coalesce into regional depressions within each exploited aquifer unit. The cones can reach 80 m deep and the seasonal fluctuation caused by agricultural consumption exceeds 30 m.

Land subsidence was first noted in the 1950s when rates of 2 mm/yr were recorded. This coincided with the start of increasing demand for groundwater due to industrial development. Subsidence increased with increasing groundwater production rates until in 1980 subsidence was recorded at 80-90 mm/yr (abstraction 50 MCM/yr). Following the campaign for increased efficiency of water usage, in 1986/87 subsidence was reduced to 40-50 mm/yr (abstraction 33 MCM/yr). General subsidence rates are currently at 40-60 mm/yr. However at the main focus of groundwater exploitation, the chemical factory, recent subsidence rates of 84-88 mm/yr were thought to be the highest in China. Unit 2, with 41% of groundwater production is reportedly responsible for 50% of the total subsidence (Fig. 3).

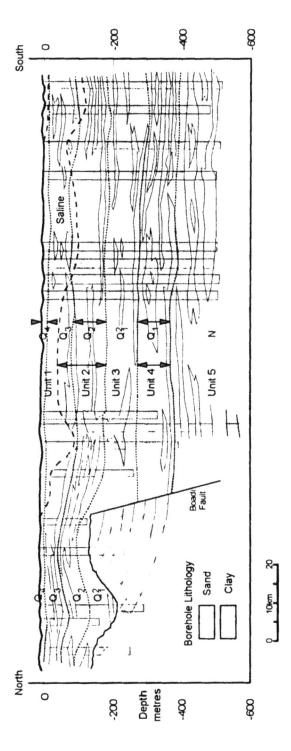

Figure 2: Geological section of Tianjin Municipality (See Fig. 1 for alignment).

In 1985 a subsidence monitoring network was established around Hangu. It centres on the "layered benchmark": a nest of extensometer wells situated at the chemical works. The deepest extensometer is 300 m and is correlated to a stable bedrock benchmark 70 km away. The layered benchmark enables the measurement of the compaction of individual aquifer layers.

The response of the rate of subsidence to changes in groundwater abstraction confirmed the belief that the abstraction was the underlying cause. It also indicated that the subsidence could be controlled by limiting abstraction. The local authorities wish to restrict the subsidence to a maximum of 25 mm pa.. They are considering replacing the local abstraction of groundwater with river water piped from neighbouring catchments.

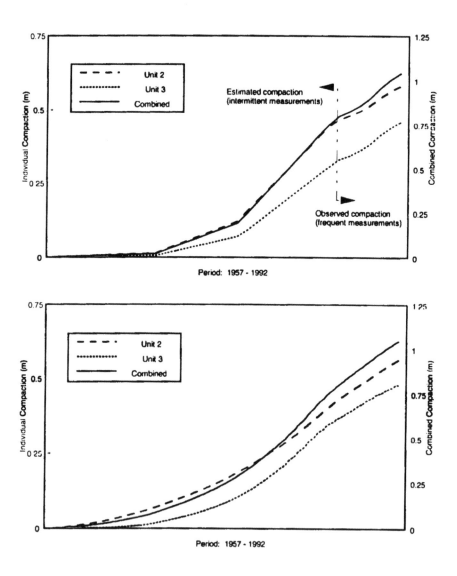

Fig. 3 Observed (upper) and computed (lower) compaction at Hangu benchmark

AQUIFER COMPACTION

It has long been recognised that the abstraction of groundwater can cause land subsidence. The scientific basis of current understanding was established during the 1920s:

Meinzer [7] observed that artesian aquifers compress when the piezometric head decreases. He concluded that water was released both by the elastic expansion of the water and by compression of the aquifer structure. He recognised that the reduction of storage following compression may be permanent (inelastic) and that these effects were much greater in low permeability aquifer systems.

Terzaghi [11] developed the theory of one dimensional consolidation of clays. He reasoned that the matrix of an aquifer was subjected to an effective stress equal to the total geostatic stress minus the pore fluid pressure.

$$P' = P - U_w$$

where P' = effective intergranular stress, P = total geostatic stress and u_w = pore fluid pressure.

If the fluid pressure is reduced, a greater proportion of the total stress is transferred to the matrix which compacts until intergranular forces increase to establish a new equilibrium.

Jacob [3] recognised that clay interbeds and adjacent confining clay layers could be the major source of water in an aquifer and that the low permeability of the clay might introduce a time lag between the lowering of the head in the aquifer and the release of water from the clay.

Riley [9] highlighted the time-dependency of the compaction on the rate of equalisation of the pore pressures within the aquitards to those within the aquifer. He considered a single homogeneous aquitard, bounded above and below by aquifers and concluded that if the head in the aquifers were instantaneously and equally lowered, the time required to attain any specified dissipation of average excess pore pressure within the aquitard would be a direct function of: (1) the volume of water that must be squeezed out of the aquitard in order to establish the denser structure required to withstand the increased stress, and (2) the impedance to the escape of this water. The product of these two parameters constitutes the aquitard time constant.

The volume of water is determined by the volumetric compressibility, m_v, of the aquitard, the compressibility, S_w, of the water, and the thickness, b', of the aquitard. The impedance is determined by the vertical permeability, K', and thickness of the aquitard.
The time constant, J, is

$$\tau = \frac{S'_s (b'/2)^2}{K'}$$

and where S'_s, the specific storage of the aquitard, is the sum of S'_{sk}, the storage component due to compressibility of the aquitard, and S_{sw}, the component due to the compressibility of water. For consolidating aquitards S'_{sk} is much greater than S_{sw}".

The inelastic compression of the aquifer/aquitard system represents a reduction in void space: there is a permanent loss of storage potential and a decrease in formation permeability.

A comprehensive discussion of land subsidence due to groundwater abstraction is given in Poland [8].

THE SUBSIDENCE MODEL

In order to better understand the complex relationship between abstraction and subsidence a computer model of the aquifer system has been developed. The simulation was based on the widely-used USGS code MODFLOW [6]. MODFLOW itself does not calculate aquifer compaction but a supplementary module known as the Interbed Storage Package (IBS) [5] does. This module is based on a simple conceptual model where compaction occurs when the previous maximum intergranular stress is exceeded which is equivalent to the piezometric pressure falling below previous minima.

A limitation of the IBS package is that pressures within the clayey interbeds is assumed to equalise with the pressures in the aquifer within each stress period. At Hangu there is a large annual fluctuation in piezometric pressures and the extreme stress occurs for a relatively short period of time. As the profile contains a high proportion of heavy clay and the interbeds are often very thick (see Fig. 2), the assumption that equalisation would occur was not considered reasonable. Consequently an alternative module for calculating compaction was developed for this application. Known as the Interbed Drainage Package or IDP [4] this also assumes that the spatial pressure variations within the interbeds can be characterised by a single value but permits pressure differentials to decay very slowly if required. This enables the simulation of compaction to respond to time-weighted stress changes rather than the momentary extreme.

The model was calibrated against the best available estimates of subsidence over the past 35 years (Fig. 3) and then used to predict the subsidence that would result from various pumping scenarios up to the year 2020. For the scenarios the model domain was divided into three zones: A, the main industrial district, B, the outer urban area, and C, the rest of the domain (mainly agricultural). Individual predictions were made assuming various augmentation schemes would be implemented in the years 2000 and 2010 which would result in changes in abstraction based on proportions of the 1992 rate (Table 1). The predictions inform resource managers of the quantity of groundwater they have at their disposal and the requirement for augmentation of supply from outside the locality within the constraints of acceptable rates of subsidence. See Shearer and Kitching [10] for details.

All of the scenarios fulfilled the objective of restricting the rate of subsidence to less than the target 25 mm/yr (see Table 2), indeed maintaining the present abstraction would result in only 18 mm/yr in 2020. However the total additional subsidence would be over 600 mm which is significant given that much of the land surface in the urban area is only about two metres above the high tide level and therefore at risk of flooding.

THE SOCIO-ECONOMIC STUDY

The socio-economic study [2] established a framework of physical impacts and associated costs.

Within the limitations of the available data, it concluded that increased operational expenses of water abstraction are the major economic consequence of the over-exploitation: the large drawdowns incur additional pumping costs. Surprisingly, the direct costs of land subsidence do not figure prominently. This may reflect data limitations to costing the impact of land subsidence.

Table 1   Hypothetical abstraction plans

Area	Production 1992 (MCM)	Multiplication factors for 1992 production							
		Plan 1 2000	Plan 1 2010	Plan 2 2000	Plan 2 2010	Plan 3 2000	Plan 3 2010	Plan 4 2000	Plan 4 2010
A	15.4	1.40	2.87	0.27	0.99	0.29	0.24	0.80	0.80
B	14.8	0.88	0.94	0.88	0.94	0.88	0.94	0.88	0.94
C	147.5	1.15	1.27	1.15	1.27	1.15	1.27	1.15	1.27

Table 2:   Predicted compaction at layered benchmark

Plan (Year)	Total Compaction (mm)	Rate of Compaction (mm/yr)
Observed (1992)	1040	37
Calibration (1992)	1045	35
Plan 1 (2020)	2054	22
Plan 2 (2020)	1365	17
Plan 3 (2020)	1092	2
Plan 4 (2020)	1598	17
Current rate (2020)	1699	18

The economic model indicates how the unit cost of water can be expected to increase over time. It also identified the low benefit value of water to the bulk consumers; the chemical industry and particularly agriculture. However because users do not bear the full economic costs of the water they consume (the value of water used for irrigation is lower than the actual cost of supply), they have little incentive to economise. This suggests that present water supply and pricing policies are not sustainable in the long-term. It is argued that a demand-side strategy should be considered alongside the supply augmentation option. The principal means of reducing the increase in demand would be the use of economic pricing with incentives also being offered for increased use of recycling and water-saving devices.

CONCLUSIONS

!   The subsidence calculation module developed for this application simulates the delayed response of compaction to groundwater abstraction caused by the slow drainage of the comparatively thick clayey interbeds found at Hangu.

!   The model predictions indicate that the rate of subsidence can be controlled by management of the quantity of groundwater pumped.

!   There is scope to achieve some of the necessary reduction in water usage by demand management including volumetric pricing and water saving.

ACKNOWLEDGEMENTS

The study was funded by the Overseas Development Agency and the Tianjin Bureau of Geology and Mineral Resources. The assistance of TBGMR in providing data for the study is gratefully acknowledged.

This paper is published by permission of the Director of the British Geological Survey (NERC).

REFERENCES

1. Adams B., Grimble R., Shearer T. R., Kitching R., Calow R., Chen D. J., Cui X. D. And Yu Z. M. *Aquifer over-exploitation in the Hangu region of Tianjin, PRC*. British Geological Survey Technical Report WC/94/42R (1994).

2. Grimble R. and Calow R. *A socio-economic analysis of aquifer over-exploitation and management options, Hangu District, Tianjin PR China.* Natural Resources Institute, Project CO484 Report (1994).

3. Jacob C. E. On the flow of water in an elastic artesian aquifer. *American Geophysical Union Transactions,* pt **2,** pp. 574-586 (1940).

4. Kitching R. and Shearer T. R. *Modelling of Subsidence due to Groundwater Extraction.* British Geological Survey Technical Report WC/95/42 (1995).

5. Leake S. A., and Prudic D. E. *Documentation of a Computer Program to Simulate Aquifer-System Compaction using the Modular Finite-Difference Groundwater Flow Model.* US Geological Survey Techniques of Water Resources Investigation, Book 6, Chapter A2 (1988).

6. Mcdonald M. G. and Harbaugh A.W. *A Modular Three-Dimensional Finite-Difference Groundwater Flow Model.* US Geological Survey Techniques of Water Resources Investigation, Book 6, Chapter A1 (1988).

7. Meinzer O. E. Compressibility and elasticity of artesian aquifers. *Economic. Geology*, 23, pp. 263-291 (1928).

8. Poland J. F., (ed) *Guidebook to studies of land subsidence due to groundwater withdrawal*, UNESCO, Paris (1984).

9. Riley F. S. Analysis of borehole extensometer data from central California. *Land subsidence, Vol. 2,* IAHS Publication 89 (1969).

10. Shearer T. R. and Kitching R. *A numerical model of land subsidence caused by groundwater abstraction at Hangu, PR China.* British Geological Survey Technical Report WC/94/46 (1994).

11. Terzaghi K. Principles of soil mechanics: IV Settlement and consolidation of clay. *Eng. News-Rec*, pp. 874-878 (1925).

Proc. 30th Int'l Geol. Congr., Vol.22, pp. 213-223
Fei Jin and N.C. Krothe (Eds.)
© VSP 1997

Optimization of Pump-and-treat Technology for Aquifer Remediation

YUNWEI SUN
Dept. of Civil Engineering, Technion-Israel Institute of Technology, Haifa 32000.
BINGCHEN WANG
Comprehensive Institute of Geotechnical Investigation & Surveying (CIGIS), Ministry of Construction, P. R. China.

Abstract

Pump-and-treat (PAT) is an important technique for remediation of contaminated aquifers. In this technique, contaminants are removed from the aquifer with the pumped water, and then treated in treatment plants. The treated water may be supplied to some appropriate consumers or injected back into the aquifer. In order to design an PAT system efficiently, we have to use optimization approaches, since the installation and operation of an PAT system is a long and costly process. In designing of a *pump-and-treat* system, with optimization approaches, for groundwater and aquifer remediation, the mass removal of contaminants by pumping, cleanup time, and total cost serve as three principal system performance indices for evaluating pumping/injection alternatives. When budget and cleanup time are given, the mass removal of contaminants should be maximized as an objective. Therefore, a mass removal model is developed and solved here. Sensitivity analysis for changing budget and cleanup time are conducted to generate a flexible decision basis.

Keywords: optimization, remediation, pump-and-treat, aquifer, contamination

INTRODUCTION

Since contamination problems have damaged and/or threatened aquifers which serve as sources of water supply, groundwater/aquifer remediation becomes an important issue. The most popular technology for groundwater/aquifer remediation is the *pump-and-treat* (PAT). In this technology, pumping wells are introduced at selected locations into the contaminated aquifer, and contaminants are removed from the aquifer with the pumped water. The contaminants are then removed from the pumped water by various treatment technologies. The treated water may then be supplied to appropriate consumers or injected back into the aquifer.

In the construction of an optimization model for *pump-and-treat* system design, Three important factors: mass removal of contaminants (or cleanup standard, MCL-Maximum contaminant level., as a constraint), resources (available budget), C_S , and cleanup time, t_f, are considered as principal objectives or constraints, as shown in Table 1, in making optimal strategies for groundwater remediation. They describe the shape of the feasible region and optimal directions. Decision makers want to use less resources and shorter cleanup time to reach maximum mass removal from an aquifer, and thus meet the water quality requirements (MCL). When two of the three factors are constrained, another one can be optimized. For example, given the required MCL standard and cleanup time, one may want to maximize the mass removal of contaminants. Various models are presently used for producing pumping/injection strategies (Sun, 1995, Wagner, 1995). Concerning the different combinations of objective(s) and constraints, which are referred here as decision situations, three general optimization models can be derived as shown in Table 1.

Table 1 Classification of PAT optimization models

Model	Decision Situation
1. Mass removal model	MCL is flexible, but the aim is to remove more mass of contaminant. Budget is given. There is cleanup time requirement.
2. Economic Model	No budget limitation, but the aim is to save money. There are MCL requirements There is cleanup time requirement.
3. Cleanup time model	Terminate cleanup as soon as possible. Budget is given. There are MCL requirements.

Much effort has been made for solving the economic model (Wagner, 1995). The combined optimization-simulation techniques have been used to design cost-effective PAT systems (Gorelick *et al.*, 1984, Wagner and Gorelick, 1987, Gorelick, 1990). The finite difference technique was used to approximate derivatives of the objective function and the constraints. Because most developments of optimization approaches have been stimulated by economic considerations, most of the previous studies (Gorelick *et al.*, 1984, Wagner and Gorelick, 1987, Ahlfeld *et al.*, 1988, Wang and Ahlfeld, 1994) on aquifer remediation took minimization of costs as an objective in an optimization model.

Since simulated solute concentrations (by numerical methods) are nonsmooth and nonconvex function of well locations when linear basis function is used in numerical simulators (Wang and Ahlfeld, 1994, Sun, 1995), mass removal of contaminants, which is a function of the calculated concentrations, has never been used as an objective in the optimization of groundwater remediation. In this paper, we extend the optimization of *pump-and-treat* technology from an economic model to a mass removal model, in which the mass removal of contaminants serves as the objective function. Then, the model is solved for the maximization of mass removal of contaminants by pumping, taking cleanup time, t_f, and available budget, C_S, as major constraints.

The optimal solution, including pumping/injection rates, well locations, and the number of wells, is sought by solving the mass removal model with a given budget in a prescribed cleanup time. The mass removal model serves as a basis for decision-making. Tradeoff among cost, cleanup time, and mass removal of contaminants, can be made on sensitivity analysis.

OPTIMIZATION MODEL

Decision variables

In order to design a PAT system, the following questions have to be answered:

1. How many wells are needed?
2. How much should be pumped and/or injected in each well?
3. Where should wells be placed?

All these mentioned items constitute the decision variables for a groundwater remediation project.

State Variables

Governing equations (Bear and Verruijt, 1987), which describes groundwater flow and contaminant transport in a heterogeneous, isotropic, confined aquifer, is expressed by the balance equations:

$$S\frac{\partial h}{\partial t} = \nabla \cdot (\mathbf{T}\nabla h) - \sum_{j=1}^{M} Q_{pj}\delta(\mathbf{x} - \mathbf{x}_j) + \sum_{j=1}^{N} Q_{rj}\delta(\mathbf{x} - \mathbf{x}_j) \tag{1}$$

where S = aquifer storage coefficient [1]; $\mathbf{x}$ = position vector [L]; $\mathbf{T}$ = tensor of aquifer transmissivity [$L^2\ T^{-1}$]; Q_{pj} = pumping rate at the jth well [$L\ T^{-1}$]; Q_{rj} = injection rate at the jth well [$L\ T^{-1}$]; δ = Dirac delta function at $\mathbf{x}_j$, and

$$R\phi\, B\frac{\partial c}{\partial t} = \nabla \cdot (\phi\, BD \cdot \nabla c) - \nabla \cdot (\phi\, BVc) - \sum_{j=1}^{M} cQ_{pj}\delta(\mathbf{x} - \mathbf{x}_j) \tag{2}$$
$$+ \sum_{j=1}^{N} c_r Q_{rj}\delta(\mathbf{x} - \mathbf{x}_j)$$

where c = solute concentration [$M\ L^{-3}$]; c_r = concentration of injected water [$M\ L^{-3}$]; ϕ = porosity [1], with $\partial\phi/\partial t = 0$, B = thickness of aquifer [L]; R = retardation factor [1], defined as $1 + \rho_b K_d \phi$; ρ = soil bulk density [$M\ L^{-3}$]; K_d = solute partitioning coefficient [$L^3\ M^{-1}$]. D = coefficient of hydrodynamic dispersion tensor [$L^2\ T^{-1}$] with components (Bear, 1979):

$$D_{xx} = (a_L V_x^2 + a_T V_y^2)/|V|$$
$$D_{yy} = (a_L V_y^2 + a_T V_x^2)/|V|$$
$$D_{xy} = D_{yx} = (a_L - a_T)V_x V_y/|V|$$

$|V|$ = magnitude of fluid velocity [LT^{-1}]; V_x and V_y = magnitudes of velocity components in x and y directions [LT^{-1}]; a_L and a_T = longitudinal and transversal dispersivities [L]. Here, dispersivity means field (or macro-) dispersivity.

The piezometric heads and concentrations are state variables and equations (1) and (2) are state equations. Given the system parameter values (transmissivities, storage coefficients, initial and boundary conditions, etc.), and the decision variables (well locations and pumping/injection rates, as well as the number of wells), equations   (1) and (2) are solved by the computer code SUTRA (Voss, 1984),   the   numerical simulator, which forecasts piezometric heads and concentrations. These forecasts, in turn, can be used to evaluate objective function and constraints.

Mass removal model
The mass removal model takes the form:

Minimize $\qquad\qquad\qquad\qquad\qquad\qquad$ (4)
X ∈ R^n
$$f(\mathbf{X}) = - \sum_{k=1}^{M} \sum_{l} \overline{c(k)} Q_{pl} \Delta t_k$$

Subject to
$$\Pr\{c_{t_f}(\mathbf{x}_i) \le c^{MCL}\} \ge \varsigma, \quad \forall i \in \Omega \qquad (5)$$

$$\Pr\{h_{\min} \le h(\mathbf{x}_i, k) \le h_{\max}\} \ge \varsigma, \quad \forall i \in \Omega \qquad (6)$$

$$\sum_{k=1} \Delta t \le t_f \qquad (7)$$

$$Q_p \le Q_p^u \quad Q_r \le Q_r^u \qquad (8)$$

$$\mathbf{x}_p \le \mathbf{x}_p^u \quad \mathbf{x}_r \le \mathbf{x}_r^u \qquad (9)$$

$$\mathbf{e}_p^T Q_p + \mathbf{e}_r^T Q_r \le C_S \qquad (10)$$

Since decision makers are most interested in removing more mass of a contaminant from the contaminated aquifer when the budget and cleanup time are given,   the objective (4) is expressed as the product of the concentrations and the volume of pumped water. Because many uncertainties, such as those from hydraulic conductivities and porosities, are involved in the groundwater system, the mean value of concentrations, $\overline{c}$, is used in the objective. In (4), M stands for the number of pumping wells, Q_{pl} is the pumping rate (As a decision in a remediation design, Q_{pl}, cannot be changed from time to time. In fact, such a pumping   rate means the capacity of the well.) of lth pumping well, and Δt_k is the kth time interval in the numerical simulation. $\mathbf{X}$ stands for the vector of decision variables including pumping/injection rates, Q_{pl}, Q_{rl} , $\forall\ l$, and their locations, $\mathbf{x}_p$, $\mathbf{x}_r$.

We intend to reduce the final contaminant concentrations to lower than the ***maximum contaminant level***, c^{MCL}, (We may refer US EPA water standards for drinking water or other purpose.),

$$c(\mathbf{x}_i, t_f) \le c^{MCL}, \qquad \forall i \in \Omega \qquad (11)$$

where i is any node of interest in the simulation. Because of the uncertainties inherent in the considered aquifer parameters, deterministic optimization models are often inadequate for effective groundwater system   planning and design. Thus, the MCL constraint should be expressed in the probabilistic form as (5). c_{t_f} stands for a final concentration and ζ denotes the reliability.

During the entire cleanup period, the piezometric head at any point of interest must remain above some minimum level and lower some specified maximum one. Similar to (5), the chance-constrained piezometric heads at points of interest should be held as (6).

The duration of numerical simulation must be equal or less than the given cleanup time, t_f. A model is usually solved when t_f has a fixed specified value. It is used as an input to the numerical simulator in order to evaluate system's response to different decision alternatives.

(8) expresses the pumping/injection capacity limitation. (9) gives possible locations for installing pumping/injection wells.

The total cost of a PAT system is expressed by the capital costs for well installation and treatment plant construction and annual costs for system operation, which is discounted over time. It determines the number of wells and pumping/injection rates. When capital costs are dominant, the total cost is approximated as the left side of (10), where e_p is the vector of cost for pumping and treating a unit volume of contaminated water, e_r is the vector of cost for injecting a unit volume of treated water, and C_S is the budget limitation.

SOLUTION METHOD: Exponential Penalty and Gradient Search

Constrained optimization models are often solved as unconstrained ones by introducing penalty function into the objective functions. The solution of a constrained problem can be obtained by arbitrarily increasing the penalty parameters, or *Lagrange multipliers*, as the algorithm gets closer to convergence. This advantage becomes much more effective in the applications where objectives and constraints are *implicit functions* of the design variables. For example, contaminant concentrations at all nodes of interest, which are implicit functions[1] of pumping/injection rates and well locations, serve as constraints in the optimization procedure. This implicit nature of the constraints makes the computation of such functions and their gradients very expensive. In optimization methods, the gradient of each constraint is required in order to calculate the search direction. This requires large computational efforts, especially, when we have a large number of constrained nodes, for example, for concentration. However, in the penalty or multiplier methods, all of implicit constraints are summed up. Therefore, the gradient of only one functional is needed in search direction calculation.

In order to solve the proposed mass removal model, we introduced the exponential penalty method, which was derived by Templeman and Li (1987) from maximum entropy formalism. It makes the constrained problems unconstrained,

---

[1] Implicit functions denote functions which cannot be explicitly and analytically described. Very often, those are described by numerical simulators.

$$\Phi = f(\mathbf{X}) + \frac{1}{\rho} \ln \left\{ \sum_{j=1}^{J} \exp\left[\rho\alpha\, C_j(\mathbf{X})\right] \right\} \qquad (12)$$

where $f(\mathbf{X})$ is an objective, $C_j(\mathbf{X})$ is jth constraint, ρ is an arbitrary positive constant, α is the Lagrange multiplier.

This function is sequentially minimized with increasing value α and some fixed positive ρ to get a solution of the problem. The second term of (12), which takes an exponential penalty form, represents a cumulative constraint for J inequality constraints $C_j(\mathbf{X})$, $\forall j=1,...,J$. It is suggested by Arora *et al.* (1991) that ρ be chosen as

$$\rho = \frac{\ln(J)}{\varepsilon}, \qquad (13)$$

where ε is the constraint tolerance at optimum. Such a solution method can be performed in following steps:

Step 1	Set k=0. Estimate ρ and α^0.
Step 2	Minimize $\Phi(\mathbf{X}, \alpha^k, \rho)$. Let $\mathbf{X}^k$ be the solution.
Step 3	If convergence criteria are satisfied, stop the iteration process.
Step 4	Update α^k and increase ρ if necessary.
Step 5	Set k=k+1 and go to Step 2.

In step 2, when α^k and ρ are fixed, common unconstrained optimization methods can be used. We use steep descent method in this paper. For every $\mathbf{X}$ in the iteration, the numerical simulator, here we use SUTRA, is run to evaluate $f(\mathbf{X})$ and $C_j(\mathbf{X})$, $\forall j=1,...,J$ (Sun, 1995). For simplifying the numerical simulation, we limit well locations only nodes in the mesh.

APPLICATION: A Numerical Example of Aquifer Remediation

A hypothetical, rectangular, heterogeneous, isotropic, and confined aquifer with distributed hydraulic conductivities, porosities, and thicknesses has been contaminated. For the numerical simulation, it is discretized by a finite element mesh with *260* nodes and *228* elements, as shown in Fig. 1. Each element has the size of $\Delta x = \Delta y = 152.5$ m (*500* ft). Zero-flow boundary conditions are assumed along upper and bottom sides. Constant piezometric heads are maintained along the left and right boundaries such that regional flow and the initial contaminant plume move from right to left. If no remediation action is conducted in this domain, the plume will be developed as shown in Fig. 2 in ten years.

The objective here is to produce the strategy of well locations to remove the contaminant as much as possible. Pumping rates are simply assumed to be constant as the capacity (*489.24* m^3/d). Thus, the number of pumping wells is obtained, $n=4$, from the given budget (4.0×10^5 dollars), if the total cost of one well is assumed to be 10^5 dollars. Then, the work of PAT design is only the optimization of the four well locations.

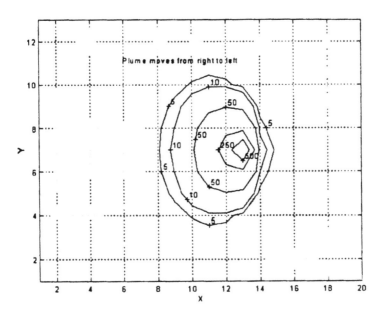

Figure 1 A numerical example of synthetic contaminant plume. The plume moves from right to left under the regional flow regime.

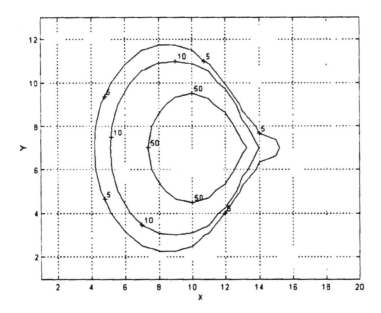

Figure 2 Contaminant plume in ten years without remediation.

By solving the optimization model, the four wells are, respectively, located at (7,7), (9,7), (10,7), and (10, 8), as shown in Fig. 3. It is understood from the solution that pumping wells should be located at downstream of the centroid of the contaminant plume rather than the centroid. The well locations are mainly determined by regional velocity of groundwater flow, pumping rates (which are assumed to be a constant in this case), and cleanup time. After 10 year operation of the designed PAT system under this derived strategy, the corresponding final concentrations are shown in Fig. 3. The contaminant concentration at 91.5% nodes in the study domain will reach the MCL requirement (≤ 50 ppm) in 10 years. 59.2% of total contaminant mass will be removed in this alternative.

Often, the feasible region of an optimal solution may not exist when constraints are too strict. To ensure a solution, some constraints should be relaxed. For example, one may consider the increase of budget or cleanup time, or relaxing the MCL requirement. Therefore, on the basis of mass removal model, tradeoffs among the three principal factors, mass removal of a contaminant, budget, and cleanup time, should be made before an PAT installation.

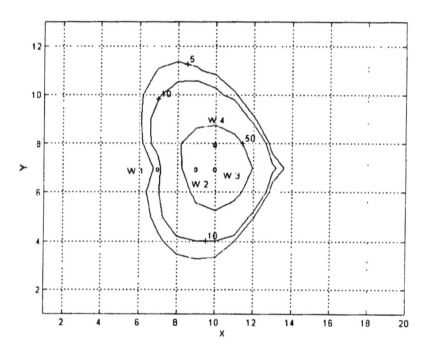

Figure 3 Optimal well locations and final contaminant plume

SENSITIVITY ANALYSIS

In order to know the information about the sensitivity of optimal solution to various aspects of the problem, we wish to estimate the effect on the objective, mass removal of a contaminant, of perturbing the constraints of budget and cleanup time. Such an information, obtained from the sensitivity analysis, will be helpful for decision makers to make tradeoff among the mass removal of a contaminant, cost, and cleanup time. One may wish to determine the changes of constraints that produces the largest improvement of the objective.

In the mass removal model, we eliminated the constraints of MCL and piezometric heads from the optimization model to assure that a feasible region exists under different constraints of budget and cleanup time. In parallel to solving the optimization model, the chance-constraint of MCL is regarded as a system performance and calculated as a function of the budget and cleanup time. For simplifying the illustration the pumping and injection rates are, respectively, assumed to be constant and zero. Once a budget and a cleanup time are given, the objective of mass removal is obtained by solving the optimization model. When the constraints of cost and cleanup time are relaxed, the model is solved again for the new decision. For the range of budget from 10^5 to 7×10^5 dollars and of cleanup time from *4* to *28* years, the tradeoff among the three factors, mass removal of a contaminant, cost, and cleanup time, are shown Table 2.

Table 2. Percentage of mass removal for different constraints of budget and cleanup time.

Cost	Cleanup time (years)						
	4	8	12	16	20	24	28
10^5	9.4235	24.2520	32.2694	38.1865	42.4579	46.0619	50.2482
2×10^5	20.1042	41.2021	51.2225	60.7507	67.6777	72.6416	76.1590
3×10^5	21.0386	42.9845	57.8646	69.2273	81.7029	87.7589	92.0838
4×10^5	22.2211	45.3093	64.7937	77.1077	86.1925	92.8625	97.7706
5×10^5	19.0408	49.3598	62.6906	75.3069	84.9975	95.0863	97.4274
6×10^5	19.4391	46.8830	63.4001	76.1787	89.0237	95.0863	98.8370
7×10^5	19.6756	47.1407	63.7562	76.6161	89.4273	96.4527	99.5918

Correspondingly, the penalty of MCL constraint violation, as another system performance is evaluated in Table 3.

Table 3 Penalty of MCL constraint violation

Cost	Cleanup time (years)						
	4	8	12	16	20	24	28
10^5	6301.7	4036.0	2991.8	2698.5	1886.7	1198.0	767.2
2×10^5	5137.6	3045.5	1791.1	828.0	293.8	247.6	7.1
3×10^5	5042.9	2891.5	1623.2	635.1	105.8	0.0	0.0
4×10^5	4967.6	2794.9	1329.3	392.2	65.5	0.0	0.0
5×10^5	5350.0	2582.3	1232.1	329.5	49.3	0.0	0.0
6×10^5	5296.9	2586.1	1166.0	282.9	0.0	0.0	0.0
7×10^5	5258.1	2551.0	1118.5	250.1	0.0	0.0	0.0

Table 2 and 3 tell us that for different values of cleanup time the objective function of mass removal increases with the investment, when the investment is small. After the budget reaches a certain value, the objective does not increase significantly. Decision makers may consider such a tradeoff among mass removal of contaminants, cost, and cleanup time, in the final design of a groundwater remediation project. The information in Tables 2 and 3 provides a basis for decision makers to determine what the combination of the three factors is favorite.

CONCLUSION AND DISCUSSION

A mass removal model has been proposed and solved for groundwater remediation. It is suitable for the design of remediation with a limited budget and a given cleanup time. If the given budget is too tight, or cleanup time is too short, or the MCL standard is high, the feasible region may not exist. Thus, the sensitivity analysis of the mass removal model may be conducted for making the tradeoff among the three basic factors (mass removal, cost, and cleanup time). It is noted that the global optimum is not guaranteed by the solution method. The results from above analysis suggest that investment for designing a PAT system should be coordinated to the cleanup time.

Dilution is not a solution for aquifer remediation. The most important performance index of *pump-and-treat* remediation is the mass removal of contaminants. Although natural dispersion may cause contaminant concentration to be lower than arbitrary MCL standard, the contamination problem is not solved. It is suggested, by the mass removal model, that we have to use prescribed budget and cleanup time to sufficiently remove contaminants from aquifers.

In the optimization procedure, the numerical simulator may be run hundreds or thousands of times to find optimal solutions, especially in a hierarchical optimization. Previous numerical simulators, such as SUTRA (Voss, 1984), MT3D (Zheng, 1990), RT3D (author's recent work), *etc.* are based on serial calculation which take a rather long time in searching optimal solutions. Of course, such simulation techniques limit the power of the PAT design. We only used a simple and hypothetical aquifer remediation case in this paper. Parallel algorithm will be good direction to further solve both optimization and simulation of PAT design.

REFERENCES

1. Ahlfeld, D. P., Mulvey, J. M., Pinder, J. F. and Wood, E. F., Contaminated groundwater remediation design using simulation, optimization, and sensitivity theory. *Water Resources Research*, 24(3), 431-441, 1988.
2. Andricevic R. and Kitanidis P. K., Optimization of the pumping schedule in aquifer remediation under uncertainty, *Water Resources Research*, 26(5), 875-885, 1990.
3. Arora, J. S., A. I. Chahande and J. K. Paeng, Multiplier methods for engineering optimization, *International journal for numerical methods in engineering*, 32, 1485-1552, 1991.
4. Atwood, D. F. and Gorelick, S. M., Hydraulic gradient control for groundwater contaminants removal, *J. Hydrology*, 76, 85-106, 1985.
5. Bear, J., *Groundwater hydraulics*, McGraw-Hill, New York,1979.
6. Bear, J. and A. Verruijt, *Modeling groundwater flow and pollution*, Kluwer Academic Publishers, 1987.
7. Culver, T. B. and C. A. Shoemaker, Dynamic optimal control for groundwater remediation with flexible management periods, *Water Resources Research*, 28(3), 629-641, 1992.
8. Culver, T. B. and C. A. Shoemaker, Optimal control for groundwater remediation by differential dynamic programming with quasi-Newton approximation, *Water Resources Research*, 29(4), 823-831, 1993.
9. Dettinger, M. D. and J. L. Wilson, First order analysis of uncertainty in numerical models of groundwater flow, 1. Mathematical development, *Water Resources Research*, 17(1), 149-161, 1981.

10. Dougherty, D. E. and R. A. Marryott, Optimal groundwater management: 1 Simulation annealing, *Water Resources Research*, 27(10), 2493-2508, 1991.

11. Gorelick S. M., S. M. Remson and C. Rechard, Management model of a groundwater system with transient pollutant source, *Water Resources Research*, 15(5), 1243-1249, 1979.

12. Gorelick S. M., A review of distributed parameter groundwater management modeling methods, *Water Resources Research*, 13(2), 305-309, 1983.

13. Gorelick S. M. *et al*, Aquifer reclamation design: The use of contaminant transport simulation combined with nonlinear programming, *Water Resources Research*, 20(4), 415-427, 1984.

14. Gorelick S. M., Large scale nonlinear deterministic and stochastic optimization: Formulations involving simulation of subsurface contamination, *Mathematical Programming*, 48, 19-39, 1990.

15. Gorelick S. M., R. A. Freeze, D. Donohue, and J. F. Keely, *Groundwater Contamination-Optimal Capture and Contaminant*, LEWIS publishers, 1993.

16. Greenwald, R. M. and S. M. Gorelick, Particle travel times of contaminants incorporated into a planning model for groundwater plume capture, *J. of Hydrology*, 107, 73-98, 1989.

17. Jones, L., R. Willis and W. W-G Yeh, Optimal control of nonlinear groundwater Hydraulics using differential dynamic programming, *Water Resources Research*, 32(11), 2097-2106, 1987.

18. Sun, Y., *Optimization of pump-treat-inject technology for remediation of contaminated aquifers*, Ph.D. dissertation, Technion, Israel, 1995.

19. Voss, C. I., SUTRA (Saturated-Unsaturated Transport) two-dimensional, density-dependent flow and transport of either dissolved solute or thermal energy, USGS, 1984.

20. Wagner, B. J. and S. M. Gorelick, Optimal groundwater quality management under parameter uncertainty, *Water Resources Research*, 23(7), 1162-1174, 1987.

21. Wagner, B. J., Recent advances in simulation-optimization groundwater management modeling, *Review of Geophysics*, Supplement, 1021-1028, 1995.

22. Wang, W. and D. P. Ahlfeld, Optimal groundwater remediation with well location as a decision variable: Model development, *Water Resources Research*, 30(5), 1605-1618, 1994.

23. Zheng, C., MT3D--*A module three-dimensional transport model for simulation of advection, dispersion, and chemical reactions of contaminants in groundwater system*, S. S. Paradopulos & Associate, Inc.,1990.

Proc. 30th Int'l Geol. Congr., Vol.22, pp. 224-228
Fei Jin and N.C. Krothe (Eds.)
© VSP 1997

A Discussion of Scientific Issues in Hydrogeology of Low Permeability Strata

CHIN-FU TSANG

Earth Sciences Division, Ernest Orlando Lawrence Berkeley National Laboratory
University of California, Berkeley, California 94720

Abstract

Current outstanding scientific issues in hydrogeology of low-permeability crystalline rocks are discussed according to five topics: measurement methods; interpretation of field tests; physical processes; modeling techniques and predictive modeling strategies.

INTRODUCTION

Low permeability strata have become a subject of intensive study in recent years because of interest in evaluating their isolation properties related to potential groundwater contaminant migration in geologic systems. In this paper we shall focus our discussions on low-permeability crystalline rocks, where much of recent search has centered around flow and transport in fractures. Flow in fractures is several orders of magnitude larger than that in the rock matrix part of the medium, thus controlling the characteristics of these low-permeability crystalline rocks.

A number of recent reviews cover various aspects of flow and transport in rock fractures (e.g., Bear et al., 1993; Tsang and Neretnieks, 1996; Tsang and Stephansson, 1996), and extensive references are given in these review papers. Below we shall present and discuss a number of key scientific issues in the hydrogeology of these low-permeability rocks. They may be grouped under several topics, (a) measurement methods, (b) interpretation of field tests, (c) physical processes, (d) modeling techniques and (e) predictive modeling strategies. The goal of these discussions is to outline outstanding scientific problems and potential directions for future research.

MEASUREMENT METHODS

Because of the low permeability, conventional pumping tests cannot be used due to the resulting large draw-downs. Injection tests and pressure fall-off tests are often employed, and techniques of measuring low flow rates have been developed. This raised the interesting scientific issue of hydromechanical effect of rock fractures during injection tests. This was studied in detail by Rutqvist et al. (1992) who performed field experiments and model simulations on this subject. They pointed out that even below the lithostatic pressure, the injection pressure may reduce the effective stress across the rock fractures sufficiently that the fracture opens up with the result of increased aperture and increased hydraulic conductivity. Much work remains to be done in the coupling of hydraulic injection pressures and mechanical opening or closing of rock fractures.

Because of the importance of rock fractures in hydrogeology of low permeability crystalline rocks, it would be extremely useful to be able to measure efficiently the hydraulic conductivity of each of the fractures intercepted by a well. Conventional hydraulic testing usually involves well testing over an extended thickness of the formation corresponding to the entire section of the well below the casing. This interval may well contain many fractures or sometimes more than one stratagraphic layer. Considerable advances have been made in pressure testing with packers in the well, which allows testing with a packer-bracketed interval shorter than 1 meter. Another method involves the use of well logging techniques, which measure the transient physical and chemical properties of fluids in the well as a function of depth, as flow is being produced from the fractures into the well (Tsang et al., 1990). This method can determine the hydraulic conductivity of a fracture at an inflow location with a resolution of the same order as the well diameter (5–15 cm).

Advanced methods are still very much needed to determine fracture hydraulic properties, especially because they vary widely even among the fractures intercepted by one well. It is estimated that about 80% of the fractures are not hydraulically active. Without proper measurement techniques, it would be difficult to understand the hydrogeology of fractured rocks.

FIELD TEST INTERPRETATION TECHNIQUES

In conventional aquifers, the entire stratum is hydraulically conducting. Even though there may be variations in conductivity values, the range is relatively small for a specified region. Thus a mean conductivity can be defined and evaluated from well tests. For low-permeability fractured media, hydraulic conductivity varies widely in space, thus requiring a very careful interpretation of well test data. The measurement of the effective hydraulic conductivity derived from type curve matching, for example, will have to be reviewed. Is it the average value over the region of influence to be calculated from the test duration? Is it the appropriate value to be used for numerical modeling of flow and transport in the medium? Is it valid only for radial flow (which is the flow pattern for the single well injection test), or is it valid for general flow patterns? These are some of the questions that do not yet have clear and practical answers.

PHYSICAL PROCESSES

Many of the physical processes in low-permeability media are similar to those in well-known aquifers. However, four additional effects may be defined special to low-permeability media. The first is the variability of the aperture in a single fracture giving rise to an important observation in field experiments. If one were to perform a flow experiment under a pressure drop, the effective permeability can be obtained which is related to the small-aperture (or constriction) part of the fracture, since this is the part that offers the most resistance to the flow. On the other hand, if one were to perform a tracer migration experiment and calculate the aperture from the mean arrival time and the flow rate, the aperture value obtained would be the arithmetic mean. The two aperture values derived from these two experiments differ by a factor of two or more. This was first pointed out by Abelin et al. (1985), subsequently discussed in a number of papers (see e.g., Tsang, 1993).

The second physical process found in fractured media is flow channeling or focused flow. Because the media is so heterogeneous, flow seeks paths of least resistance and these tend to be focused on a few main channels in the media. This was first pointed out for a single fracture (Tsang and Tsang, 1989) and later confirmed for a 3-D strongly heterogeneous medium (Moreno and Tsang, 1994).

The third physical process of interest is matrix diffusion of tracers (Neretnieks, 1983). In fractured rocks, tracer migration is dominated by flow through fractures, which is relatively fast. The matrix, though having very little flow, contains pores where tracers can diffuse into and be retarded. Matrix diffusion is a slow process; however, it is an important one in considering the isolation characteristics of low permeability rocks where isolation is expected for thousands of years.

The fourth process of particular interest for fractured media is the coupled thermo-hydro-mechanical processes. Both thermal expansion and hydraulic pressure may close or open fractures mechanically causing a significant change in their hydraulic conductivity values and hence modifying the total flow field. Details of these processes with information of recent studies may be found in Stephansson et al. (1996). These four and other processes in low-permeability fractured media are in different degrees of understanding, and techniques of including them in modeling studies are still being developed.

MODELING TECHNIQUES

In general these are two main modeling approaches: fracture network models and stochastic continuum models. The double-permeability models may be useful for high-permeability fractured porous media with large fracture density. They are not used for cases where a few fractures dominate the flow in the region of interest.

The fracture network models assume that the matrix permeability is negligible and flow is mainly through a network of fractures. Though the conceptual framework of these models are reasonable, they suffer from the fact that it is almost impossible to obtain the actual structure of the network from field data. Thus one resorts to stochastic concepts that fracture lengths, orientations, and apertures are described by distributions with the respective mean and standard deviation values. This makes the application of modeling results to field application very difficult. Cacas et al. (1990) proposed a way of probability distributions to consider field results. Their method and general fracture network models are surveyed in Tsang and Neretnieks (1996).

An alternative is to consider the low-permeability stratum to be a continuum with hydraulic conductivity varies by orders of magnitude over space. By a stochastic generation technique, large conductivity points are made to lie in prescribed planes at orientations corresponding to those of observed fracture sets. The generated stochastic continuum can be conditioned to point field data and used for flow and transport calculations (e.g., Tsang et al., 1996).

Both approaches are stochastic in the sense that sufficient data are not available to deterministically define the system and the models are possible realizations which represent a real physical system. How to use the stochastic models to understand field tests and calibrate parameters and how to use them to simulate physical phenomena remain open questions. A number of attempts (e.g., Cacas et al., 1990) were made, but clear and practical answer is yet to be found.

PREDICTIVE MODELING STRATEGIES

Much research has been done on many hydrologic processes related to flow and transport of solutes in geologic formations. Examples of these processes include dispersion and its scale dependence, matrix diffusion, flow channeling, and density driven flow. However, with the recent interest in flow and transport of contaminants of radionuclides in the subsurface as related to environmental remediation and nuclear waste disposal problems, the need is not only for an advanced understanding of these individual processes, but also for the capability of predicting such transport in the geological medium at a given site as a system. Processes like dispersion have been studied from the early days of hydrogeology while density driven flow is a clearly defined problem whose main difficulties may be technical, i.e., how to numerically calculate the process in three-dimensional complex media with sufficient accuracy. Other processes, such as flow channeling and matrix diffusion, have been relatively more recently addressed specifically for low-permeability fractured media.

Just to understand the individual processes, however, is not the goal. Recent interest has focused how to predict flow and transport in geologic systems. To go from processes to the prediction of system behavior opens up a new series of issues that have not yet been adequately addressed. Understanding a process is a well-defined problem in which the equations and experiments are restricted within the problem definition. To predict system behavior, on the other hand, requires a "sufficient" level of knowledge about the system—which is often difficult to attain for geologic systems. Even when we have a level of knowledge of the system, it is still not simple to translate that knowledge to characteristic parameters that can be used in model calculations of system behavior.

Examples of the scientific issues associated with predictive modeling include how to obtain appropriate boundary conditions and long term scenarios on the system (both of which having temperal and spatial variations); how to handle heterogeneity of the geologic system whose details cannot be found deterministically; how to develop the conceptual model of the system; and how to combine results of alternative conceptual models which are all consistent with available data. Many of these issues are discussed in Tsang et al. (1994) and Tsang (1996).

Experience in predicting flow and transport characteristics of a low permeability stratum is still limited. One would need more experience with case studies where such prediction is made for tens of thousands of years. Since there will be no data that can verify these predictions, cross comparison between two or more independent groups making predictions need to be done. These groups will need not only to present their predictions but also to state the estimated confidence level or uncertainty ranges of these predictions. In this area much work remains to be done.

CONCLUDING REMARKS

Hydrogeology of low-permeability crystalline rocks has been intensively studied over the past ten to twenty years. However, much remains to be done. One reason is the society's demand for prediction of the isolation characteristics of these strata over tens of thousands of years. This demand emphasizes the need for careful science and understanding of physical processes and their uncertainties, which were never required before to such a degree. The present paper outlines various scientific issues that need to be further addressed and investigated in our endeavor to advance the state-of-the-art of hydrogeology of low-permeability systems.

Acknowledgment

The paper is prepared with the joint support of Nuclear Power and Fuel Development Corporation (PNC), Japan, and Office of Energy Research, Office of Basic Energy Sciences, Engineering and Geosciences Division, of the U.S. Department of Energy under Contract No. DE-AC03-76SF00098.

REFERENCES

1.   H. Abelin, I. Neretnieks, S. Tunbrant, and L. Moreno. Final report of the migration in a single fracture, experimental results and evaluation, Stripa Project. Tech. Rep. 85–103, Nucl. Field Safety Proj., Stockholm, Sweden (1985).

2.   J. Bear, C.-F. Tsang, and C. de Marsily (Eds.). Flow and Contaminant Transport in Fractured Rocks. Published by Academic Press, 1993.

3.   M.C. Cacas, E. Ledoux, G. de Marsily, A. Barbreau, P. Clamels, B. Gaillard, and R. Margritta, Modeling Fracture Flow with a Stochastic Discrete Fracture Network; Calibration and Validation, 2, The Transport Model, Water Resour. Res. 26(3), 491–500, 1990.

4.   L. Moreno, and C.F. Tsang, Flow Channeling in Strongly Heterogeneous Porous Media; A Numerical Study, Water Resour. Res. 30(5), 1421–1430, 1994.

5.   J. Rutqvist, J. Noorishad, O. Stephansson, C.F. Tsang, Theoretical and Field Studies of Coupled Hydromechanical Behaviour of Fractured Rocks—2, Field Experiment and Modeling, Int. J. Rock Mech. Min. Sci. and Geomech. Abstr. 29(4), 411–419, 1992.

6.   O. Stephansson, L. Jing, C.-F. Tsang (Eds). Coupled thermo-hydro-mechanical processes of fractured media: Mathematical and Experimental Studies. Elsevier Science Publishers, December 1996.

7.   C.-F. Tsang. From processes to systems: A discussion of predictive modelling of flow and transport in geologic systems. In Groundwater and Subsurface Remediation: Research Strategies for In-situ Technologies. H. Kobus, B. Barczewski, and H.-P. Koschitzky (Eds.), Springer Verlag, Germany, 207–218, 1996.

8.   C.-F. Tsang, L. Gelhar, G. de Marsily, and J. Andersson, Solute Transport in Heterogeneous Media: A Discussion of Technical Issues Coupling Site Characterization and Predictive Assessment, Advances in Water Resources 17, 259–264, 1994.

9.   C.-F. Tsang and I. Neretnieks. A review of flow channeling in heterogeneous fractured rocks. Submitted to Reviews of Geophysics, December 1996.

10. C.-F. Tsang, P. Hufschmied and F.V. Hale, "Determination of fracture inflow parameters with a borehole fluidconductivity logging method," Water Resources Research 26(4), 561–578, 1990.

11. C.-F. Tsang and O. Stephansson. A conceptual introduction to coupled thermo-hydro-mechanical processes infractured rocks. In Coupled thermo-hydro-mechanical processes of fractured media: Mathematical and Experimental Studies. O. Stephansson, L. Jing, C.-F. Tsang (Eds), Elsevier Science Publishers, 1996.

12. Y.W. Tsang, Usage of "Equivalent Apertures" for Rock Fractures as Derived From Hydraulic and Tracer Tests, Water Resour. Res. 28(5), 1451–1455, 1992.

13. Y.W. Tsang, Y.W., and C.-F. Tsang, Flow Channeling in a Single Fracture as a Two-Dimensional Strongly Heterogeneous Permeable Medium, Water Resour. Res. 25(9), 2076–2080, 1989.

14. Y.W. Tsang, C.-F. Tsang, F.V. Hale, and B. Dverstorp. Tracer transport in a stochastic continuum model of fractured media, Water Resources Research 32(10), 3077–3092, 1996.

Proc. 30th Int'l Geol. Congr., Vol.22, pp. 229-234
Fei Jin and N.C. Krothe (Eds.)
© VSP 1997

THE PALEOHYDROGEOLOGY OF THE SIBERIAN PLATFORM

EUGENE V. PINNEKER, A.A.DZYUBA

Institute of the Earth's Crust, Russian Academy of Science, Siberian Branch,
664033 Irkutsk, Russia

Abstract

The knowledge of the problems of paleohydrogeology provides a principally new approach to understanding of the character of deep groundwater formation and accompanying oil-gas deposits. The methods of paleohydrogeological reconstruction are very different. Conventional methods of investigation of the relationship of paleohydrogeological reconstruction, the role of variability of the thickness and quality of water-bearing and screen rock mass, discontinuity of sedimentation have been supplemented by the development of the analysis of continental drift, the water-rock balance, isotope of a wide range of elements. The retrospective review of the hydrogeological history evidences that the evolution of the underground hydrosphere is attributed primarily by its dynamics within fault zones which control the accumulation and discharge rate of groundwaters, specificity of their chemical composition as well as the oil and gas migration. At the old Siberian platform the stages of hydrogeodynamical activity existed in Proterozoic, Wend and Phanerozoic times. Paleohydrogeological reconstructions hypothesize that preservation of the oil and gas deposits are associated primarily with the presence of Mesozoic-Cenozoic centers of hydrogeodynamical activity. Among those are, for example, the oil-gas deposits in the fault zones along the rivers Vilui, Angara, Oka and Upper-Lena. The former hydrogeodynamical activity can be proved also by the presence of concentrated brines with high contents of hydrogen sulphide ($1\text{-}2$ g/dm^3) in the productive series, as well as by the organic matter rich ($C_{org.}$ up to $30\text{-}40$ mg/dm^3) freshwater in suprasalt formations.

Key words: *paleohydrogeology, oil-gas deposits, methods of reconstruction, water-rock balance, paleohydrogeological schemes, isotopic diagnostics.*

INTRODUCTION

The study of problems of paleohydrogeology enables to form a principally new understanding of the character of groundwater formation as well as the accompanying oil and gas accumulation. Methods of the paleohydrogeological reconstruction are different. In addition to the well known ways of the paleohydrogeological reconstruction we used the analysis of the continent drifts, material balance within the "water-rock system" and the isotopic diagnostics. Retrospective review of the hydrogeological history shows that the evolution of the underground hydrosphere is determined primarily by its dynamics in fault zones, especially within tectono-magmatic cycles, determined primarily by its dynamics in fault zones, especially within tectono-magmatic cycles, which control the accumulation and discharge of groundwaters, their specific composition, as well as migration of hydrocarbons and other fluids.

The Siberian platform (Fig.1) is a typical example for the paleohydrogeological study; particularly, it represents the old tectonic structures   (Russian, African, Canadian etc.) of the Earth. The vast area of the Siberian platform (about 3.6 million km^2), its prolonged evolution (approximately one billion years), diverse material composition of the sedimentary cover, repeating magmatic events of considerable intensity facilitate the estimation of global And particular tendencies in the hydrogeological history.

Three megachrons marked by different hydrosphere types may be distinguished at the Siberian platform in Phanerozoic (Fig.2).

The Siberian platform in the megachron of the Early and Middle Paleozoic was situated near the Equator. The megachron is marked by the highly arid climate and accumulation of the concentrated brines. Under the present conditions the brines of this kind show the chloride calcium composition and very high salinity (500 g/l and more). During this time a maximum amount of salts has been accumulated. Within the megachron of Late Paleozoic, Early and Middle Mesozoic the Siberian platform gradually turned about the Equator. This megachron was marked by the predominantly humid climate and less mineralized groundwaters. At present the water-bearing formations which date from that stage of the geological history contain the sodium chloride brines as well as saline waters of the sodium chloride sulphate and   hydrocarbonate composition. This megachron was characterized by accumulation of subsaline and fresh groundwaters.

Up to now the areas of the Siberian platform occupied by sea basins were predominantly investigated, which was quite reasonable because the deposits and syngenetic groundwaters provide the fundamental   principles for the interpretation. Generally the traces of the groundwater activity after sedimentation are feebly marked. In this context, a package of measures and methods of analysis can be proposed to gain an efficient understanding of the paleohydrogeology under the continental conditions.

On the Siberian platform (as well as on some other older platforms) the occurrences of the rock salts are widely developed. The thickness of salt formations   (Fig.3)   accounts for about 1 km within the 3 km thickness of the sedimentary cover. Leaching of this salt by groundwaters influences the chlorine and sodium ion contents in river water. Hence, the mass balance factor should be considered. The paleohydrogeological reconstructions may be also effectively traced by the isotopic data.

The paleohydrogeological schemes of different periods demonstrate the change of hydrogeological situation at the south of Siberian platform.

The subsidence of this territory and advance of the sea started at the Late Pre-Cambrian. The scheme of Motskian time (beginning of the Low Cambrian) shows the saline waters and brines being formed in terrigene and gypsiferous-carbonate formations. In the later Usolskian and Lenskian times the brines originated mainly in halogene formations, practically everywhere on the internal field of the south of Siberian platform. However, during the Middle and Upper Cambrian (Verkholenskian time) the salt-forming formation reduced and displaced northwards.

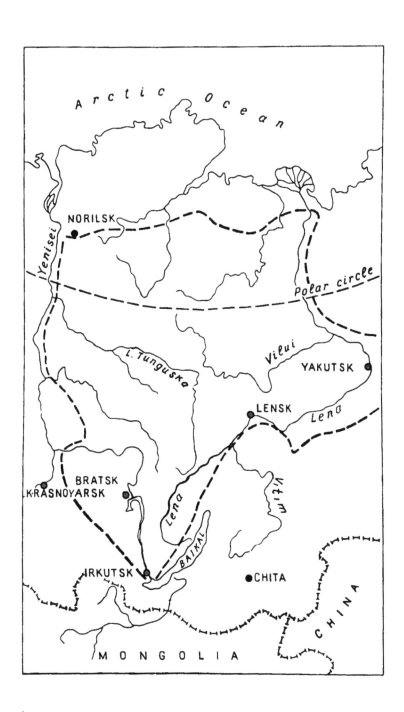

Figure 1. The countour of the Sibirian platforme

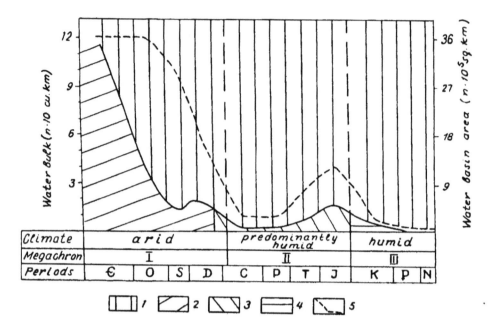

Figure 2. Plot of climatic changes, salt contents, mineralization of groundwaters, and catchment areas at the Siberian platform in Phanerozoic. 1- occurrence of hydrocarbonate magnesium/calcium waters within the near-surface area of sedimentary cover, 2- accumulation of chloride calcium/sodium groundwaters of highest mineralization (320-350 g/l), 3- accumulation of chloride sodium and hydrocarbonate sodium low mineralized (5-20 g/l) groundwaters, 4-accumulation of hydrocarbonate sodium and calcium containing fresh and subsaline (1-5 g/l) groundwaters , 5- diagram for the catchment areas.

Considerable changes in dynamics and chemical composition of groundwaters arose from the tectonic-magmatic activization in Triassic. Redistribution of groundwaters occurred through all parts of the geological sections with their intensive discharge; the abundance of hot   springs at the south of Siberian platform at that time   is similar to that of present-day Iceland. The radical changes came over the hydrogeochemical situation, which appeared particularly in the form of freshwater intrusion into the deep of the geological section and probable eruption of waters of magmatic genesis, as well as substitution of the older connate for infiltrational waters.

During the Jurassic time the gradual sinking of a considerable part of the territory (to the depth of about 450-600 m) occurred. Bulks of water of Jurassic basins have been intaken by karsted rocks of Lower Cambrian. Very important in the further geological history was the neogenic activization along the fold framework at the south of the Siberian platform, accompanied by formation of the Lake Baikal trough (depth about 6,000 m) and the East Sayan mountain range being 1,000-1,500 m uprisen. This all initiates the intrusion of the "fresh" water in the deeper part of the geological section.

The hydrogeological history has naturally affected the isotope composition of brines. As the isotope diagram shows (Fig4), the brines of highest mineralization (up to   500 g/l) are presented by the mixture of both connate and older infiltrational waters.

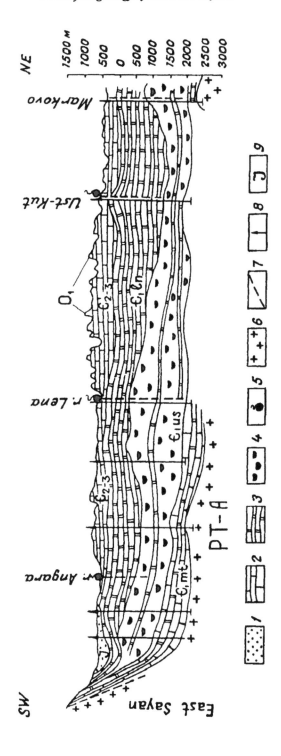

Figure 3. Geological section of the south of Siberian platform. 1-Sandy-argillaceous coal-bearing deposits; 2- dolomite; 3- carbonaceous-anhydrous series; 4- salt rock; 5- spring; 6- crystalline base; 7- brine removing fracture; 8- borehole; 9- geological index of water-bearing rocks of the south of Siberian platform.

CONCLUSIONS

The trappean magmatism in Triassic and tectonic activization   in Cenozoic resulted in the partial
or full destruction of the oil and gas deposits, which caused the hydrogen sulphide increase (up to
1,000-2,000 mg/l). Considerable oil and gas accumulation survive   primarily   within the middle
area of the south   of the Siberian platform (the Angara-Lena interstream and Nepa arch), where
the trappean magmatism in Triassic and Cenozoic activization did not have any effect.

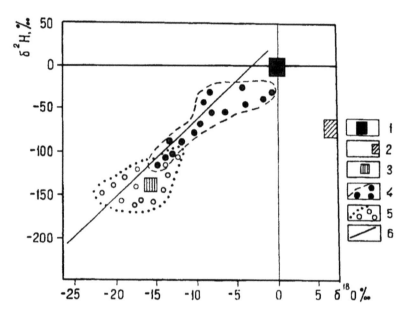

Figure 4. Isotope diagram of hydrogen and oxygen composition in groundwaters of different genesis.   1- SMOW,
2- hypothetical specimen of juvenile water, 3- water of the Lake Baikal, 4- calcium chloride brines, 5- sodium
chloride brines, 6- the Craig lines of surface waters.

Proc. 30th Int'l Geol. Congr., Vol.22, pp. 235-257
Fei Jin and N.C. Krothe (Eds.)
© VSP 1997

Exploiting Ground Water from Sand Rivers in Botswana using Collector Wells

HERBERT, R[1], BARKER, J A[1], DAVIES, J[1] and KATAI, O T[2]

[1] (BGS, UK), [2] (DWA, Botswana)

Abstract

Ephemeral sand rivers are common drainage features in north eastern Botswana where they form major sources of under-utilised groundwater.   These systems are recharged by unpredictable and infrequent wet-season storm rainfall events that result in short lived flash floods or mere trickle flows along sand river channels.   The sand rivers have well-defined, steep-sided channels, with flat floors, infilled with thick alluvial sands.

Studies of the sedimentology of the top few metres of the sand rivers show them to have been laid down in the last 100 years or so.   One model for such river deposition suggests there can be a peat marker layer at the base of these recent deposits, which, if dated, will indicate the age of the last major flood and recharge event.   It is possible the denudation of vegetation that is occurring now will enhance erosion and contribute to the above process.   Little is known about the deeper sediments in the sand rivers.

Various systems are employed to extract water from sand rivers.   Temporary abstraction points include hand excavated shallow pits for abstraction of small quantities for cattle watering purposes, mechanically excavated large pits for abstraction of large quantities of water for road construction purposes as from the Ntshe, and well points for village domestic supply.   More permanent structures include large diameter wells installed behind sand dams as at Mahalapye and Shashe.   At the latter site the well construction shows some indication of undermining and ultimate failure.

Recently, a technique for extracting water efficiently from thin shallow aquifers, collector wells, has been developed by the British Geological Survey, BGS, using the Overseas Development Administration of UK, ODA, funding.   As part of an ongoing village water supply project the Department of Water Affairs, Botswana have already begun constructing collector wells at eight village sites.   The DWA sink 2 m diameter shafts into the weathered crystalline rocks of the river embankments and then with the assistance of a specially designed drilling rig, horizontal adits are drilled out into the sand rivers to allow exploitation of the groundwater within them.

All eight collector wells have been tested for yield by pumping.   A novel means for interpretation of these tests has been developed.   In addition, salt dilution tests have been carried out within the sand river sediments at distance from the collector wells and estimates have been made of the naturally occurring downstream flow in the sand river aquifers.   These estimates were made at the end of the dry season and so assist with estimates of sustainable yields at the abstraction points. Further work is still required if sustainable development of the sand river system of Botswana is to be attained.

INTRODUCTION

Collector wells are large diameter dug wells with adits drilled horizontally from their base. These wells are suitable for exploiting thin shallow aquifers. They enhance the supply and provide large storage for low yielding aquifers. Sand rivers are ephemeral rivers which are incised into bedrock are filled with alluviums and have significant groundwater stored within.

This paper describes the work done at eight sites where collector wells have been constructed on sand river banks and which tap the water in the sand river deposits.

EARLIER WORK WITH COLLECTOR WELLS IN UNCONSOLIDATED DEPOSITS

A project funded by the Overseas Development Administration, ODA, and entitled Development of a Horizontal Rig for Alluvial Aquifers was undertaken between 1989 and 1992. The project s main aim was to develop a drilling rig, string and technique (a drilling system), which would be suitable for the construction of collector wells in alluvial valleys, plains or sand rivers. The system was to be relatively cheap and suitable for developing countries. Two reports summarise work which used telescoped-jetting as the drilling technique in alluvial valleys in Malaysia (Allen, 1988) and the alluvial plains of Zimbabwe (Herbert and Rastall, 1991). A third report, Morris *et al* (1991) describes work done in the UK using a moling technique. A final report Herbert 1992, describes the construction of 2 collector wells on the banks of sand rivers in Botswana. It also reviews the findings of the entire project and makes recommendations for the use of collector wells in alluviums. The unique problem posed by the work done in Botswana related to the fact that the shaft of the collector well was dug in weathered hard rock. The sand river cuts into the hard rock and the collector s adits have to be drilled laterally out through the hard rock, using air hammer and then through the non-cohesive sands of the sand river.

Construction of collector well shafts: Collector wells were constructed in the UK, Zimbabwe, Malaysia and Botswana using a variety of techniques; a) shafts were constructed using concrete rings, precast concrete keystone-shaped units, reinforced brickwork and galvanised steel (Armco) lining. In semi-consolidated deposits unskilled labourers could construct a 2 m diameter shaft to 15 m in one month given full logistical support i.e. the ARMCO lining, a winch to raise the lining, digging tools, sustenance and transport. This was the method preferred for the subsequent work in Botswana.

Collector wells adits: After much experimentation two drilling techniques, telescoped jetting and moling, were developed. Both are capable of emplacing screens in such material. In the wells sunk in hard rock adjacent to the sand rivers of Botswana, air-hammer techniques had to be used before break-out into the unconsolidated sand.

Jetting: In jetting, water is pumped through a hollow drill string and bit, and returns via the annular space between the string and the hole. The water removes sand distributed by the drill bit thus allowing steady advance of the string into the sand. A variant of this, telescoped-jetting, was used in the coarse sand of Malaysia, in the sand river beds of Botswana and in the alluvial plains of Zimbabwe. With this technique, four-inch diameter casing was jetted out to about 12 metres. Then, two-inch screen, encircling one-inch jetting pipe fitted with a loosely-connected sacrificial bit was jetted through the casing and into the sands beyond. Eventually the jetting pipe and casing were removed, leaving about 20 metres of screen permanently in place in the aquifer.

Moling: Moling is in regular use in the West for laying telephone lines etc, under roads. A hydraulic jack is used to push out a solid large diameter head in front of a smaller diameter drill string. The head creates a cavity by lateral compression of the material through which it is pushing. For this project a lightweight jacking rig pushed out a four-inch head using 1.5 inch push rods. These were surrounded by plastic-mesh-wrapped 1.5 inch screen and temporary, blank casing three inches in diameter. After reaching 20 or 30 metres, the three-inch casing and push rods were extracted leaving the screen behind.

Air hammer drilling: A hammer at the end of the drill string is driven by air which also blows the cuttings from the hole. This technique is ideal for hard rocks and was used in this project to drill through the hard rock banks until the hammer drill entered the sands of the river. The technique used is routine and is in regular use worldwide.

Selecting the best horizontal drilling technique: Adits created by moling are of smaller diameter than those created by jetting and consequently they can transmit far less water. The criteria for choosing which is best suited to a particular situation are outlined in the diagram below (Fig. 1). The thickness of two metres in the first decision-diamond is somewhat arbitrary. A widespread aquifer one metre thick would be sufficient for drilling of laterals. If the bedrock surface undulates however, the length of adit that can be drilled will be limited.

Key results of the 1989-92 project: a) Collector wells have an advantage in shallow, thin aquifers or where fresh water lenses must be skimmed from above deeper saline water; b) Moling cannot easily penetrate medium to coarse sands and telescoped jetting or other drilling methods cannot penetrate far into cobbly, coarse sands; c) When sand rivers are underlain by weathered hard rock, it is possible to air-hammer adits and their yield will depend on the type of weathered rock penetrated.

SAND RIVERS IN BOTSWANA

Occurrence

In Botswana, outside the eastern margin of the Kalahari Beds which cover almost 85% of the country lies a relatively small but highly populated, area drained by a large number of ephemeral rivers. The water courses are actually wide (up to 100 m) sandy beds which flow only for short periods in the wet season and in some years not at all. These sandy water courses are rather shallow with depths normally ranging between 3 and 10 metres, are generally called sand rivers, and contain valuable resources of groundwater of various volumes during different seasons.

Sand rivers in Botswana are classified in three groups; major, intermediate and smaller sand rivers (Nord, 1985). The first group comprises the Lower Shashe, Motloutse and Mahalapye rivers capable of yields exceeding 75 m^3 $d^{-1}km^{-1}$. The border river of Ramokgwebane also belongs to this group. The Upper Shashe, Tati and the Tutume rivers form the second group with expected yields of 30 to 45 m^3 $d^{-1}km^{-1}$, while the smaller river of Lotsand, Thune and Metsemotlhaba have yields estimated at 20 m^3 d^{-1}km^{-1}.

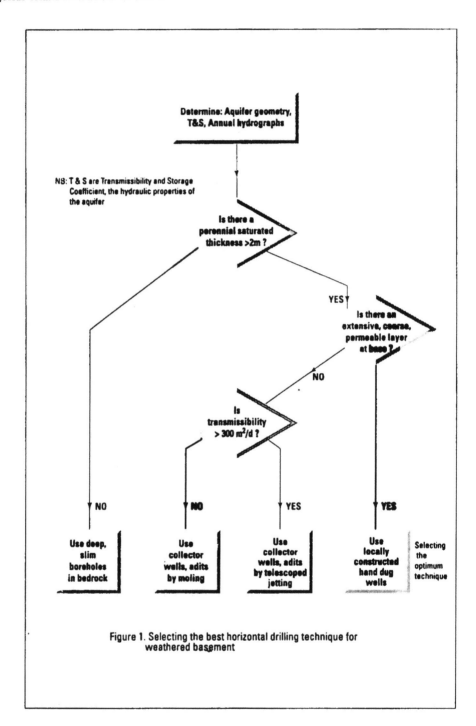

Figure 1. Selecting the best horizontal drilling technique for weathered basement

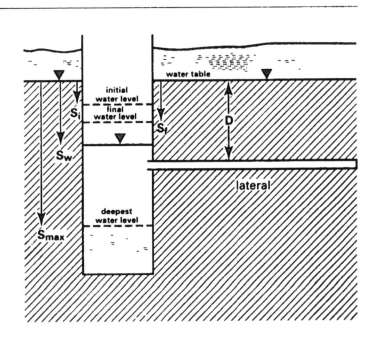

Figure A1. Geometry for Collector Well Tests with nearby recharge source

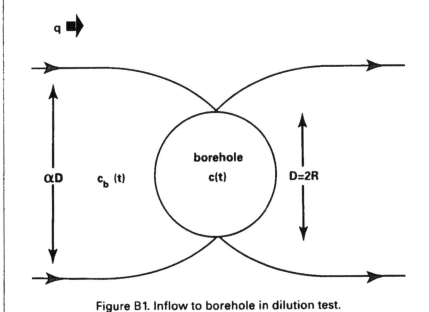

Figure B1. Inflow to borehole in dilution test.

The sand rivers in the north are well developed and show a more regular flow pattern and dry periods are shorter than the rivers further to the south. Common to the rivers in Botswana is the gentle slope of about 0.15% which tend to increase up to 0.5% higher up in the catchment areas. A perennial flow is considered to occur in the major sand river beds although rather slow due to the slight slopes. Because of intense, highly variable thunderstorms that fall with consequent high runoff rates, the rivers can occasionally carry large quantities of water.

Sedimentology
Earlier work: The hydrogeological potential of ephemeral sand rivers in north-eastern and eastern Botswana was studied by Thomas and Hyde (1972), Wikner (1980) and Nord (1985). They investigated river channel gradients and occurrence of country rock barriers to groundwater flow and sediment filled basins. Augering equipment was used to determine sand river sediment thickness and texture, both across and along channels.

Picard and High (1973) and Williams (1971) describe depositional structures within ephemeral stream sediments in semi-arid areas of the USA and Australia. There, as in Botswana, large scale ripples and dune sets are common sand river surface bedforms . These are composed of cross-bedded medium to coarse sands and gravels deposited during low stage flow. Thomas and Hyde (1972) noted that the upper 1-1.5 metres of sand river sediments are typically coarse sand and gravel that fine with depth; coarse sediments occur along the middle of straight river channels; banks of silt and clay occur adjacent to river banks; and coarse sediments occur along river channel bottoms. This general description of the deposits agrees well with the findings of NORD in Botswana and the recent work described in this paper.

The deeper layers: Horizontal wells have been drilled into the base of the river deposits at eight sites. These drillings confirm the general description given above. Near bank clayey deposits are common and high yielding channels occur in the middle of the sand rivers. Further, the sand rivers are often underlain by weathered semi-consolidated bedrock which is a potential lowyielding source of water. More work needs to be done to develop a model of these deeper layers.

The superficial layers: Sand pits dug down to the water table were available for inspection. The exposure at one site Masunga is given on Fig. 2. Two distinct fining upwards cycles of coarse sand deposition were recognised, separated by a coarsening upwards clayey fine to medium sand cycle. These sequences appear to have been deposited under differing climatic conditions, as indicated by the variation in grain size, cross bedding, and peaty clay content present. The lower cycle of sedimentation between 2.3-2.6 m is composed of well graded coarse to medium sands (Fig. 2). These are abruptly overlain by the middle coarsening upwards cycle of sedimentation. The lower part of the latter cycle is characterised by thin peaty clay layers interbedded with thin cross-bedded fine to coarse sands. These pass upwards initially into well graded medium to fine sands, that are succeeded by medium sands and thence upwards into medium to coarse sands. The peaty clay layer, although relatively thin was fairly persistent along the entire sand pit exposure. The upper-most cycle is dominated by poorly graded, cross bedded, sub-angular coarse sands and gravels that gradually fine upwards through a series of stacked cross bedded units.

Crowley (1983) describes macro-bedforms deposited during high stage flow under bank-full flood conditions within the ephemeral Platte Rivers of Colorado, USA. In their simplest form such bedforms are composed of a coarsening upward sequence with apron laminae - foreset laminae -

and topset laminae (see Fig. 3). This distinctive coarsening upward sequence within a predominantly coarse grained fluvial system correlates well with the findings in the pit of Fig. 2. C^{14} radio-carbon dating of the organics at the base of this type of macroform described in the Masunga sand pit sequence are 80-100 years old.

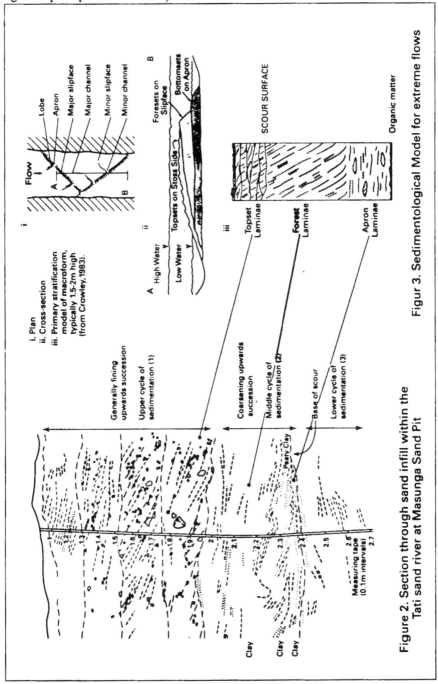

Figure 2. Section through sand infill within the Tati sand river at Masunga Sand Pit

Figur 3. Sedimentological Model for extreme flows

Thus, it is proposed the layers above the peaty clay layer were all deposited in the last 100 years or so and that any horizontal pipework below this level is unlikely to be disturbed.

Demand for supplies

The government of Botswana has two main policy themes in the water sector. Firstly to provide the entire population with reasonable access to a safe water supply and secondly to provide water supplies to facilitate the achievement of the broad objectives of rural development and employment creation. In pursuance of this, all recognised villages are now supplied with clean piped water. But with growing village population and increasing development in villages, water demand has outgrown most water sources resulting in a need for identifying new sources, rehabilitation and expansion of original schemes. The growth in population also results in new villages included in the programme.

While the above is try for villages, the remainign 25% residing in lands areas and cattle posts rely most on traditional water sources or water sources owned by syndicates or private individuals. To supply these with the same standards as villages is prohibitively expensive as the potential beneficiaries at one location are so few, hence the need to identify inexpensive and simple means of providing safe water for domestic use.

It is common practice for people to establish settlements along the main rivers as they draw water from shallow wells dug into river bed or adjacent river banks. The traditional methods of abstracting water from sand rivers tend to be temporal during dry season and the use of collector wells will go a long way in providing permanent structures if the technique proves worthwhile.

The river bed aquifers are considered to form reliable sources of water for domestic supplies and small scale irrigation schemes in eastern Botswana. This can clearly be confirmed by previous schemes which used to supply some of the major villages (Mahalapye, Tonota, Mmadinare) but are now abandoned due to availability of other sources. As a result of low yielding boreholes drilled in the basement complex, the Department of Water Affairs has diversified resource investigations of exploiting sand river storage by means of collector wells to augment water supplies in Mathangwane, Chadibe, Borolong, Masunga, Tobane and Matshelagabedi. The water demand for the above listed villages which are shown in Table 1 could possibly be met from sand river abstractions in the near future. The quality of water is generally acceptable, so the need to guard these valuable resources against pollution which may arise as a result of rapid development.

Table 1.

Village	Source (m^3/d)-1996	Demand (m^3/d)-1996	Demand (m^3/d)-2006
Chadibe	65	47	66
Borolong	102	78	110
Mathangwane	164	160	225
Masunga	370	200	350
Tobane[1]	40	64	95
Matshelagabedi[2]	40	78	110

[1] - still to connect the two collector wells

[2] - laterals still to be constructed

The groundwater system

Sand river geometry: Sand rivers present a unique set of problems to the hydrogeologist. Most techniques available for the analysis of groundwater systems are designed to cope with horizontally bedded deposits of infinite horizontal extent. A sand river is a narrow (20-100 m wide) deposit which has a natural slope to its base, which mirrors the fall of the land surface. It is thin (5-10 m), it is therefore a narrow, shallow sinuous deposit on a continuous slope. It is more like a pipe than an extensive aquifer. Furthermore, the deposits within it are variable. There are significant thicknesses of clays and silts as well as productive fine to coarse sands. In places these deposits overlay weathered basement, which are difficult to distinguish from alluvial clays and silts and it is likely the sand river s location is controlled to some degree by the faulting in the basement. The hydraulic situation is therefore complex.

Standard pumping tests: Ordinary pumping tests have been run at various locations in the middle of the sand rivers. This investigation carried out such a test at Borolong. A 2 inch diameter well was installed to bedrock in the centre of the river and observations were made of change in water level at 5 m and 10 m distance. The results were of a standard nature and transmissivity (T) was 1670 m^2/d with a 50% confidence of lying between 1535 and 808. The storage coefficient(s) was 0.16 (0.13-0.2 at 50% confidence). Similar tests run by earlier investigators Nord (1985) gave T&S values on the Shashe River (Fig. 4) as:

	T (m^2/d)	S(-)
Shashe (Sabina)	600	0.15 - 0.2
Shashe (Tonota)	2700-4900	0.03-0.04

ONGOING WORK

New testing methods used: *Pumping tests for collector wells with adits in low permeability deposits of sand rivers*

At the time of writing, the greater majority of the horizontally drilled adits do not penetrate the coarser, higher yielding deposits of the sand river. (Plans are underway to change this situation.)

Thus, almost all the adits are in relatively low permeability layers, which are in proximity to high permeability deposits. Thus, when abstraction begins from the collector wells, there will be a short period of time while head changes propogate outwards from the adit towards the nearby high permeability layers but then a quasi-steady-state will be established and the flow into the adit will be governed by the head difference between the undisturbed water level in the high permeability layers and the level in the adit. Appendix A, provides an analytical solution to this behaviour for the case of a single adit.

Pumping tests on these wells can at best only yield approximate values for the permeabilities of the low permeability layer. The main use is in a diagnostic role. Thus, if the pumping test behaves as described in Appendix A we can deduce the adits do not penetrate the higher yielding layers and the yields from the adits will be proportional to the drawdown relative to the groundwater level in the coarse deposits and will have a maximum steady yield proportional to the head difference between the adit and the groundwater level in the river.

Table A1. Drawdowns and times for a collector well with a single lateral

phase	Range of drawdowns	Drawdown formula, $s_w(t)$	Time period	Time formula
1	$s_i \leq s \leq D$	$s_Q + (s_i\text{-}s_Q)e^{-\beta t}$	$0 \leq t \leq t_1$	$t_1 = \frac{1}{\beta}\ln\left(\frac{s_i - s_Q}{D - s_Q}\right)$
2	$D \leq s \leq s_{max}$	$D + \beta\,(s_Q\text{-}D)(t\text{-}t_1)$	$t_1 \leq t \leq t_2$	$t_2 = t_1 + \frac{s_{max} - D}{\beta(s_Q - D)}$
3	$D \leq s \leq s_{max}$	$s_{max} - \beta\,D(t\text{-}t_2)$	$t_2 \leq t \leq t_3$	$t_3 = t_2 + \frac{s_{max} - D}{\beta D}$
4	$D \leq s \leq s_r$	$De^{-\theta\,(t\text{-}t_3)}$	$t_3 \leq t \leq t_4$	$t_4 = t_3 + \frac{1}{\beta}\ln\frac{s_r}{D}$

Pumping test procedure: In this situation it is recommended the collector well is pumped dry or below the level of the lowest adit. (If this cannot be done relatively quickly (<6 hours) it is likely the adits tap into the main aquifer, and can be treated as a normal pumping test on a large diameter well in a homogeneous extensive aquifer.) If the decline and rise in water level is (i) linear with time when water levels are below the adit, and (ii) exponential with time when above the adit, then it can be assumed the adits do not penetrate to the most permeable deposits of the sand river.

Point dilution tests: In sloping aquifers, the minimum amount that can be continuously abstracted at any one location is equivalent to the groundwater flow downslope at the end of the dry season. This assumes there is no interference from close-by wells upstream. For this reason point dilution tests were developed to allow estimation of this downstream flow.

Procedure: A hole is drilled to bedrock within the sand river. A tracer, sodium chloride, is dded instantaneously, and the fall in concentration of the tracer is monitored with time. Appendix B presents the theory behind this test. A plot of log $[(C-C_b)/(C_i-C_b)]$ vs linear time should be linear and the darcy velocity, v_D, is calculated using:

$$v_D = 0.9 \, D.A$$

Where C_b is natural background concentration of tracer
            C_i is start of test concentration of tracer
            C is monitored concentration of tracer
            D is hole diameter
            A is slope of plot

and v_D x cross sectional area of aquifer is equal to the downstream flow rate.

Results: Fig. 4 shows the location of the collector wells constructed to date. Point dilution tests were carried out at Masunga 1 and 2, Mathangwane and Tobane 1 and 2.

The results are summarised in Table 2 below.

Table 2. Results of Point Dilution Tests

Location	No of test holes per cross section	Total flow m^3/day
Masunga 1	1	61
Masunga 2	1	43
Mathangwane	3	101
TTobane 1	3	230
Tobane 2	3	230

It is clear, from the table and inspection of Fig. 4, that there is a correlation of downstream dry season flow, with distance downstream from river source.

Detailed results from Masunga collector well

The location of the 2 collector wells at Masunga and the Tati sand river reach between are shown on Fig. 5a. Detailed levelling of the sand river surface and the water table, the latter via a series of dug pits, was undertaken between the two Masunga collector wells and a longitudinal section along the sand river was produced (Fig. 5b). The river surface gradient was determined to be 0.005 and the river water table gradient to be 0.0013. The damming effect of the gneissic rocks that crop out upstream of the second collector well site was demonstrated.

The **Masunga No.** 1 collector well is located on the west bank of the Tati River, east of Masunga Village. Before well digging, probing of the river cross-section proved some 4 m of sandy sediment within the channel. This thickness was also predicted by the results of a hammer seismic survey. On well completion the water table in the sand river was 3.1 m below the sand infill surface whereas in the well the water level was 7.9 m below a difference of 4.8 m, indicating that the sand river aquifer is perched above the weathered basement aquifer.

Six laterals were drilled from this collector well (Fig. 6). Lateral 1 was drilled 28.5 m using a 100 mm drag bit with air flush through weathered basement, very weathered biscuit saprock and compact basement. Lateral 2 was drilled 1 m above and in the same direction as lateral 1, through clayey basement and into hard quartz bands at 21 m and 23 m before alluvial gravels. Lateral 3, drilled 1.25 m above lateral 2 in the same direction, passed through clayey basement into hard quartz at 21 m, then into alluvial gravels. Similar lithologies were met by laterals 4 and 5 drilled each side and at the same depth as lateral 3. All these laterals were dry on drilling but lateral 3 flowed during test. Lateral 6 (flowing) was drilled at 5.3 m below top of casing through 13.5 m of weathered basement into saturated alluvial sand. Laterals 3 and 6 were screened, lateral 3 to 22 m with 2" diameter 1 mm slot steel pipe, and lateral 6 with 19 m of 2" diameter pvc 1 mm slot pipe with 5 m of 2" diameter 1 mm slot steel pipe to 24 m.

Lithological data obtained from laterals and observation boreholes indicate a river channel infilled with sand down to 6 m underlain by weathered basement 2-10 m thick.

Data from a salt dilution flow test conducted upon one observation well indicated groundwater flow through the saturated basal layer, of cross section area c. 30 m^2, during the dry season to be of the order of 0.33 l/sec. The aquifer has a permeability of c. 700 m/day and a transmissivity of 700 to 1000 m^2/day. The results of this test are shown in Fig. 7.

Two yield tests were undertaken.

1.   On 8/2/95, after drilling of laterals, the collector-well was pumped at 10 l/s for one hour for a drawdown of 6.4 m, 36 m^3 of water being pumped. More importantly, recovery of 75% of drawdown took only 75 minutes.

2.   On 15/8/95 the collector-well was pumped at 4-3.25l/sec for 154 minutes for a drawdown of 4.65 m. 32.6 m^3 of water were pumped of which 14.6 m^3 came from storage and 18 m^3 as inflow from the aquifer. Recovery of 75% of drawdown took 3.5 hours, this being relatively slow compared to the earlier test.

The Masunga collector well has been used for public water supply since mid June. Fig. 8 shows the amount pumped each day for the first 10 months. Initially the well yields c. 100 m³/day. This eventually fell to c. 50 m³/day which mirrors the groundwater flow downstream estimated from the salt dilution test. Following recharge of the system during late November pumping rates exceeded 200 m³/day. These have since declined to about 100 m³/day by the end of March.

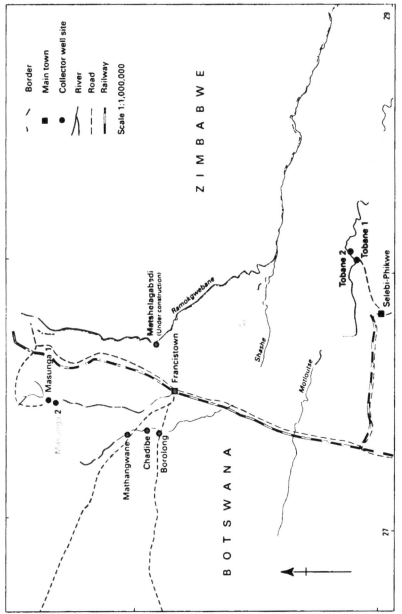

Figure 4. Collector well sites, location map.

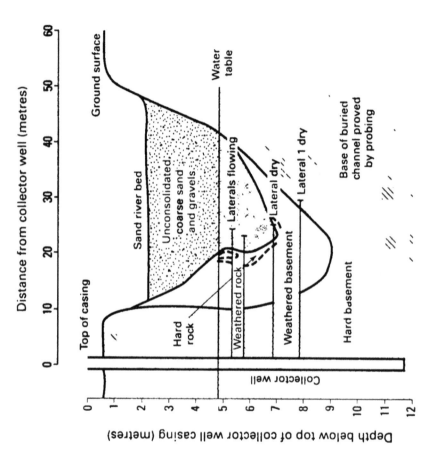

Figure 6. Buried channel cross-section at Masunga collector well 1 site.

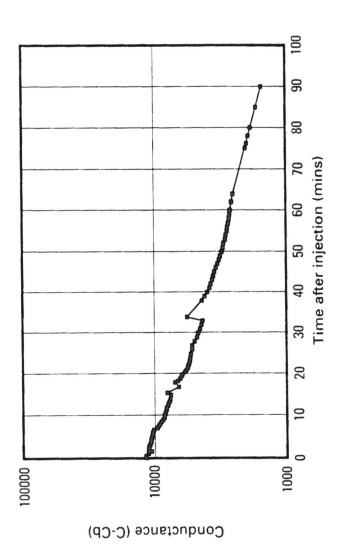

Figure 7. Masunga Collector Well 1 - Salt Dilution Test.

The achievements at 7 sites

As part of an ongoing village water supply project of the Government of Botswana seven 2 m diameter wells, were sunk by the Department of Water Affairs (DWA) (Fig. 4). One other site is under construction. All seven completed wells were converted into collector wells by the drilling of lateral boreholes into the saturated basal sands of adjacent sand-river channels.

During this programme the following were undertaken:-

1. The depth to bedrock across a river section at each site was determine using probing equipment.
2. 2 m diameter wells were dug to below the base of the adjacent sand-river channel.
3. Yield tests were run before and after conversion of wells to collector wells.
4. Laterals were drilled   into three target horizons: the weathered rock below the channel base, the coarse sediments at the river base and the main sands of the sand river.
5. Some seismic refraction surveys were run.
6. A number of salt dilution tests were run to determine the presence or absence of groundwater flow through sand rivers at the end of the dry season, to quantify the rate of flow, and derive aquifer parameters.
7. Pumping tests to determine inflow rates into collector wells, and determine possible well-pumping regimes as well as determine aquifer parameters from drawdown data measured in observation boreholes.
8. Survey of water table and sand surface levels along a sand river channel, to determine aquifer parameters from drawdown data measured in observation boreholes.
9. Sediment samples were obtained for grain size analysis from various sand pits, to determine modes of sediment deposition.
10. C^{14} dating of organics to determine age of shallow sediment deposition.

These results are published in a series of internal reports written for the DWA (see Davies *et al*, 1995).

Collector well performance: The best way to interpret the behaviour of large diameter wells with high storage in low permeability strata is to compare rates of recovery of well water level after pumping. This is because normal rates of pumping greatly exceed the rates at which the aquifer can replenish the well during the drawdown phase. It is only during the recovery phase that the contribution of the aquifer can be accurately determined.

The recovery tests on the seven collector wells are mixed, however they can be interpreted qualitatively as follows:

·   At Borolong the shaft was dry before drilling. Afterwards, the collector well was pumped dry and 50% recovery to the undisturbed well water level was achieved in only 70 minutes. It is clear this well could be pumped dry and fully recover more than once during a single day.

·   At Chadibe the shaft was dry before drilling. Afterwards laterals were drilled into weathered bedrock and recovery of well water levels was achieved in 90 minutes. Again, a good result.

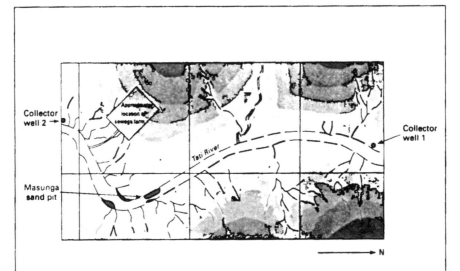

Figure 5a. Plan of Masunga Collector Wells.

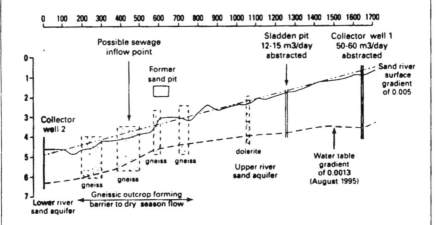

Figure 5b. The Tati River between collector wells 1 and 2 at Masunga

- At Tobane (1 and 2) the shafts were filled with water. After drilling the rates of recovery increased in both cases. At Tobane 1, 75% recovery took 700 minutes before drilling and 450 minutes afterwards. At Tobane 2, 10% recovery took 790 minutes before and 70 minutes after drilling. Recovery is generally slow at Tobane and deeper penetration of the laterals into the sand river is desirable.

- At Masunga the results were different. The rates of recovery fall after drilling. However, the post drilling test was carried out well into the dry season. It is thought continuous abstraction at the collector well site had caused a significant recession in background levels and hence caused a reduction in yield and rate of recovery. The main text records the results for Masunga 1 and the results were similar for Masunga 2 where 75% recovery defore drilling (12/2/95) took 158 minutes and afterwards (12/8/95) took 175 minutes.

Similar results to those at Borolong were obtained at Mathangwane.

In summary it is clear that collector wells work. They can meet a different set of drilling conditions and careful site investigation of the sand river prior to drilling will help to set drilling targets more precisely.

THE FUTURE

Eight collector wells have been constructed prior to June 1996 and sustainable yields of 40 to 230 m³/day are attainable. Improvements are still possible, however, and suggestions for the future are given below.

Site investigation
In order to determine the maximum yields at each collector well site, the geometry and lithology of the sand river must be determined as the collector well adits must be drilled into the most water-productive layers of the sand river. The DWA utilise a penetration probe and sampling device which is adequate for this purpose. A full cross-section with sieving analysis to determine sand properties would take up to 3 days work at any one section. Ground penetrating radar or perhaps automatic resistivity imagery techniques could probably do the same job quicker, in more detail and more cheaply. This option could be investigated.

Drilling
The adits are usually constructed by drilling through the weathered basement rocks of the sand river bank and out into the sand river deposits, which may be clays, silts or sands. Drilling through weathered basement and into semi-consolidated beds like silty sands is no problem, air hammer or rotary drilling can be used. To date success in drilling into the unconsolidated material has been mixed. Some success has been had using telescoped jetting. However, further work is needed to fully determine the capabilities of the many different drilling methods. Two new drill strings are being considered. A special GRP screened well strong enough to be jetted directly in place and a Duplex system consisting of PVC screen emplaced within temporary steel casing are both strong candidates for better results.

Determining sustainable yields by modelling studies

The sustainable yield of sand river system will depend on:

- the recharge it receives and its distribution in time
- the geometry of the sand river deposits
- the geometry of the collector well, shaft and adits
- the hydraulic properties of the sand river
- the upstream use of the sand river groundwater resources

It is recommended a generalised digital model be constructed of the groundwater system of a sand river. A subroutine for simulating different patterns of recharge should be developed. The model could be used to examine the issues, which decide the long term sustainable yield and the factors which affect efficient design of the collector well adits. The model will need to be proven against the observations of long term behaviour of an existing collector-well site, like Masunga.

REFERENCES

1. G.E. Williams. Flood deposits of the sand-bed ephemeral streams of central Australia. Sedimentology, vol 17, pp 1-40 (1971).

2. D. Koltz, h. Moser and w. Rauert. Model tests to study groundwater flows using radioisotopes and dye tracers. In: Fundamentals of Transport Phenomena in Porous Media, IAHR, 356-367. Amsterdam: Elsevier (1972).

3. E.G. Thomas and L.W. Hyde. Water storage in the sand rivers of Eastern Botswana with particular reference to storage in the Mahalapswe River. UNDP/SF/FAO Project Bot. 1, Tech Note 33 (1972).

4. M.D. Picard and L.R. High. Sedimentary Structures of Ephemeral Streams. Developments in Sedimentology 17. Elsevier (1973).

5. T. Wikner. Sand Rivers of Botswana. Results from Phase 1 of the Sand Rivers Project sponsored by the Swedish International Development Authority (SIDA) (1980).

6. K.D. Crowley. Large scale bed configurations (macroforms), Platte River Basin, Colorado and Nebraska: Primary structures and formative processes. Geological Society of America Bulletin, vol. 94, pp 117-133 (1983).

7. M. Nord. Sand rivers of Botswana. Results from Phase 2 of the Sand Rivers Project sponsored by the Swedish International Development Authority (SIDA) (1985).

8. S.J. Allen. Construction and testing of two collector wells at Tampin, Malaysia, April-July 1988. BGS Technical Report WD/88/20, Wallingford, UK (1988).

9. R. Herbert and p. Rastall P. Development of a horizontal drilling rig for alluvial aquifers of high permeability. ODA/BGS R&D Project (91/7): Final Report on Work in Zimbabwe. BGS Technical Report WD/91/50 (1991).

10. B.L. Morris, J.C. Talbot and D.M.J. Macdonald. Radial collector wells in alluvium project: Final Report (No. 3) on trenchless moling trials at Carmer Wood, Laughton, Lincolnshire. BGS Technical Report WD/91/69 (1991).

11. R. Herbert. Final report on ODA/BGS R&D Project 91/7: Development of Horizontal Drilling Rig for Alluvial Aquifers of High Permeability. Work on Sand Rivers in Botswana and Summary of work in Malaysia, Zimbabwe and UK. British Geological Survey Technical Report WD/92/34 (1992).

12. J. Davies, R. Herbert and P. Rastall. Collector well systems for sand rivers: Report on Visit to Botswana, 4 August - 10 September 1995. Technical report WD/95/53C, Overseas Geology Series, British Geological

APPENDIX A: COLLECTOR WELLS TAPPING SAND RIVERS: A quasi-steady-state approach to analysis

Consider the situation depicted in Fig. A1 where a collector well is constructed   with a single lateral extending into a sand river.   A simple model is presented here for analysing the water level data from the collector well in terms of the hydraulics of the lateral.

The assumptions made in this analysis are:-
1)   The water table in the sand river is negligibly affected by the flow to the lateral and into the well.
2)   Flow to the lateral is proportional to the head difference between the lateral and the water table.
3)   The head loss along the lateral is negligible.
4)   When the water level in the well falls below the lateral, the lateral remains saturated and at a head equal to its elevation at the well.
5)   There is no direct flow into the well.

Pumping starts and the water level in the well falls to the lateral (Phase 1) and beyond (Phase 2). Pumping then ceases and the water level recovers to the elevation of the lateral (Phase 3) and above that level (Phase 4).   The drawdowns and times for these phases are given in Table A1. Where:

s_w is the drawdown in the well
s_i is the initial drawdown in the well
s_{max} is the maximum drawdown achieved

$$s_Q = Q_p / \left(\pi r_w^2 \beta \right),$$ the drawdown at which pumping, Q_p,, balances flow from the lateral

β is an empirical   parameter characterising the the lateral
D is the depth of the lateral below the water table.

Analysis of pumping tests in the well can be used to determine β from the drawdown formulae in Table A1.   In a homogeneous system with a horizontal water table and uniform hydraulic conductivity K,   the parameter β would be expected (from application of Darcy's law) to be related to other parameters through:

$$\pi r_w^2 \beta = \frac{\gamma L K}{\ln(2D / R)} \tag{A1}$$

where L and R are the length of penetration of the lateral and its radius, respectively.   The parameter   is a geometrical factor which probably has a value in the range 1 to 2.

If the above scenario is regarded as a period of normal well use, the total time for recovery to drawdown s_f is

$$t_{rec} = \frac{S_{max} - D}{\beta D} + \frac{1}{\beta} \ln \frac{S_f}{D} \tag{A2}$$

For several laterals at the same depth, the effective total value should be the sum of the values of the separate laterals provided the laterals do not interfere hydraulically with one another (this is likely to be the case if they are closer to the water table than to one another). If there are laterals at several depths, the number of phases of behaviour will be greater than the four described above. The data will be amenable to analysis by very similar means but the effects of leakage from all of the laterals will need to be considered simultaneously.

Once the lateral parameters have been determined, the drawdown and recovery of the water level in the well can be predicted for alternative pumping regimes and water table elevations. However, it cannot be assumed that the model will remain valid or the lateral parameters will not change as the water-table elevation varies.

Equation (A1) might be used to estimate how lateral performance would vary with changes in length or diameter. Complications arise, however, because of the heterogeneity of sand rivers which will change the effectiveness of a lateral.

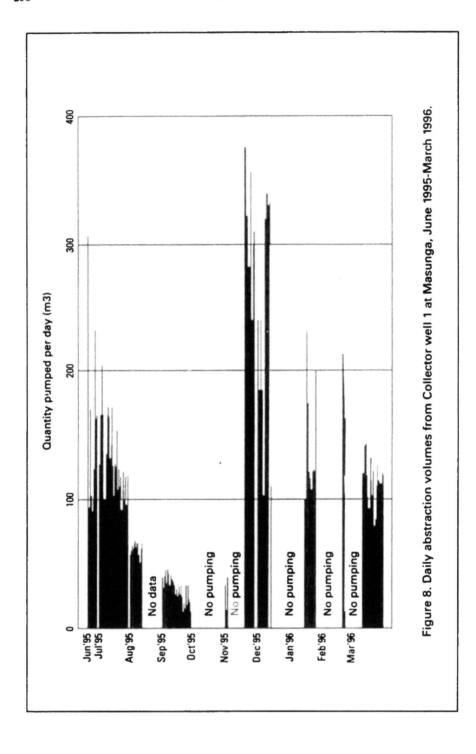

Figure 8. Daily abstraction volumes from Collector well 1 at Masunga, June 1995-March 1996.

APPENDIX B: POINT DILUTION METHOD

The point dilution method is used to estimate groundwater flow velocities.   A tracer, such as salt, is injected into a borehole and the concentration within the borehole measured over time.   Water flowing into the open section of the borehole will dilute the water within the borehole giving a falling tracer concentration; the rate of fall is used to determine the groundwater velocity.

The following assumptions are made:-
·   The concentration within the borehole remains uniform and equal to the concentration leaving the borehole.
·   The concentration at time zero is instantaneously raised to c_i .
·   Water enters the borehole from an aquifer thickness equal to the screened length of the borehole (i.e. there is no vertical flow in the aquifer).
·   Water upstream of the borehole is at a uniform background concentration of c_b .
·   The flow is steady-state.

It follows that the flow velocity, v, *is given by*:

$$v = \frac{\pi D L_{sat}}{4 n_e L_{scm} \alpha t} \ln\left(\frac{c - c_b}{c_i - c_b}\right)$$ (B1)

where:
D is the borehole diameter
L_{sat} is the saturated depth of the borehole
c_{sa} is the time varying borehole concentration
c_b is the background aquifer concentration of the tracer (often zero)
c_{ib} is the initial concentration in the borehole (at $t=0$)
n_e is the effective (or kinematic) porosity
L_{scm} is the open length of the borehole (often the screened length)
α_s is the ratio of the width of the aquifer contributing flow to the borehole to the borehole diameter (see Fig. B.1)
t is time since the injection of the tracer.

The quantity   can have a value anywhere in the range 0 to 8 and depends on: the hydraulic characteristics of the aquifer, the gravel pack (if any) and the well screen (Klotz *et al.* 1972). When there is no gravel pack $\alpha = 2$, which will suffice in many cases.

Field data are often plotted on semi-logarithmic paper: the aquifer flow velocity is determined from the slope, A, of the plot of $\log[(c - c_b)/(c_i - c_b)]$ versus t :

$$v = \frac{\ln(10) \pi D L_{sat} A}{4 n_e L_{scm} \alpha} \approx \frac{1.8 D L_{sat} A}{4 n_e L_{scm} \alpha}$$ (B2)

Proc. 30th Int'l Geol. Congr., Vol.22, pp. 258-269
Fei Jin and N.C. Krothe (Eds.)
© VSP 1997

The Groundwater Assessment for the Western Jamahiriya System Wellfield, Libya

J.W. LLOYD, *University of Birmingham, UK*. A. BINSARITI, *University of Benghazi, Libya*. O. SALEM AND A. EL SUNNI, *General Water Authority, Libya*. A.S. KWAIRI, *Great Man-Made River Project, Libya*. G. PIZZI, *Geomath, Italy and* H. MOORWOOD, *Brown and Root, UK*.

Abstract

Cambro-Ordovician sandstones in western Libya form an extensive groundwater system. Studies of the groundwater distribution and hydraulic characteristics using numerical modelling techniques show that it is feasible to make major abstractions from unconfined areas.

Keywords: aridity, groundwater abstraction, specific yield.

INTRODUCTION

In the regional sedimentary basins of Saharan North Africa good quality groundwaters occur related to pluvial recharge preceding the present aridity. In Libya and Egypt these groundwaters are utilised for *in situ* agriculture, but unfortunately, because of the hard climatic conditions, poor soil quality and their relative remoteness such developments are not proving socio-economically attractive.

To utilise the resources the Libyan Government in 1980 conceived the Great Man-made River Project (GMRP). In the project groundwater is to be conveyed from the interior desert areas to coastal areas, thus avoiding the difficulties of *in situ* development, but taking advantage of the coastal area infrastructure, population centres and better quality farm land. Additionally, to the agricultural use, the project will provide essential water for a number of population centres.

The layout for the project is shown in Fig. 1 and consists of three phases. Phase I, the Sarir-Sirt and Tazerbo-Benghazi system is largely completed with water now being conveyed to Benghazi from Sarir. Phase II, known locally as the Western Jamhariya System (WJS), is under construction and is the subject of this paper. Phase III, which will abstract water from the Kufra Basin, is in the investigatory stage. The anticipated groundwater abstractions for the GMRP are 2.10^6 m^3/day each for Phases I and II and $1.68 \ 10^6$ m^3/day for Phase III.

The hydrogeological assessment carried out for the proposed $2 \ 10^6$ m^3/day Phase II abstraction to the south of Tripoli is described below. The area, of 864,000 km^2, is shown on Fig. 2.

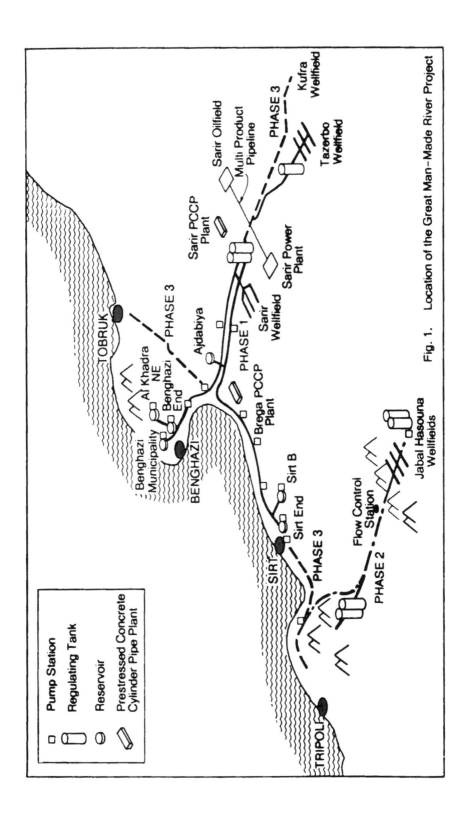

Fig. 1.   Location of the Great Man-Made River Project

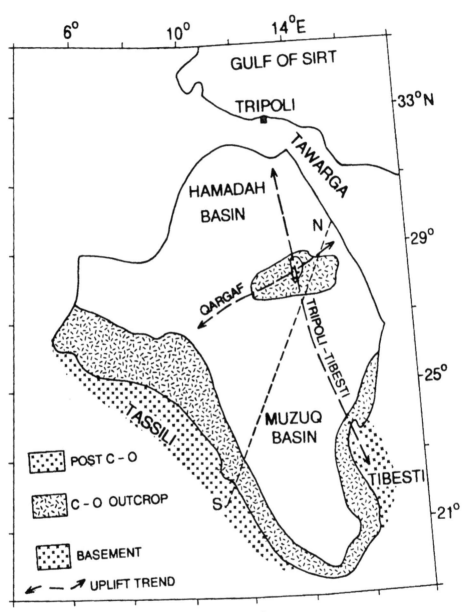

Fig. 2. Regional geology

GEOLOGY

The geological succession is given in Table 1 with the distribution, pertinent to the eventual wellfield area discussed below, shown in simplified form on Fig. 2.

Cambro-Ordovician sandstones define the limits of the study area and underlie the two sedimentary basins of which it is comprised; the Murzuq Basin in the south and the Hamadah Basin in the north. Only in a small central area, where the basement outcrops at Jabal Hasouna, are Cambro-Ordovician units absent. The lithologies range from coarse to fine grained sandstones with minor siltstones. The sandstones are variably, poorly to well cemented, with quartzite layers and friable horizons.

Immediately overlying the sandstones, Palaeozoic shales with interbedded sandstones are present (Table 1). In the Murzuq Basin the Palaeozoic sequence is overlain by Jurassic to Cretaceous sandstones. Northwards, in the Hamadah Basin, the sandstone sequences give way to laterally equivalent mudstones, limestones and sandstones. Pleistocene to Recent valley alluvium, sand dunes and sebkhas are present.

The Cambro-Ordovician rocks were structurally disturbed by the Hercynian orogony which imparted a significant NNW-SSE discontinuity trend comprising the Tripoli-Tibesti uplift and many small scale faults. Late Palaeozoic to Mesozoic movements imparted an ENE-WSW discontinuity trend comprising the Al Qargaf uplift with associated faulting. The latter uplift initiated the definition of the Murzuq and Hamadah Basins and influenced their subsequent sedimentation. Locally, Tertiary volcanism has occurred related to the main structural trends.

HYDROGEOLOGY

Aquifers

Exploitable groundwater has been identified [1,2] in three main groups of rocks (Table 1).

The GMRP investigations have concentrated upon the Cambro-Ordovician sandstones, although, because of erosion processes at various stages through to the Tertiary, there are some local hydraulic connections between the Cambro-Ordovician sandstones and aquifers higher in the sequence. An indication of this is given in Fig 3. The characteristics of the higher aquifers are outside of the scope of this paper, although they were accounted for as part of the overall study.

Groundwater Head

The Cambro-Ordovician sandstones form the most extensive continuous aquifer unit. As seen on Fig. 4 the groundwater head distribution shows mound conditions in the Tassili-Tibesti outcrop areas, where unconfined conditions exist, with flows north-eastwards towards the Tawarga sebkha, north-westwards into Algeria and south-eastwards into Niger. Away from the Tassili-Tibesti outcrops the aquifer is confined, apart from the area surrounding Jabal Hasouna where the structural uplift has raised the elevation of the sandstones and their subsequent exposure has resulted in an unconfined window.

Table 1. Hydrogeological Succession

Age	Lithology	Aquifer
Recent- Pleistocene	Sand Dunes Sebkhas Basalts	-
Palaeocene- U. Cretaceous	Limestones	Mizda
U. -L. Cretaceous	Marls with Gypsum	
L. Cretaceous - Triassic	Sandstones Limestones	Murzuq\ Kikla
Carboniferous -Silurian	Shales with Sandstones	
Ordovician- -Cambrian	Sandstones	Palaeozoic

Table 2. Hydraulic Properties
Hydraulic Parameter Ranges from
Pumping Tests

Transmissivity
           500-2500 m^2/day
Storage $2.2\ 10^{-5}$ - $6.4\ 10^{-2}$
           (confined) (unconfined)

Parameter Ranges from
Laboratory Tests

       Intergranular Permeability
           1-4000 mD
       Intergranular Porosity
           $3.0\ 10^{-2}$ - $2.6\ 10^{-1}$
       Intergranular Specific Yield
           $6.0\ 10^{-3}$ - $1.4\ 10^{-1}$

Table 3. Steady State Flow Balance
(10^6 m^3 year)

Tassili Depletion	+ 389
Jabal Hasouna Depletion	+ 26
Murzuq Inflow	+ 5
INFLOW TOTAL	+ 420
Sebkha Losses	- 74
Western Boundary Flows	- 142
South Eastern Boundary Flows	- 117
Losses to Higher Aquifers	- 34
North Eastern Outflow	- 53
OUTFLOW TOTAL	- 420

Table 4. Project Details

4m Dia. Conveyance Pipeline	799km
0.3m to 3.6m Dia. Collector Pipelines	724km
Transmission & Distribion Line System (220KV/66KV/ 33KV) and Substations	1645km
Permanent Control and Communication System	
Degassing Plants	3
Pumping Stations	3
Regulating Tanks	2
Flow Control Stations	2
Wells	440 + 10% standby
Minimum Well Spacing	1.5km
Well Yields (Modelled)	45-56l/s
Total Abstraction	$2.10\ ^6 m^3$/day
Projected Life	50 years

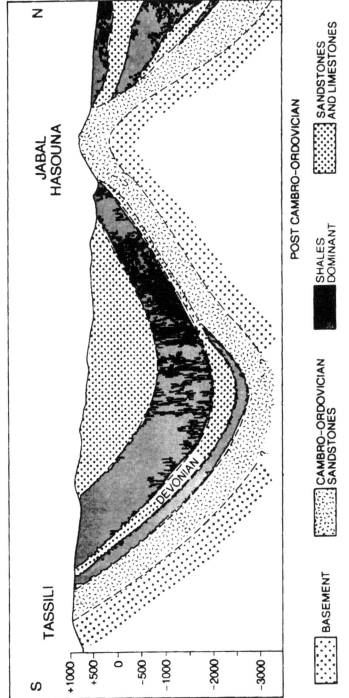

Fig. 3. Geological section (location on Fig. 2) Elevations m O.D.

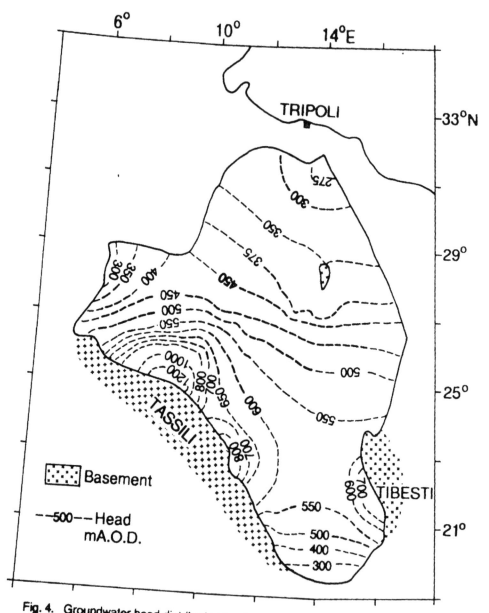

Fig. 4. Groundwater head distribution for the Cambro-Ordovician aquifer

To the immediate south of Jabal Hasouna sebkha discharges occur where groundwater flows emerge moving from the Tassili. In the north-east some of the groundwater flows into the large coastal sebkha of the Tawarga but some moves upwards into the overlying Kikla aquifer through a structural dislocation.

Aquifer Parameters

The parameter data for the study were obtained from both water wells and oil wells. Pumping tests were carried out on the water wells using observation wells, and cores for laboratory testing were obtained from the unconfined area of Jabal Hasouna. The range of parameters determined from the various tests are shown in Table 2 and indicate a mixed porosity domain.

Hydrogeological Concepts

The main Tassili-Tibesti outcrop area lies in the central Saharan region where rainfall is extremely sparse and sporadic. Recharge, significant in resources terms, is not believed to occur so that the groundwater gradients away from the unconfined mound are considered to be artefacts of past pluvial events maintained by groundwater storage (specific yield) depletion. In consequence, current groundwater throughflow must be small in relation to the aquifer extent. Because of the perceived small throughflow, abstraction development was targeted at the unconfined area in the sandstones adjacent to Jabal Hasouna in order that the long-term abstraction could utilize the specific yield of the aquifer. Access constraints and limited data dictated that any development should be concentrated to the east and north-east of Jabal Hasouna.

Regional Groundwater Modelling

Conventional groundwater modelling was adopted to assess the potential resources. A steady state calibration followed by a transient calibration covering 1970-1990 was carried out. Various abstraction scenarios were simulated for a fifty year period.

A quasi three-dimensional finite element model [3] was used consisting of 12048 aquifer elements for the three aquifers identified (see above). As appropriate, 10536 linear elements were used to connect aquifer elements vertically between aquifers. For the potential wellfield areas, a 6.5 km^2 element size was used with up to 20 km^2 elements in the overall area of the wellfields. The grid automatically generated from these areas to the model boundaries.

The steady state groundwater flow balance obtained from the modelling is given in Table 3 and confirms the small throughflow of the system relative to the required abstraction. The transient calibration was made using historical head data, from abstractions in the confined area chiefly to the south and north east of Jabal Hasouna. An example of the calibration is shown on Fig. 5.

Wellfield Selection

Adopting the principle of utilising specific yield the unconfined area east of Jabal Hasouna shown on Fig. 6 was examined through abstraction simulation scenarios, initially by simulating elemental area abstraction. Abstractions were also extended into the confined area to the immediate north of the unconfined area because of the economic advantage of reducing the length of the conveyance and of locating wells close to the conveyance line. Following acceptable long-term drawdown results, individual well abstractions were simulated within the model. 440 wells pumping between 45 and 56 l/s were modelled, as shown on Fig. 6, with drawdowns of up to 90 m in the unconfined area and 210 m in the confined area predicted. An arbitrary 1.5 km minimum well spacing was selected with 10 km between lines.

In the simulations drawdown was found to be sensitive to regional specific yield, a parameter which cannot be satisfactorily determined for major wellfield abstraction from laboratory tests of individual well pumping tests. In the absence of any significant historical abstractions in the unconfined area, which would have helped to define the parameter, the specific yield was examined using confined area historical pumping data. Head measurements in the unconfined area showed that no recognisable variations had occurred over the period 1980-90 so that it was concluded that the specific yield in the unconfined area must be sufficient to offset any impact of confined area abstraction. Using this zero impact criterion, the limiting specific yield value obtained was 0.05 which has been taken to represent the minimum regional value for the unconfined area.

To provide a final wellfield distribution design an hydraulics model and cost program were linked to the hydrogeological model accounting for well locations determined by local topography, collector line distributions, power requirements and hydraulic aspects of the conveyance line. The overall project details are shown in Table 4.

Hydrochemistry

As would be expected for such a large study area the hydrochemistry is very varied [4]. Much of the water, even in the outcrop areas, is thought to be old (>4000 years). Calculations from the regional model suggest that groundwaters entering the Jabal Hasouna area from the Tassili may be up to 0.5 million years old.

In the wellfield area the hydrochemistry has been complicated by the mixing of flows from the Tassili, local recharge in the pluvials and some salinity inflow thought to occur from Devonion sandstones at the southern confined-unconfined interface (see Fig. 3).

The bulk water quality for the conveyance is shown in Table 5, assessed from pumped samples with account taken for CO_2 degassing. The water is generally satisfactory for agricultural purposes but may require some nitrate treatment for potable use.

Different water qualities have been found within the wellfield area. As a result of microbiological array measurements and corrosivity meter readings, corrosivity coupons of various metals were installed in selected wells at three depths. The results have shown potential metal corrosion. Glass reinforced plastic has been chosen for the well casing and stainless steel 318 for riser pipe.

Table 5. Predicted Quality of Water to be Conveyed

Ca	77 mg/l	Mg	32 mg/l
Na	198 mg/l	K	9 mg/l
HCO_3	58 mg/l	SO_4	220 mg/l
Cl	346 mg/l	NO_3	54 mg/l
TDS	994 mg/l	EC	1.55 μS/cm
pH	8.2		

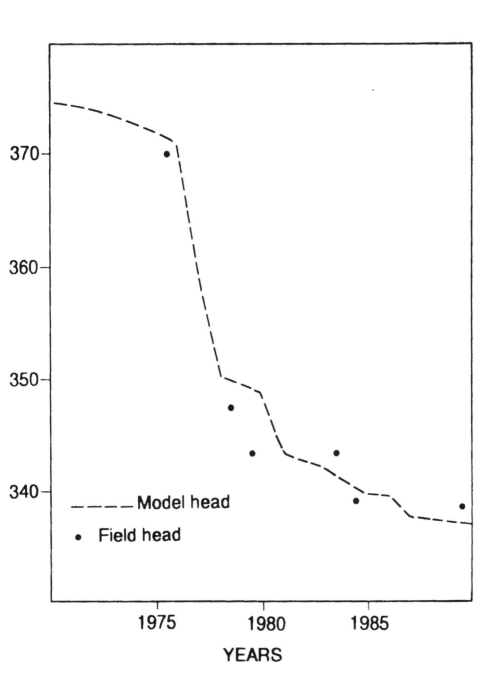

Fig. 5 Example of calibration for transient model

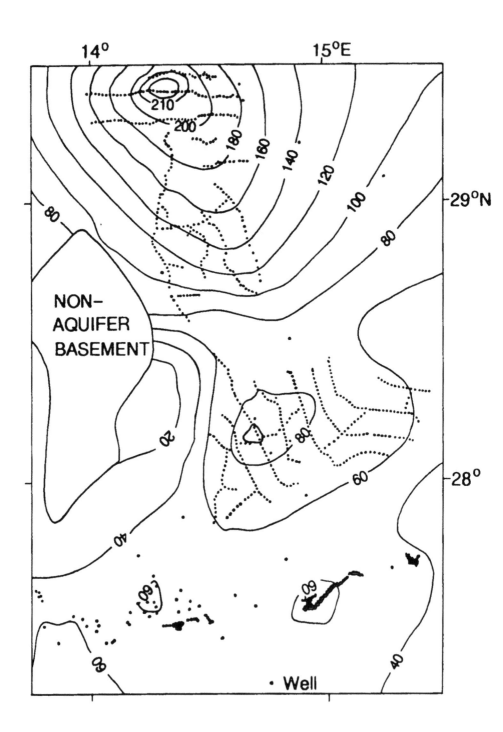

Fig. 6 Modelled 50 year drawdown (m) for the optimal wellfield (lines of wells) in Jebel Hasouna area

CONCLUSIONS

The hydrogeological assessment shows that, in regional terms, the Cambro-Ordovician sandstone aquifer can support a 50 year abstraction of $2 \ 10^6$ m^3/day from the area. east and north east of Jabal Hasouna. Drawdown is acceptable and the scheme is practical in its hydraulics. The groundwater quality is generally suitable for irrigation and for potable use with minor treatment. Materials of construction and degassing requirements have been selected to reflect the aggressivity of the water.

REFERENCES

1. Pallas, P. 1980. Water resources of the Socialist People's Libyan Arab Jamahiriya. [In the Geology of Libya, Vol. II. [Salem. A. and Buswreil, P] Academic Press, London, 539-594.

2. Idrotecneco. 1982. Hydrogeological study of Wadi Ash Shati, Al Jufrah and Jabal Fezzan. Rept. to Government of Libya.

3. Pizzi, G. and Sartori, L. 1984. Interconnected groundwater system simulation (IGROSS) - Description of the system and a case history application. Jl. Hydrol., 75, 255-285.

4. Salem, O.M. and Lloyd, J.W. 1990. Groundwater chemistry of the Fezzan area of Libya. Proc. Int. Conf. Groundwater in Large Sedimentary Basins. Australian Water Resources Council, Perth, Western Australia. 10p.

Proc. 30th Int'l Geol. Congr., Vol.22, pp. 270-280
Fei Jin and N.C. Krothe (Eds.)
© VSP 1997

Optimal Allocation of Water Rsources with Quality Consideration in Kaifeng City, Henan Province, China

CHITSAZAN Manouchehr[1], LINXueyu[2]. SHU Longcang[3]
(1) *shahid Chamran University,Ahwaz,iran*
(2)(3) *Institute of Applied Hydrogeology,Changchun University of Earth ciences,China*

Abstraces

For establishing optimal allocation model of water resources,the authors analyzed that how many sources of water could be used to supply a number of users at different places and time.In Kaifeng city,the water users are municipal, industrial, agricultural demands or other uses. The water supply sources are as follow : (1)Yellow river ; (2) shallow aquifer, including potable and non potable quantities ; (3) intermedium aquifuer,including potable and non potable quantities.

The nature of water resources allocation closely parallel the "transportation or transshipment" problem. As we konw, every user has certain water quality requirement, but the quality constraints cannot be considered in transportation programming. Linear progreamming can overcome the shortcoming. For this reason, the authors have established an optimal model of water resources subject to quality constaints for the study area by using the combination technique of transportation and linear programming.

The main task of the model is to determine the amouzts Xij to be allocated over all selected user's routes so as to minimize costs. Owing to the allocation cost changes with time and other influence factors, so the relative costs of various water supply were chosen based on the cost of fresh water supply. three plans were designed for the study area. According to plan 1, Yellow river was allocated to all users, shallow aquifer was allocated to agriculture and intermedium aquifer was allocated to the industrial and domestic users. In plan 2, Yellow river and potable quantities of shallow aquifer was allocated to all users; non potable quantities of shallow aquifer to industrial and agricultural users; potable quantities of intermedium aquifer to domestic and industrial users; non potable quantities of intermedium aquifer only to industrial user. Finally in plan 3, Yellow river was allocated for domestic users, shallow aquifer for agriculture and intermedium aquifer was allocated to industrial demands.

The allocation model was solved by linear programming. The results show that from economic point of view, plan 2 is the best plan. Eventhough plan 1 has 10 percentage more total cost than plan 2, from quality point of view, plan 1 is the best plan.

If detailed data are available, it is possible to apply reuse concept in the allocation model proposed in the paper.

INTRODUCTION

Optimal allocation of water resources in a region requires an analysis of how various sources water could be used to supply a number of users at different places and times. The nature of the problem closely parallel the "transportation or transshipment" porblem of linear programming.

The trans-shipment formulation is a powerfull analytical tool in that relatively large number of alternatives can be examind effieiently and throughly to determine that combination which minimize or maximize the chosen objeetive function. However. it does suffer form some drawbacks. The most severe shortcomings of transportation techniques are:

1) Quality constraints cannot be considered in trnsprotation programing.

2) Optimization of the amount of water to be desalinated and hence blended cannot be achieved in either of transportation techniques.

Linear programing technique provide means where by both of these shortcomings can be overcome. In linear progreming both quality and quantity constraints, and any other linear constraints, can be manipulated to yield an optimun solution. For this reason, an optimal allocation model of water resources subject to quality constraints was developed for the study area that can be solved by linear programing.

PREVIOUS WORK

Considerable research and work has been directed to the concept of optimal allocation of water resources in recebt years. Characteristic of the significant work were studied by Bishop and Hendricks (1971) and Bishop et al.(1974 and 1975). work by these investigators has been extended by Pingry and Shaftel (1979) and Ocanas and Mays (1981) in terms of addressing econmics of scale and development of more efficient solution of algorithms, still lacking in considering the water quality dircectly as constraint in their models.

In this paper an application of transportation is demonstrated. with suitable modification to include water quality directly as a constraint in the allocation model, Stephenson, D (1982). The model is comprehensive and it can handle all kinds of water supply and demends with their special water quality characteristics, The model can be easily solved by linear progreming. In addition, the model can be presented graphically and it can be used as a sopport tool for decision makers involve in water resouree management. Although, in this paper only application of model for allocation of primary sources of water and imported water is demonestrated, according to needs and availability of data, the other sources of water including reused water can be easily included in the model.

SYSTEM DEFINITION

The major categories of supply and demand for water resouce system are showm in he matrix of Table 1 as sources and demands, respectively. Requirement for water, the column headings or demands are grouped into the broad sectors of municipal, industrial, agricltural demands or other usese. Each of them could be further subdivided to show specific water using entities. The water supply soerces are shown by the row headings, and each source of water is classified as: (1) primary or base supply: (2) secondary or effluent supply: or (3) supplementary or imported supply. The matrix of water supply source and user demands depicts all possible commbination, by which the total available supply may satisfy the aggregate system.

BASIC PRINCIPLES

The problem of allocating water from various sources to points of use or demands is closely related to the classical transportation problem. In the general transportation problem, a homogeneous product is available in the amounts $a_1, ..._m$ from each of m shipping origins and is required in amounts $b_1, b_2, ..._n$ by each of n shipping destination. The term X_{il} represents the amount shipped from the ith origins to the jth destination. The cost of shipping unit amount from the jth origin to the jth destination is C_{ij} and must be known for all combinations.

MODEL FORMULATION

The problem is to detetmine the amounts X_{ij} to be shipped over all routes so as to minimze costs. The mathematical statement of the transportation problem is to find values for the variables X_{ij} which minimize the total cost (TC):

$$\text{Min} \quad Tc = \sum_{i=1}^{m} \sum_{j=1}^{n} C_{ij} X_{ij} \qquad (5-1)$$

subject to the constraints:

$$\sum_{i=1}^{m} X_{ij} = b_j \qquad\qquad j = 1, 2, \cdots, n \qquad (5\text{-}2a)$$

$$\sum_{j=1}^{n} X_{ij} \le a_i \qquad\qquad i = 1, 2, \cdots, m \qquad (5\text{-}2b)$$

$$\sum_{i=1}^{m} X_{ij} a_i \le a_i b_j \qquad\qquad j = 1, 2, \cdots, n \qquad (5\text{-}2c)$$

$$X_{ij} \ge 0 \qquad\qquad (5\text{-}2d)$$

Where:
m = Total number of sources of supply
n = Total number of demands or water users
a_i = Water quantity available at source i , $i = 1, 2, \cdots$, for users
b_j = Water quantity required by demands (users)j ,$j = 1, 2, \cdots$, n
C_{ij} = The unit transportation cost form source i to demand j
a = Total dissolve solid or ion concentration of source i or demand j
X_{ij} = Water allocated from source i to demand j

Table 1 Allocation alternative for water as system

Demands Supply Sources	Domestic	Industrial	Agricultural	Recreation Wild life hydropower
Primary Supply Surface Water Groundwater	-------------------	---------- initial	allocation ------	-------------------
Secondary supply Domestic effluent	recycle reuse	sequential reuse	sequential reuse	sequential reuse
Industrial waste	sequential reuse	recycle reuse	sequential reuse	sequential reuse
Agricultural return flow	Agricultural returen flow	sequential reuse	recycle reuse	sequential reuse
Supplementary supply imported water	-------------------	----- allocation of supplementary	supply ------	-------------------
Use sector requirement	Domestic diversion requirement	Industrial diversion requirement	agricaltaural diversion requirement	mincellaneous diversion requirement

MODEL APPLICATION TO THE STUDY AREA

Geographic discription and hydrogological attributes of the study area

The model developed in the preceding section was applied to the Kaifeny city that is 767 Kilometers south of Beijin, north of Henun province. It is bunded to the south by Yellow rivers and Long Hai Rillway passes through the south of study area (Fig1). The economy and location of this city make it very important city in China.

Development of industry and agriculture also modernization of Kaifeng city has increased ground water extraction for irrigation, industry and domestic. The intense extraction of groundwater in the study area that is part of Kaifeng city has caused groundwater levels to decline.

Water resource system in the study area is very complex. Ground water resoures occurs in three layers. The top layer is uncofined (shallow aquifer). The second layer is semiconfined (intermedium aquifer) and third layer is confined (Deep aquifer). Even though shallow and intermedium aquifers are hydraulically connected to each other, each one had its own characteristic. Degradation of shallow aquifer by municipal effluent and high floride content of intermedium aquifer need special attention. Also, yellow river has high silt content.

As a result, optimal allocation of water resources for the study area should be a comprehensive one that can deal with these contradictory problems. From ground water management point of view, ground water problems can be avoided or at least minimizes by implementing approproiate groundwater management strategies. But, previous work (Chitsazan, 1995) has shown that these strategies cannot solve the problems completely. On the other hand, from water resource management point of view water resources should be allocated in the most economic pattern subject to quality and quantity constrains. Thus, optinal allocation model was chosen as a best solving methodology and was applied to the study area.

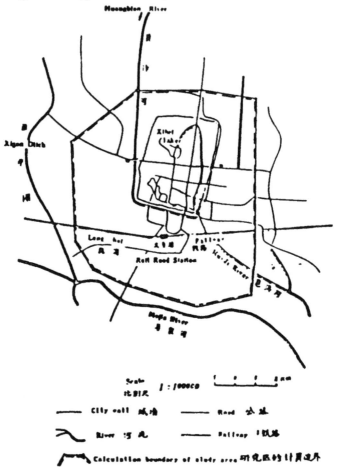

Fig 1:   Location of the study area

WATER RESOURCE AWAILABILITY AND WATER USE REQUIREMENT

The primary souces of water supply in the study area are as follow:

1) Yellow rive water, 2) shallow aquifer, potable and nonpotable quantities, 3) Intermedium aquifer, potable and nonpotable quantity. The availability of groundwater as a water supply source is based on optimal value of shallow and intermedium aquifer. Potable part of shallow and intermedium aquifer is considered that portion of groundwater that is free from pollution and ion concentrations in groundwater does not exceed the standard for drinking water.

Water use sectors for the study area are as follow: 1) Domestic demand; 2) Industrial demand, and 3) Agricultural demand supply requirements and alteratives are depicated in Fig 2. Sources of water are shown as (T, U, X, Y, Z) and they each have limited resources indicated inside the circle, first row, in terms of Megaliters perday (ml/day). Destinations water users, also are shown as L, M, N which require some quantity of water, the first row inside rectangular, in erms of ml/day.

WATER ALLOCATION COSTS

Cost functions for supplying water from available sources to various users are in terms of Yuan (RMB) per cubic meter of water. The costs of water are based on good sources but they would require further refinement if the results were to be used in actual planning and decision making rather than simply as demonstration. However, the results obtained do indicate trends and orders of magnituds. The cost of ransportation along each is indicated in Fig 2. further refinement if the results were to be used in actual planning and decision making rather than simply as demonstration. However,the results obtained do indicate trends and orders of magnitude. The cost of ransportation along each is indicated in Fig 2.

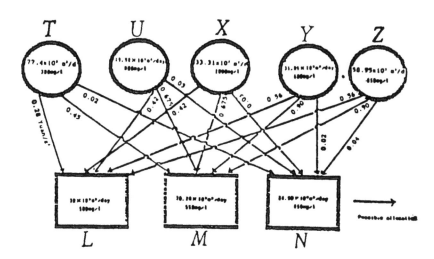

Fig . 2 Water supply requirements and alternatives

WATER QUALITY CONSIDERATION

The situation of water allocation becomes complicated by the fact that every user has certain water quality requirment. The measurements of relevant impurity e.g. TDS (total dissolved solids) is in mg/l and the requirements of L. M. and N are that the TDS shall not exceed certain quantity (Fig 2 second rows inside circles and rectangulers). The multiplication of ml/day by mg/l will be kg/day of salts, a mass flow rate. Because of data problem and time, water reuse analysis was not considered in this study. If data are available about secondary supply, muncipial effluent, industrial waste, agricultural return flow, they can be easily extended into exiting model as extra sources with their relevant quantities and qualities.

WATER RESOURCES MANAGEMENT PLANS

Three plans were designed according to water allocation possibilities. A general flow diagran of possible alternative for each plan is shown in Fig (3).

Destination	plan1			plan2			plan3		
Sources	L	M	N	L	M	N	L	M	N
T	*	*	*	*	*	*	*		
U			*	*	*	*			*
X			*		*	*			*
Y	*	*		*	*			*	
Z	*	*			*			*	

Fig . 3 Alternatives of water allocation plans   T: Yellow river   U: shallow aquifer (potable)   X: shallow aquifer (nonpotable)   Y:Intermedium aquifer   M: Industrial demand   N: Agricultural demand   * Allowed water allocation according to plans

Plan 1

According to plan 1, Yellow river is allocated to users, shallow aquifer is allocated to agricultural demand and intermedium aquifer is allocated to domestic and industrial users. The supply requirement and alternatives of plan 1, along with transportation cost, are depicted in Fig 4. The unallowed destinations for example from shallow aquifer to domestic and industrial users are eliminated by a high cost in the objective function. The optimum allocation subject to constraints is indicated in Fig 5.

Plan 2

According to plan 2: 1) Yellow river is allocated to all users, 2) potable shallow aquifer is allocated to domestic, industrial and agricaltural demands,   3) non potable shallow aquifer is allocated to industrial and agricultural demands, 4) potable intremedium aquifer is allocated to domestic and industial demands and 4) non potable intermedium aquifer is allocated to industrial demand as depicated in Fig 6. The optimal allocation of plan 2 is indicated in Fig 7.

Plan 3

According to plan 3: 1) Yellow river is allocated to domestic demand, 2) shallow aquifer,   potable and nonpotable, is allocated to agricultural demand and 3) intermedium aquifer is allocated to industrial demand as depicated in Fig 8. The optimum allocation of watew resources, according to plan 3, is indicated in Fig 9.

RESULTS ANALYSIS

Table 2 shows the surplus of water supply sources according to optimal allocation of each plan. According to optimun allocation of plan 1. all water supply of Yellow river. all potable portion of shallow aquifer and almost all potable portion of intermedium are used completly but unpotable portion of intermedium aquifer remains untouched. Also more than 86 percentage of unpotable portion of shallow aquifer remains unused. Advantages of this plan is that unpotable portion of shallow aquifer and intermedium aquifer that generally have high TDS and harmful elements are not allocated to the users, and can be used under emergeney situation in future.

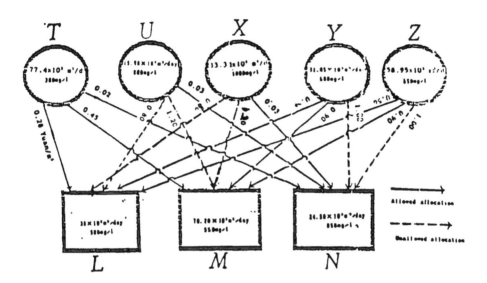

Fig 4: Water supply requirement and alternatives

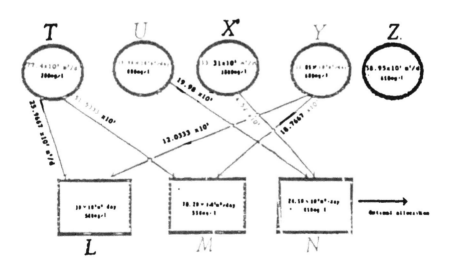

Fig 5: Water optimal allocation according to plan 1

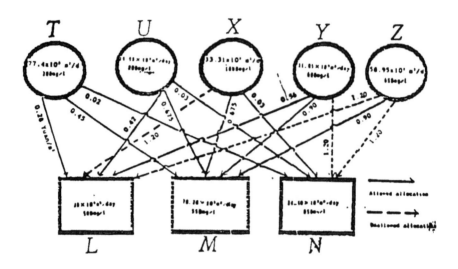

Fig 6:   Water supply requirement and alternative according to plan 2

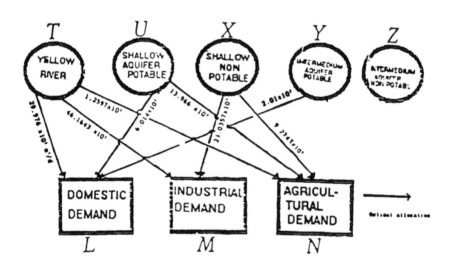

Fig 7:   Water optimal allocation according to plan 2

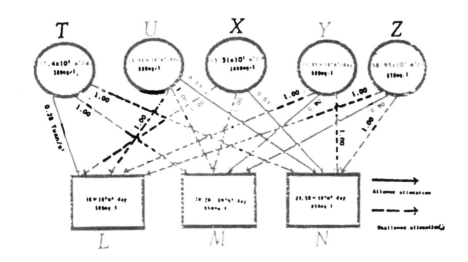

Fig 8: Water supply requirements and alternatives according to plan 3

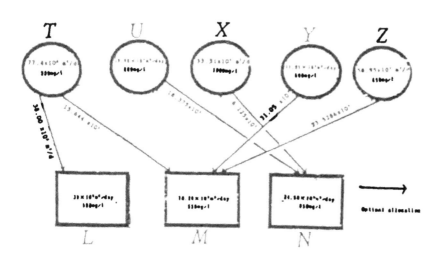

Fig 9: Water optimal allocation according to plan 3

Table 2 Surplus of water supply source according to optimal allocation (ml/day)

Plan	Source Number	T	U	X	Y	Z
	1	0	0	28.79	0.25	58.95
	2	0	0	0	29.01	58.95
	3	23.7786	1.605	27.185	0	35.12311

Note: T: Yellow river   U: shallow aquifer (potable)   X: shallow aquifer (unpotable)   Y: Intermedium aquifer
(potable)   Z: Intermedium aquifer (unpotable)

According to plan 2 all water supply of Yellow river all portions of potable and unpotable of shallow aquifer are allocated completly. Also according to this plan, more than 93 percentage of potable portion of intermedium is allocated but all unpotable portion of intermedium aquifer remains untouched. Advantage of this plan is that this plan has the lowset total cost among three plans (11765. 1990 Yuan/day) but from quality point of view. because unpotable portion of shallow aquifer is allocated to industry and agriculture is not atractive as plan 1.

According to plan 3. all potable portion of intermedium aquifer and more than 94 percentage of potable of shallow aquifer are allocated to users but 30 percentage of water supply from Yellow river, 81 percentage of unpotable protion of shallow aquifer and 60 Percentage of unpotable protion of intermedium aquifer are not allocated to the users. This plan has the highest total cost (79241.4220 Yuan/day) among three plans because the bigest users (industries)use the most expensive source (intermedium aquifer). Another cause of high total cost of plan 3 is related to the high costs along unallowed destinations. From quality point of view, plan 3 is not attractive because almost 40 percentage of unpotable portion of intermedium aquifer is allocated to the industry and more than 18 percentage of unpotable portion shallow aquifer is allocated to the agriculture.

It can concluded that from economic point of view, plan 2 is the best plan. Eventhough plan 1 has 10 percentage more total cost than plan 2, from quality point of view, plan 1 is the best plan.

REFERENCES

1. Bishop, A. B. and D. W. Hendricks 1971. Water Reuse systems analysis Journal of sanitary Engingeering Division, ASCE, Vol.97. No SAL. New York. NY.

2. Bishop, A. B. 1974. Evaluating water Reuse Alternative in water Resoureces planning. Utah state University, publication No. PRWG 123-1. Logan Utah.

3. Chitsazan, M. , optimal Regulation of water Resources and water allocation, ph.D. Theisis, chang chun university of Earth sciences, china, 1995.

4. Ocanans and W. Mays, 1981. A model for water Reuse planning water Resources Research 17 (1): 25-32.

5. Pingry, D.E. and T.L. shaftel. 1979. Integrated water
   Management with Reuse: A programing Approach, Water Resources Research Vol. 15 No.1. American Geophysical union, Washing ton D.C.

6. Stephenson, D. (1982) optimum allocation of water resources subject to quality constrnints (proceeding of symposium, July 1982) IAHS publ no 135.

Proc. 30th Int'l Geol. Congr., Vol.22, pp. 281-290
Fei Jin and N.C. Krothe (Eds.)
© VSP 1997

Annual Variation of δ¹³C Infiltrating Water through a Soil Profile in a Mantled Karst Area in Southern Indiana

George Haichao   Yu and N. C.   Krothe
Department of Geological Sciences, Indiana University, Bloomington, IN 47405, U.S.A.

Abstract

Samples of infiltrating soil water through a 30- foot (~ 10m) unsaturated and saturated section of the unconsolidated material above a Karst terrain in southern Indiana were collected via an array of suction   lysimeters   and wells to study the annual variation of stable carbon isotopes. Soil waters were collected monthly from 9 suction lysimeters installed at various depths in the soil matrix to the top of the underlying bedrock. Groundwater samples from a nearby water well installed into the top portion of the limestone aquifer (epikarst) were also collected. All samples were analyzed for stable carbon isotopes, major ions, pH, Eh, conductivity, and temperature. Depth and time variations in stable carbon isotopes through the unsaturated zone were studied.

A preliminary evaluation of the data shows a clear and consistent shift towards higher C^{13}/C^{12} ratio as soil water infiltrates. The carbon isotopic composition ranged from - 35 ‰(PDB) in soil water samples collected near the surface to approximately - 14.7 ‰ in the epikarst zone. The study also shows significantly high HCO_3^- concentrations in the deeper soil zone indicating   continued interaction between soil CO_2 and the soil water. The annual variation in C^{13}/C^{12} ratio through the thick soil zone to the epikarst zone in this study is believed to be influenced by multiple factors. Among them, the soil CO_2, organic content, available $CaCO_3$ and chemical composition of the soil are thought to be the most important.

INTRODUCTION

Identifying the changes in dissolved inorganic carbon species and associated stable carbon isotopic signature is an important step in understanding the geochemical evolution of groundwater systems. This is particularly true in karst hydrogeological settings because of the ability of infiltrating water in the mantled portion of the karst water system to recharge the karst aquifer. It is important to understand how the chemical and isotopic compositions of the soil water would change as it infiltrates and how they would affect chemical and isotopic characteristics of water entering the karst water system. This study was designed to investigate how the stable carbon isotopic composition and major ion chemistry change as water infiltrates   through the thick soil mantle to the upper part of the karst aquifer known as the   epikarst   zone. The results of this study can be used in developing geochemical reaction path models   by providing the initial chemical and isotopic composition of the recharging water.

Stable carbon isotopes are important source indicators of dissolved inorganic carbon in a karst water system. This is due to the production of δ ¹³C values that differ in karst environments depending on the hydrogeological conditions. Many case studies documented in the literature have indicated that different geological materials have different isotopic signatures due to natural processes such as dissolution, precipitation, and fractionation. Marine carbonate minerals generally have δ ^{13}C values between -3 ‰ and +3 ‰ (Carothers and Kharaka 1980, Drever , 1982) which compares to much lighter δ ^{13}C of soil- gas carbon dioxide (CO_2) between -21 ‰ and - 31 ‰ (Rightmire and anshaw, 1973). Atmospheric CO_2 is characterized by δ ^{13}C values between -7 ‰ and -8.5 ‰ (Drever , 1982, Mook , 1980)

DESCRIPTION OF THE STUDY AREA

The study area is located in the eastern portion of the Mitchell Sinkhole Plain in southern Indiana as shown in Figure 1. The Mitchell Sinkhole Plain, covering an area of 1,125 square miles (699 sq K), is a classical example of karst topography (Schneider, 1966) and is considered to be one of the best developed Karst regions in the United States (Malott , 1945). The geology, geomorphology, and karst hydrogeology of this area have been described by a number of researchers (Malott , 1922, 1945; Powell, 1966; Sunderman , 1968, Palmer and Palmer, 1975; Ruhe , 1975, Bassett , 1976; Krothe and Libra, 1981).

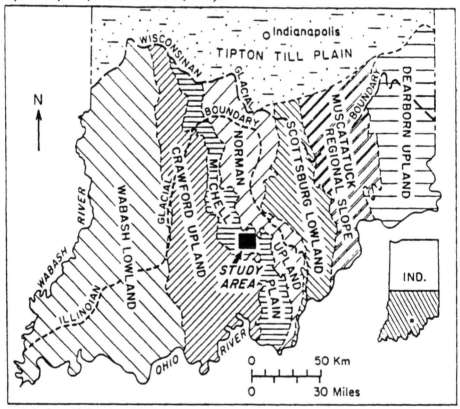

Figure 1 physiographic Location of the study Area

Developed on Mississippian carbonate rocks (Blue River Group) through open dissolution, the Mitchell Plain is characterized by two distinct hydrogeologic areas; the open sinkhole plain to the west with thin soil cover and a myriad of sinkholes and the mantled portion to the east with up to 40 feet of silt- loam soil cover formed in loess or bedrock residuum. The thickness of the soil mantle increases from west to east. The extensive sinkhole plain is developed along a gentle southwest regional dip of the carbonate bedrock and is bounded to the west by the Crawford Upland. Subsurface drainage systems under the Mitchell Plain have been defined by dye tracing tests (Murdock and Powell, 1968; Bassett , 1976). The head of the Lost River is located in the mantled part of the Mitchell Plain in western Washington County.

The lysimeter experiment site was located in the town of Campbellburg of Washington County, about 100 miles south of Indianapolis (Figure 2).

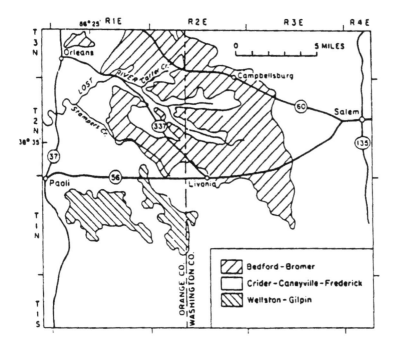

Figure 2 Major Soil Types in the Study Area

The site was selected because of the thickness of the soil cover allowing installation of lysimeters at various depths in the unsaturated zone and the relative flat topography allowing for uniform infiltration of rainwater. In addition, because the site is in the vicinity of the head of the Lost River, the chemical and isotopic compositions of the infiltrating water at the site are likely to be representative of the water sustaining the base flow of both the Lost River and the karst aquifer system, and thus provide valuable information about initial composition of recharge water in developing a reaction- path model. Therefore, this location is believed to be suitable for the purpose of this study.

The soils in the study area belong to the Bedford-Bromer group which are nearly level or gently sloping, moderately well-drained to poorly- drained silt and clay soils (Wingard , 1984). The sources of these soils include: 1) those derived from weathering of the westward retreating Crawford Upland, which is composed of clastic formations as sandstone, siltstone£, shale, limestone, and dolostone ; 2) erosional residuum of underlying carbonate rocks with limestone fragments. This type of soils predominantly covers a large portion of the Lost River drainage basin.

The soils at and adjacent to the lysimeter site are silt loams with high clay content and a thickness of approximately 30 feet. During the installation of lysimeters , soil core samples were visually inspected for texture and clay content. It is noted that both clay content and limestone rock fragments increase with depth. This observation is consistent with more detailed soil surveys conducted in the same general area (Wingard , 1984). According to the soil survey, in the top 10 feet of the profile, the material is an organic- rich brown silty clay to clay which gradually grades to a reddish brown clay with trace rock fragments. The materials in the lower part of the profile consist of reddish silty clay with increasing rock fragments. Soil pH ranges from less than 4 in the shallow part to nearly neutral in the deeper soil. Recent studies by (Iqbal , 1994, Iqbal and Krothe 1995, 1996) in the same area have shown that soil water infiltration and environmental contaminant migration (nitrate) were affected by preferential flow through macropores.

SAMPLING AND ANALYTICAL METHOD

An array of suction lysimeters were installed at various depths in a small lot (8 x 8 feet) in a flat pasture adjacent to a corn field (Figure 3). Four boreholes were drilled for installation of lysimeters . Two shallow boreholes were drilled using stainless steel hand augers with one for installation of 1 and 2 feet deep samplers and the other for 3 and 5 feet samplers. Two deep boreholes were drilled using a drill rig for installation of two groups of lysimeters One group consists of 10, 20, and 30 feet samplers and the other consists of 7.5, 15, and 25 feet samplers. After the boreholes were advanced to the desired depths, a small amount of clay slurry was poured to the bottom of the hole to seal off possible water coming from below (Iqbal , 1994).

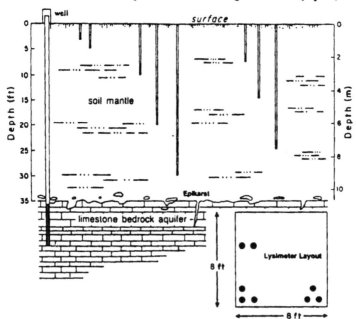

Figure 3   Lysimeter Experiment

Soil water samples were collected on a monthly basis from May 1993 through June 1994. To eliminate atmospheric interaction with soil water, pressurized inert gas (argon) was released into the lysimeter via a PVC tubing and soil water inside the lysimeter cup was driven upward through another PVC tubing which was connected to the sampling bottle as demonstrated by Figure 4.

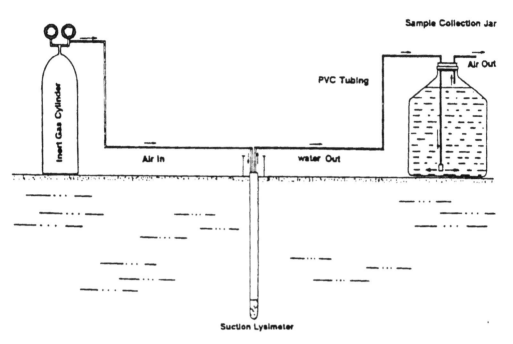

Figure 4 Schematic Diagram Shoeing Soil Water Sampling Procedures

Groundwater samples from a drinking well about 300 feet (~100 meters) south of the lysimeter site were also collected concurrently with the soil water samples. Samples from this well, which was installed at an approximate depth of 45 feet (13.7 m) below ground surface and about 15 feet (4.6m) into limestone bedrock representing the phreatic water stored in the epikarst zone.

Field collection and laboratory treatment of soil water samples for δ ^{13}C analysis generally followed the procedures described by Bishop, 1990. Water samples were collected in 500 ml amber glass bottles and about 30- 50 ml of saturated $SrCl_2$ -NH_4OH reagent solution was injected via a syringe into the bottle immediately after the sample was collected to precipitate all dissolved inorganic carbonate as $SrCO_3$. Sample bottles were then sealed with PVC tape and transported to the Hydrogeology Research Laboratory at Indiana University. The samples were allowed to set for at least 48 hours in the lab to insure a complete reaction and to eliminate possible isotopic fractionation during precipitation [Bishop, 1990]. Several precautionary steps were taken to eliminate the potential for sample contamination from atmospheric CO_2 during filtration and drying of precipitates by filtering precipitates through 5 micron membrane paper in a stream of Argon gas. The $SrCO_3$ precipitates were then placed   in a vacuum desiccator until completely dry. Dry $SrCO_3$ precipitates were scraped from the filter paper into a small air tight glass vial and stored in a desiccator until analyzed. Isotopic analyses were performed on pure CO_2 gas liberated from $SrCO_3$ precipitate by 24- hour reaction with 100 percent H_3PO_4 at 50 ℃ under vacuum condition. The isotopic ratio was determined on a Finnigan MAT 252 mass spectrometer with the analytical error estimated to be smaller than 0.05 ‰ The results quoted relative to the PDB standard. The carbon isotope data represent the bulk composition of the total dissolved inorganic carbonate and the ^{13}C /^{12}C ratio is given as in ‰ delta notation:

$$\delta\ ^{13}C(\text{‰})= \{ (R_{sample}\text{-}R_{standard})/R_{standard} \} \times 1000$$

whereand $R=^{13}C/^{12}C$, the standard is CO_2 prepared from a Cretaceous belemnite from thePeedee formation of South Carolina.

Water samples for major ion chemistry analysis were collected in 250 ml high- density white polyethylene bottles to avoid direct sunlight and were sealed and placed in a cooler packed with ice. Before sample collection, field parameters of pH, dissolved oxygen, temperature, Eh, and conductivity were measured. Water samples were brought back to the laboratory usually within 8 hours. Upon arriving in the lab, titration with 0.0200 N sulfuric acid was performed immediately to determine the bicarbonate concentration of the water sample. Samples for cation analyses were also filtered through 0.45 micron membrane and acidified to pH less than 2 with concentrated nitric acid prior to being placed in the refrigerator. Samples for anion analyses were filtered but not acidified.

Major cation concentrations including Ca, Mg, Na, K, Al, Si, Mn, Sr, and Ba were analyzed by using a Jerrel -Ash Plasma Atomcomp inductively coupled plasma atomic emissions spectrometer (ICP Model 975) in the chemical laboratory of Indiana Geological Survey. Anion concentrations including Cl, SO_4 , NO_3, F, PO_4 and Br were analyzed by the Dionex 4000I aqueous ion chromatography in the Hydrogeology Research Laboratory at Indiana University. The anion column was frequently flushed with $NaCO_3$ -$NaHCO_3$ eluant to obtainmaximum analytical precision

RESULTS AND DISCUSSION

Soil water and groundwater samples were collected between 1993 and 1994 from the lysimeter site in the mantled karst area of southern Indiana. During this one year sampling period, lysimeters were sampled monthly under natural soil hydrological conditions of varying amounts of precipitation, and hence, varying amounts of infiltration through the soil profile. Given the low permeability of the silty clay soil and the low infiltration rate, seasonal groundwater level fluctuations at this site are believed to be relatively small compared to downstream of the Lost River drainage basin. Therefore, $\delta\ ^{13}C$ values and major ion chemistry characteristic of infiltrating soil water are not significantly affected by such changes in groundwater levels.

The stable carbon isotope compositions ($\delta\ ^{13}C$)of the soil water collected at different time of the year were plotted versus depth as shown in Figure 5. The changes in dissolved HCO_3^- concentrations of the soil water collected at the same time as the carbon samples were also plotted versus depth and are presented in Figure 6. The data as shown on Figure 5 suggest that there are three distinct patterns of $\delta\ ^{13}C$ change along the 30- feet (~ 10m) soil profile, corresponding to three depth intervals: shallow, intermediate, and deep soil zones. The first pattern, recognized from the surface down to about 5 feet (~ 1.53m), is characterized by a gradual decrease of $\delta\ ^{13}C$ values from an average -25.12 ‰ feet (~ 1.0m) to an average of-29.32 ‰ at 5 feet (~ 1.53m), accompanied by a slight increasing average HCO_3^- concentrations from 76 ppm at 3 feet (~ 1.0m) to about 116.4 ppm at 5 feet (~ 1.53m). Thesecond pattern shows a rapidincrease from -29.32 ‰ to an average of -16.75 ‰ between 5 (~ 1.53m) and 10 feet (~ 3.05m). Similarly, the mean concentration of HCO_3^- this depth interval also increased from 116.4 ppm at 5 feet (~ 1.53m) to

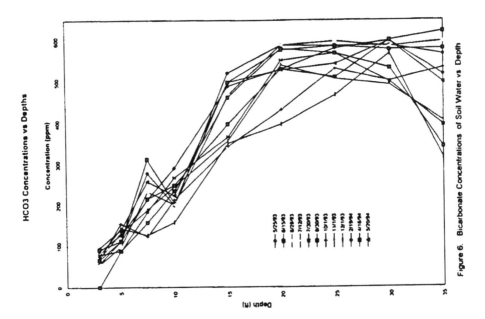

Figure 6. Bicarbonate Concentrations of Soil Water vs Depth

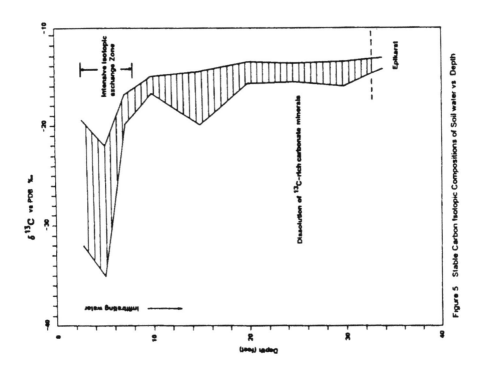

Figure 5 Stable Carbon Isotopic Compositions of Soil water vs Depth

223.6 ppm at 10 feet (~ 3.05m). The third pattern is identified by a steady and slow increase of δ ^{13}C from the average of -16.75 ‰ at 10 feet (~ 3.05m) to an average of -14.7‰ at 30 feet (~ 10m), except at depth of 15 feet (~ 4.6m) where three measurements with an average of -19.11 ‰ from the general trend. Within this depth interval, HCO_3^- concentrations increased substantially from 223.6 ppm at 10 feet (~ 3.05m) to an average of 531.6 ppm at 30 feet (~ 10m).

Shallow Soil Zone (0-5 feet, 0-1.53m).

A wide range of δ ^{13}C values was measured (-19.32 ‰ to -32.43 ‰ during the sampling period. This range is similar to the- depleted δ [13] C values characteristic of☐the soil- gascarbon dioxide (CO_2) between -21 ‰ and-31 ‰ and is probably also influenced byoxidation of the soil organic carbon with a typica δ ^{13}C lvalue of -25 ‰.The measured low HCO_3^- concentration and isotopic composition of the near surface soil water are probably modified by atmospheric exchange and photosynthesis. As water infiltrates deeper into the soil profile, intensive isotope exchange takes place between the soil water and soil CO_2 due to the elevated CO_2 partial pressure. In addition, the wide range of δ ^{13}C values observed may also be attributed to physical conditions of the shallow soil zone such as the relatively large hydraulic conductivities due to macropores which are known to exist. Less residence time for infiltrating water means isotope exchange process between soil water and soil CO 2 is unlikely to proceed to completion. Oxidation of soil organic carbon also affect the δ ^{13}C values of the soil water under aerobic conditions. There is no evidence for substantial dissolution of carbonate minerals within the shallow depth interval because of very low calcium and magnesium concentrations which indicates that most carbonate minerals have already been dissolved.

It is known that gaseous transport in the unsaturated soil zone is dominated by diffusion which allows CO_2 generated by oxidation of soil organic materials with high partial pressure to diffuse from the oxidation zone not only to the atmosphere but also into the unsaturated zone pore water (Wassenaar et al., 1990). It is likely that in the near surface soil environmental the CO_2 diffuses into the atmosphere while in the subsurface condition the diffusion process will result in dissolving CO_2 into soil water and in CO_2 build- up at depth.

Intermediate Soil Zone (5 feet , 1.53m -10 feet, 3.05m).

δ ^{13}C values of soil water collected from this depth interval showed a rapid δ ^{13}C increase from -29.32 ‰ to an average of -16.75 ‰ between 5 (~ 1.53m) and 10 feet (~ 3.05m). Similarly, the mean concentration of HCO_3^- in this depth interval also increased from 116.4 ppm (~ 1.53m) feet to 223.6 ppm at 10 feet (~ 3.05m). The increase of HCO_3^- concentration andthe sharp shift of δ ^{13}C signatures toward heavier values seem to suggest that within this depth interval soil CO_2 continues to build up and dissolution of carbonate minerals become increasingly important. At the same time, the influence from oxidation of soil organic matter is probably continuously present but its relative importance declines. The infiltration rate through the soil matrix within the soil interval is likely to be in the order of several inches per year compared to a higher rate in shallower depths, resulting in a longer residence time for isotopic exchange among soil water, soil CO_2 , organic carbon, and carbonate minerals. This is reflected by the narrow range ofvariations as soil water percolates to a deeper depth. Therefore, the observed δ ^{13}C variation in soil water is representative of a combined effect from all sources of dissolved inorganic and/or organic carbon. Given the presently available data, it is not possible to determine the specific contributions from aerobic degradation of δ [13]C- depleted organic materials, from dissolutionenriched carbonate minerals, or from the dissolution of soil CO_2 into infiltrating water in the intermediate depth interval.

Deep Soil Zone (10 feet, 3.05m-30 feet, 9.24m)

The observation that HCO_3^- concentrations increased sharply along infiltration path to an average 531.6 ppm and δ ^{13}C values become significantly heavier (30 feet, 9.24m) indicates a progressive dissolution of isotopically heavy (δ ^{13}C of -3 ‰ to +3 ‰)carbonate minerals in the presence of soil CO_2 (δ ^{13}C betweenand -21 ‰ and -31 ‰). The unusual high average HCO_3^-concentrations (513.6 ppm) is likely due to high soil CO_2 partial pressure, dissolution of carbonate material and also to the increase of diffuse infiltration from the soil matrix at deeper depth. The fact that HCO_3^- concentrations increased while δ ^{13}Cvalues of soil water indicated effect of carbonate mineral dissolution seems to implicate that even at 20 feet (6.1m) below the surface in this soil environmental the system remains open to soillt CO_2.

It is unlikely that oxidation of soil organic carbon plays a major role in changing the isotope composition of soil water at this depth interval, especially below 20 feet (6.1m) This is due to dramatic decrease in organic carbon accumulation with depth, as seen during lysimeter installation and reported in the soil survey. In fact, it has been documented that the decreasing organic carbon with depth in certain type of soils is in an exponential pattern (Arrouays and Pelissier , 1994).

SUMMARY

Data as presented and discussed above have shown that three distinct patterns eachcharacteristic of δ ^{13}C value and HCO_3^- concentrations are recognized within the 30-feet thick soil mantle covering a karstified bedrock. The shallow soil zone (0 to 5 ft, 1.53m) shows a wide variation of δ ^{13}C values (-19.32 ‰ to -32.43 ‰) and a low HCO_3^- concentration, reflecting the predominant influence of atmospheric CO_2 through precipitation/diffusion and dissolution of soil CO_2 of organic origin. No evidence was found to indicates dissolution of ^{13}C-enriched carbonate minerals within this depth. The intermediate soil zone (5, 1.53m to 10 ft, 3.05m) is characterized by a rapid increase invalues (-16.75 ‰) and HCO_3^- concentration indicating an intensive isotopic exchange and mixing among various sources of dissolved inorganic and organic carbons, especially the dissolution of isotopically heavy carbonate minerals. In the deep soil zone, δ ^{13}Csignatures of the infiltration water continue to show a steady and slow shift toward heavier values (average at -14.7 ‰), reflecting increasing control from dissolution of carbonate minerals and mixingl with dissolved soil CO_2 . An unusually high HCO_3^- concentration (averaged at 513 ppm) observed in the deeper soil depth and heavy δ ^{13}Cseem to suggest that the entire 30-feet (9.15m) soil thickness is under an open system condition.

This study shows that infiltrating water through a 30 -feet (9.15m) soil zone reaches the epikarst zone with a characteristic δ ^{13}C of about -14.7 ‰ and an average HCO_3^- concentration of 513 ppm. Because this area is believed to be part of the recharge area of the Upper Lost River drainage basin, this information is very important in modeling chemical and isotopic evolution of karst groundwater along its flow path.

Acknowledgment

This project is partially supported by the Graduate School and the Department of Geological

Sciences, Indiana University at Bloomington. We would like to thank Ms. Jane Jing Li for her assistance in field sampling and chemical analysis.

Literature Cited

1. Arrouays , D., and P. Pelissier , 1994, Modeling Carbon Storage Profiles in Temperate ForestHumic Loamy Soil of France, Soil Science, 157: 185 -192, 1994.

2. Bassett , J.L. , 1976, Hydrology and Geochemistry of the Upper Lost River Drainage Basin, Indiana. National Speleogical Society Bulletin, v. 38, no.4 October 1976.

3. Bishop, P.K. , 1990, Precipitation of Dissolved Carbonate Species from Natural Waters for δ 13C Analysis - A Critical Appraisal. Chemical Geology, v. 80, 251- 259.

4. Carothers , W. W. and Y.K. Kharaka , 1980, Stable Carbon   Isotopes of   HCO 3 - In   Oil-Field Waters-Implications for the Origin of CO2. Geochimica et Cosmochimica Acta, v.44, 323 -332

5. Drever , J. I., 1982, The Geochemistry of Natural Waters, Englewood Cliffs, Prentice -Hall, 1982.

6. Iqbal , M. Z., 1994, A Study of the Infiltration Processes Responsible for Contamination of Groundwater by Fertilizer -Derived Nitrate and other Chemical Compounds in the Karst Aquifers of Orange and Washington Counties, Indiana, Ph.D. Dissertation, Indiana University, Bloomington, 172 p., 1994.

7. Iqbal , M. Z. and N. C. Krothe , 1995, Infiltration Mechanisms Related to Agriculture Waste Transport through the Soil Mantle to Karst Aquifer of Southern Indiana, USA, Journal of Hydrology, 164, p.171 -192.

8. Iqbal ,   M. Z. and N. C. Krothe , 1996, Transport of Bromide and other Inorganic Ions by Infiltrating Storm Water beneath a Farmland Plot, Groundwater, Vol. 34, No.6, p.972- 978.

9. Krothe , N. C. and R. D. Libra, 1981, Sulfur Isotopes and   Hydrochemical   Variations in Spring   Waters of Southern Indiana, U.S.A. Journal of Hydrology. 81: p.267 -283.

10. Mallott , C. A., 1922, The Physiography of Indiana, Handbook of Indiana Geology, Indiana Department of Conservation Publication 21, p.59 -256.

11. Mallott , C.A., 1945, Significant Features of   the   Indiana Karst: Proceedings of the Indiana Academy of Sciences, v.54, p.8 -24.

12. Mook , W. G., 1980, Carbon- 14 in Hydrological Studies, in Handbook of Environmental Geochemistry, Vol.1, The Terrestrial Environmental A, edited by P. Fritz and J. C. Fontes , p 49-74, Elsevier , New York, 1980.

13. Murdock ,   S.H. and R.L. Powell, 1968, Subterrane an Drainage Routes of   Lost River, Orange County, Indiana: Indiana Academy of Sciences, Proceeding 77, p.250-255

14. Powell, R. L.,1966, Cavern Development in Northwestern Washington County, Indiana -Bloomington Indiana Grotto Newsletter, National Speleol . Soc. 6, p.33-50.

15. Palmer, M. V., and A. N. Palmer, 1975, Landform Development in the Mitchell Plain of Southern Indiana: Origin of a Partially Karsted Plain, Zeitschrift fur Geomorphologie N. F. Bd. 19, Heft 1.

16. Rightmire , C. T. and B. B. Hanshaw , 1973, Relationship between the Carbon Isotope Composition of Soil CO₂ and Dissolved Carbonate Species in Groundwater, Water Resources Research, V.9, no.4, 958-967.

17. Ruhe , 1975, Geohydrology of Karst Terrain, Lost River Watershed, Southern Indiana, Indiana University Water Resources Center, Report of Investigation, No.7, 91 pp.

18. Schneider, A. F., 1966, Physiography, in Natural Features of Indiana: Proceedings of the Indiana Academy of Sciences, Sesquicentennial Volume, p. 40 -56.

19. Sunderman , J. A., 1968, Geology and Mineral Resources of Washington County, Indiana, Indiana Department of Natural Resources, Geological Survey Bulletin, Vol.39, 90 pp., 1968.

20. Wassenaar , L. I., M. J. Hendry , R. Aravena , and P. Fritz, 1990, Organic Carbon Isotope Geochemistry of Clayey Deposits and their Associated Porewaters , Southern Alberta. Journal of Hydrology, 120 (1990) 251-270.

21. Wingard , Robert C., 1984, Soil Survey of Orange County, Indiana, Soil Conservation Service.

Proc. 30th Int'l Geol. Congr., Vol.22, pp. 291-298
Fei Jin and N.C. Krothe (Eds.)
© VSP 1997

Groundwater Protection Strategy, Policy and Management

JAROSLAV VRBA

DHV CR, Drahobejlova 48, 190 00 Prague 9, The Czech Republic

Abstract

Groundwater protection strategy, policy and management is analyzed with spatial regard to: environmentally sound development and protection of groundwater resources, technical and economic constraints within which groundwater must be protected, social policy particularly with respect to the potential impact of polluted water on human health and cultural and historical tradition of the society. The concept of groundwater protection strategy is based on a long -term multidimensional programme, managed, coordinated, subsidized and implemented by governmental authorities. The objective of groundwater protection policy is timely identification and analysis of potential conflicting issues and constraints, formulation, quantification, analysis and validation of competitive factors, hierarchical screening with the aim of finding a balance between groundwater protection, economic development and social and health implications with a view to the short term and long term prospects. The objectives of groundwater protection management is to ensure the quality, safety and sustainability of groundwater as a drinking water source and a valuable component of the environment

Keywords: groundwater protection, groundwater vulnerability, groundwater quality control programme, groundwater protection zones.

INTRODUCTION

Earlier, little attention was paid to the protection of groundwater quality, mainly because people were unaware of the treats to this invisible and hidden resource. The idea that groundwater resources are naturally well protected by geological environments, isolated from pollution and not vulnerable to uman activities have   survived a very long time. This erroneous approach to the groundwater protection led in several parts of the world to the extensive groundwater quality deterioration and pollution. Till the beginning of the sixties almost all investments tended to be channelled into groundwater resources development and only technical and economic factors were considered. In the late of 1960's there is a growing public awareness of the need to protect groundwater, as an important natural resource, as a source of drinking water and a valuable component of the ecosystem. Modern concepts of groundwater protection management are based on the assumption, that groundwater development and protection are not two separate phenomena, but mutually integrated environmental, technical, economic and social issues which must be dealt with simultaneously.

GROUNDWATER PROTECTION STRATEGY

Groundwater protection strategy should be based on environmentally sound management and the concept that prevention of pollution is always less expensive than groundwater quality rehabilitation - a costly, long- term and technically demanding task.

Groundwater protection should not be an emergency action taken only when pollution problems arise. The concept of groundwater protection strategy is based on a long- term multidimensional programme managed, coordinated, subsidized and implemented by governmental authorities. It is supported by relevant legislation, requires regular supervision and inspection, includes training of responsible technical staff, and education and information for the general public [3].

The principal criteria of aquifer protection strategy include:
·   comprehensive, aquifer analysis based on good knowledge of the groundwater system'sproperties and the philosophy that only groundwater system can be effectively protected whose properties are well researched and known;
·   the operation of groundwater quality monitoring systems, analysis and implementation ofmonitoring data;
·   the identification of the previous, existing and potential pollution sources and evaluation of their nature, extent and real or potential impacts on groundwater quality;
·   the existence of institutional structures having the necessary power and resources for the creation, coordination and implementation of a comprehensive groundwater protection strategy;
·   a legislative basis, regulatory statues and standards to regulate and control a groundwater protection and quality conservation programme;
·   an effective inspection and control system, including fines on the "polluter pays" principle; availability of experienced, qualified, trained and motivated professional personnel   responsible for   groundwater protection management;
·   a research programme supporting the development of methods and techniques for groundwater protection;
·   public information, education and involvement, particularly on the "consumer pays" principle.

Groundwater protection strategy is effective if the above criteria are applied in an consistent manner. Their partial application cannot lead to any groundwater protection strategy being successful over the long term.

GROUNDWATER PROTECTION POLICY

Not all groundwater resources need to be protected comprehensively, since not all of them are equally vulnerability or valuable. To protect all groundwater resources to the same extent would be unsustainable economically, pointless   hydrogeologically   and unrealistic in terms of management and control. Comprehensive protection of such a diffuse and vast resource as groundwater certainly   throughout the national territory would be too costly, would provoke conflicts in the exploitation of other natural resources, and could lead to an undesirable restriction of economic development in many areas. Groundwater protection policy should be constructive and not based on bans and restrictive measures only.
Groundwater protection policy should or must depend on:
·   the value of groundwater resources;

· the current and expected demands for ground water use and related protection in a given area;
· importance of groundwater for the ecosystem (e.g. wetlands);
· land use planning and economic development in a given area.

The classification of groundwater using the above criteria is always a complicated task, may be controversial and requires good knowledge of the hydrogeological , economic and social aspects of the area concerned.

A good example of groundwater protection criteria is from the U.S. in which three classes of groundwater are defined [1]. A three- level classification system for groundwater quality based on the sophisticated level of purification has been also applied in the Czech Republic.

Modern policy of groundwater protection must be
· integrated with the protection of the remaining components of the hydrological cycle and other elements of the environment (soil, vegetation, etc.);
· coordinated with economic development;
· linked with social policy;
· reflective to the cultural and historical traditions of the society.

The objective of groundwater protection policy is timely identification and analysis of potential conflicting issues and constraints; formulation, quantification, analysis and validation of competitive factors; hierarchical screening with the aim of finding a balance between groundwater protection, economic development and social and health implications with a view to both the short term and long term prospects.

GROUNDWATER PROTECTION MANAGEMENT

The objective of groundwater protection management is to ensure the quality, safety and sustainability of groundwater used as a drinking water source. Groundwater protection management is not a universal process; it is carried out and applied to particular basins having specific natural properties and therefore varied from place to place.

Groundwater protection management is based on a holistic approach and must take into account: some degree of ignorance, uncertainty and unpredictability in terms of the groundwater system's behaviour and properties, cumulative human impacts identifiable many times with difficulties, the effects of which cannot be predicted accurately over a longer term and financial and time constraints. The uncertainties in defining internal and external influences on groundwater systems are associated with certain risks. Therefore environmental impact assessments and risk analyses are usually part of groundwater protection management.

Groundwater protection management must be dealt with simultaneously with groundwater resources development as part of a Water Management Plan (Master Plan).

Two categories of groundwater protection management can be considered: general protection of groundwater resources and comprehensive protection around public water supplies.

General protection of groundwater

General groundwater protection is based on the assumption that effectively accessible groundwater resources are, or may be, used for drinking or other purposes in the future, and therefore their preventive protection is desirable. Water authorities should therefore also bear the responsibility for the protection of potentially usable groundwater resources.

Implementation of general protection of groundwater resources calls for the following activities:
· investigation of the groundwater system and determination of its vulnerability;
· identification, listing and control over the existing and potential pollution sources;
· operation of groundwater quality monitoring systems;
· control over human activities, in particular in vulnerable areas of groundwater basins and aquifers;
· stipulation and implementation of legislative measures for the protection of groundwater resources.

Of particular importance   on a general level effectively supporting groundwater quality management are: mapping and assessment of groundwater vulnerability and monitoring of groundwater quality.

Mapping and assessment of groundwater vulnerability
Groundwater vulnerability is an intristic property of a groundwater system that depends on the ability of that system to cope with natural and/or human impacts [5]. Vulnerability of groundwater is a relative, non- measurable, dimensionless property. The principal attributes used in the assessment of groundwater vulnerability are recharge, soil properties, thickness, permeability and attenuation capability of unsaturated zone and attenuation capacity of saturated aquifer.

In addition to intristic properties of a groundwater system, some users of vulnerability maps may also wish to include potential human impacts, which may prove detrimental - in space and time - to the uses of groundwater resources. For this concept, the term specific vulnerability is used.

Groundwater vulnerability is expressed numerically using indexes having relative numerical values and graphically (maps).
Indexes are a combination of the importance, weight and rating of the above mentioned attributes that involve on groundwater vulnerability.

Groundwater vulnerability maps belong to the category of special purpose environmental maps and are useful for planning, regulatory, decision making and managerial purposes. Vulnerability maps are a good tool to assess groundwater vulnerability potential, to identify areas susceptible to contamination, to evaluate risk of contamination, to design groundwater quality monitoring networks and help to define groundwater protection strategy.

Groundwater quality control programme
Groundwater quality control programme (GWQCP) is composed of three systems (Fig. 1,2): groundwater quality monitoring (1), information (2) and management (3).
GWQCP   is beneficia in terms of
· the information obtained about the current state and trends of the groundwater quality;
· clarification of the magnitude of different types of human impact on the groundwater quality;
· assessment of groundwater system vulnerability and pollution risk;
· definition of the concept of groundwater protection strategy and management.

Monitoring system strategy is based on the objectives and requested output information product. Clearly defined monitoirng objectives are the cornerstone of the monitoring system. The requested information and monitoring strategy cannot be defined clearly when to vague or inaccurately formulated objectives exist. The strategy of monitoring system must be determined before the first sample of water is taken. Choosing the correct strategy of monitoring system in relation to the information requested is the first critical point of GWQCP .

Transformation of analyzed data into the information product is the second critical point of GWQCP . It is being emphasized that however high the quality of our data are, as long as they are not transformed into quantified information, and this information product transmitted in a intelligible form to the user, the programme of groundwater quality control is not effectively utilized and the expenditures on its operation are not vindicated. In other words monitoring system must be cost- effective.

The groundwater quality management system helps address economic and social development of the area of interest in relation to the need for protection of groundwater resources quality as well as define the priorities in groundwater protection and the potential risks and conflicts. A groundwater quality management system should be supported by legislative and institutional measures [4].

Comprehensive groundwater protection
Groundwater resources utilized for public drinking water supplies are comprehensively protected by protection zones referred to as wellhead protection areas in the U.S. , usually comprising two or three levels of protection. The main purpose of groundwater protection zone delineation is to protect drinking water supply wells or wellfields from pollution and provide the population with water which meets the standards for drinking water. In many countries protection zones are an obligatory part of groundwater protection management programmes, and are based on the relevant legislation [2].

Comprehensive protection of public water supply wells requires:
· cooperation between water authorities, waterworks companies and land users (particularly farmers);
· establishment of a specific concept of management of groundwater resources and land use in protection zones, with regards to the balance between urban or rural comunity , economic development and maintenance of good quality and safety of drinking water;
· implementation of technical, institutional, legislative and control measures and regulations in protection zones.

Groundwater protection zones should be delineated on the basis of reliable data and by implementation up- to date methods and techniques to minimalize the degree of uncertainty of their definition. The extent of groundwater protection zones depend primarily on aquifer permeability (porous, fissured of karstic), complexity and vulnerability, the properties and thickness of the unsaturated zone, groundwater flow direction, distance of pollution sources from the wells or wellfield and properties of pollutant.

General approach to the groundwater protection zones delineation
In general, the level of restrictions and prohibitions in groundwater protection zones decreases with the distance from a well or wellfield . The second and third level protection zones cover significant areas of ground, frequently fertile arable land. Over- protection of water supply wells is not therefore desirable because restrictive measures or exclusion of land from farming lead to economic losses. On the other hand, under- protection of wells and wellfields may cause groundwater

pollution requiring long- term and costly remedial action.

I believe that   protection zones should be as small as possible but as large as necessary. Expressed in financial terms, higher input costs on accurate protection zone delineation will result in reduced operational costs for well and wellfield protection. Since water supply systems operate for a long time the operating costs involved on their protection should be reduced as much as possible. A sophisticated approach to protection zone definition is therefore preferable. There is no question about groundwater protection zones having a positive influence on the quality of groundwater resources.

Distribution of costs and benefits is a very sensitive and critical point in groundwater protection management, and should be dealt with a timely and responsible manner.

In spite of the two systems, the social- economic system and the physical (soil and water) system being independent of each other, their mutual relations must be coordinated and integrated with the objective of deriving benefits from utilisation of soil and water resources while conserving the quality of the environment. Management of both systems requires a specific approach in each region because the natural conditions and intensity of human activities, particularly farming, differ locally and regionally. Comprehensive protection and management of groundwater quality achieved through stringent control over agricultural activities affect unfavourably farmers' production targets. Objective evaluation and integration of the soil/water users' interests and allocation of the benefits and costs between the water and agricultural sectors is the key factor in the management strategy of effective utilisation of soil and water resources in protection zones.

Acknowledgments

I thank to DHV CR company which provided me by office facilities in preparation of the manuscript.

REFERENCES

1.   EPA. Groundwater protection strategy. United States Environmental Protection Agency - Water. pp. 55 (1984)
2.   G. Matthes ,   S.S.D. Foster and A.CH Skinner. Theoretical   background, hydrogeology   and practice of groundwater protection zones. International Contributions to Hydrogeology Volume 6. pp. 204   Verlag Heinz Heise   (1985)
3.   J. Vrba   Economic   aspects of groundwater protection. In: Groundwater Economics. E. Custodio   and A. Guargu   (Editors). pp. 153- 180. Developments in Water Science 39.   Elsevier   (1989).
4.   J. Vrba and V. Pekny. Groundwater quality monitoring-effective method of hydrogeological   system pollution prevention, Environmental Geology, Vol 17, No. 1, 9- 13. Springer- Verlag   (1991).
5.   J. Vrba and   A. Zaporozec Guidebook   on mapping groundwater vulnerability. International Contributions to Hydrogeology Vol. 16. pp.131. Verlag Heinz Heise (1994).

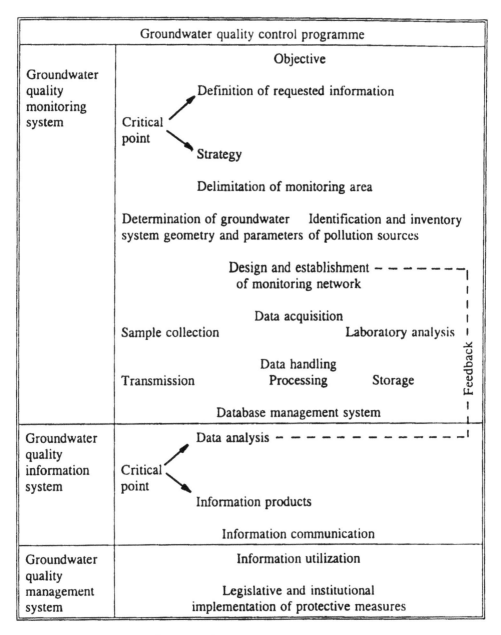

Fig. 1. Groundwater quality control programme

Monitoring programme	Category and importance of monitoring station			Station density	Sampling frequency	Variables analyzed	Monitoring stations characteristics
	Baseline	Trend	Impact				
Internaitional	D	D	LS	VL	L	B+ O-	Baseline station: natural background groundwater quality
National	D	D	LS	V	L	B+ O-	Trend station: trends in groundwater quality due to natural processes and human impacts
Regional	LS	D	C	M	M	B+ O+	
Local	LS	LS	D	H	H	O+	Impact station: changes of groundwater quality due to various impacts

Fig. 2. Categories of groundwater quality monitoring stations operating in the framework of groundwater quality monitoring programmes

Station' importance: D Dominant, C Complementary, LS Less important

Station density: 1 Station per km^2: H High - m^2 to 10, M Medium - 10 to 100, L Low - 100 to 1000, VL Very low - 10000 to 100000

Sampling frequency: H High - more than 12 times a year, M Medium - 2 to 12 times a year, L Low - 1 to 4 times a year

List of variables analyzed: B Basic - physical, chemical and biological variables included into the drinking water standards

O Optional - heavy metals, organochlorine compounds, oil hydrocarbons and other variables depending on monitoring program objectives

+ Regular analysis

- Occassional analysis

Proc. 30th Int'l Geol. Congr., Vol.22, pp. 299-307
Fei Jin and N.C. Krothe (Eds.)
© VSP 1997

Implementation of Groundwater Protection Strategies in the UK

MENGFANG CHEN AND CHRIS SOULSBY

Department of Geography, University of Aberdeen, Aberdeen AB24 3UF

Abstract

In England and Wales about 35% of public water supplies comes from groundwater abstraction, but this becomes less significant in Scotland with groundwater abstraction representing only 3% of total water supplies because of abundant surface water. Groundwater resources are vulnerable to contamination and protecting aquifers from pollution is becoming increasingly important. One solution to this problem is to delineate groundwater protection zones around public supply boreholes and springs and restricting various polluting activities within such zones.

The Environmental Agency (EA) in England and Wales has defined groundwater protection zones based on the 50-day (*Zone 1*) and 400-day (*Zone 2*) travel times and the whole catchment source (*Zone 3*) as part of its Groundwater Protection Policy. A similar approach has been advocated in the more recent development of a Groundwater Protection Strategy for Scotland.

A case study from Spey Abstraction Scheme in Scotland is presented to show how these groundwater protection zones are implemented, with emphasis on their implications in aquifer protection for the Scheme.

Keywords: Groundwater Protection Zones (GPZs), Particle Tracking,   MODFLOW   MODPATH

INTRODUCTION

In England and Wales about 35% of public water supplies comes from groundwater abstraction, and a large number of people rely on the groundwater supplies. But this becomes less significant in Scotland, with groundwater abstraction representing only 3% of public water supplies because of abundant surface water resources.

Groundwater resources are under constant pressure from human activities, and are vulnerable to contamination from various sources such as agricultural practices, landfills,   and nuclear waste disposal. In addition, remediation is very difficult and expensive once groundwater is polluted. Protecting groundwater resources from pollution is thus becoming an increasingly high priority in both developed and developing countries.

The Environment Agency (EA) in England and Wales (like its predecessor the National River Authority) is required by the Water Resources Act (1991) to monitor and protect groundwater resources used for public water supplies from a wide range of potential threats including point and diffuse sources [1]. The Scottish Environmental Protection Agency (SEPA) has a similar duty for Scotland [2]. The Groundwater Protection Policy of the EA is usually based on the definition   of groundwater protection zones, which are to be delineated from steady-state numerical groundwater flow models.

This paper is concerned with implementing UK groundwater protection strategies, particularly definitions of Groundwater Protection Zones (GPZs) and demonstrates their application to the Spey Abstraction Scheme in Scotland.

GROUNDWATER PROTECTION ZONES AND THEIR DELINEATION

Groundwater is a valuable resource and needs to be protected from polluting activities. Improper management of contamination sources resulting from human activities often causes degradation of aquifers. One solution to this problem is to prevent groundwater pollution by delineating groundwater protection zones around public supply boreholes and springs and imposing the restrictions on potentially polluting activities.

Assessing the vulnerability of groundwater resources is a difficult task, as it depends upon many factors such as presence and nature of overlying soil and drift, the nature of hydrostratigraphical units, and the depth of unsaturated zone [1]. A full assessment of groundwater resource vulnerability requires detailed hydrogeological and site investigations. Nevertheless, groundwater vulnerability mapping provides an essential framework that   serves as guidelines for strategic land use planning. In the UK, a groundwater vulnerability map at a scale of 1:1,000,000 was produced and   work for a larger scale map is currently ongoing.

Groundwater protection zones for a particular groundwater source are normally carried out by delineating a number of travel time related groundwater protection zones around a borehole or the wellfield from a steady-state groundwater model. Certain potentially polluting activities are banned or regulated within such zones. The Environmental Agency (EA) in England and Wales has defined groundwater protection zones, based on   50-day (*Zone 1*) and 400-day (*Zone 2*) travel times and the whole catchment source (*Zone 3*) as part of its Groundwater Protection Policy [1]. A similar approach has been advocated in the more recent development of a Groundwater Protection Strategy for Scotland [2].

The 50-day capture zone (*Zone 1*) is based on the time necessary for biological contaminants to decay and is designed to protect against the effects of human activities   immediately adjacent to the groundwater source. The 400-day capture zone is (*Zone 2*) defined by the time required to provide delay and attenuation of slowly degrading contaminants. It would also allow sufficient time to take any necessary remedial measures once contaminants are detected. The catchment source zone (*Zone 3*) covers the entire catchment area of a groundwater source beyond the 400-day capture zones [1].

Because of readily available commercial software, groundwater protection zones are usually delineated using reverse particle tracking techniques after groundwater flow simulations. These methods offer the most accurate delineation (given a sufficient hydrogeological and geological data) at a modest cost.

FLOWPATH and MODPATH are the most commonly used particle tracking programs used for delineating groundwater protection zones [6, 7]. FLOWPATH is a two-dimensional finite difference flow code combined with particle tracking, whereas MODPATH is three-dimensional particle tracking code, which uses a MODFLOW head solution to calculate flow velocity distribution. However, particle tracking techniques only consider contaminants being transported in groundwater by advection, while hydrodynamic dispersion and chemical reactions are ignored. Despite this assumption, the particle tracking of advective transport is an attractive alternative to solving a contaminant transport model as it does not involve the complicating effects and uncertainties in quantifying dispersion and chemical reactions [8].

APPLICATION TO SPEY ABSTRACTION SCHEME

Study Area

Particle tracking techniques have been recently been used to identify GPZs at an abstraction scheme on the River Spey in Scotland.

The River Spey, with a catchment area of some 3000 km^2, is one of the largest rivers in Scotland (Fig. 1a). The Scheme wellfield extends some 3 km along the west bank of the river. Abstraction boreholes have been drilled into floodplain alluvium, and are generally located 50-100 meters from the river and are around 100 m apart (Fig. 1b,c). The area modelled in the present study is approximately 6 km2 of the floodplain in the lower River Spey catchment. Land use is dominated by agricultural activities including arable and livestock production.

The Spey Abstraction Scheme is one of the largest developments in the UK that involves indirect river water abstraction through alluvial gravels adjacent to the river bank. This offers many advantages compared with traditional direct river abstraction, the most important being a significant improvement in water quality due to biological and chemical processes in the alluvium that result in much reduced water treatment costs. Additionally, environmental impacts on river flow and migratory salmon fisheries are minimized.

The Scheme has been designed to provide a maximum of 27000 m³/d to meet increasing demand forecasts in the Lower Moray and Banff coastal areas of Aberdeenshire by the year of 2011 (Fig. 1a).

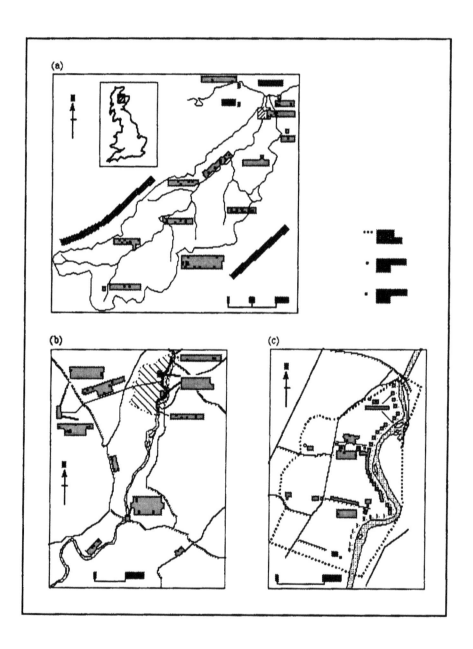

Fig. 1   The study area showing (a) the River Spey catchment; (b) the Scheme layout and (c) wellfield layout

The Scheme consists of a wellfield of 36 production boreholes (B1-B36), North and South control installations, and a twin transmission main to deliver the pumped water into the nearby Badentian Reservoir where the facility for full treatment work may be available in the future. From the reservoir, water is distributed into the northern supply destinations (Fig. 1b).

Conceptual Model and Numerical Modelling

The conceptual model for the study area is relatively simple, consisting of two hydrostratigraphical units: upper alluvium (layer 1) and lower (layer 2), separated by a aquitard, averaging 0.5 m thickness at a depth of around 7 m below the ground level. The alluvium is underlain by the Old Red Sandstone (Fig. 2). A quasi three-dimensional model was formulated, with the layer 1 representing a unconfined condition and the layer 2 a leaky confined condition. The aquitard was introduced into the model through a leakage term for the confining silt layer.

Evaluation of pumping test data from production boreholes indicates that the hydraulic conductivity of the alluvium varies considerably, between 20 and 200 m/d. The average value is 80 m/d and this reduces towards the edge of the floodplain. The storage coefficients of the alluvium range from 9 × 10^{-5} to 0.1 for the confined and unconfined alluvium respectively.

No-flow boundaries were assigned to the western and eastern boundaries of the modelled area. The northern and southern boundaries of the modelled area were assigned constant hydraulic heads, which were derived from the observed water table data for both layers. The other boundary conditions include no-flow boundaries along the limit of alluvium for both model layers and the bottom of the layer 2 below which the Old Red Sandstone is present. Fixed heads were specified for river nodes and a recharge boundary was applied to the top of layer 1.

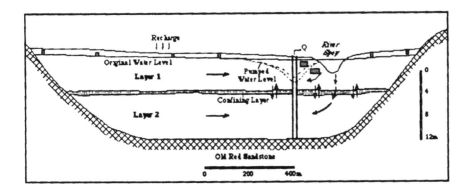

Fig. 2 Conceptual model of river-aquifer interactions at the Spey Abstraction Scheme showing potential effect of abstraction on water levels

Under natural conditions, groundwater generally flows from the southwest to northeast in the alluvium, with a hydraulic gradient of approximately 0.003. Discharged is predominantly into the river at the northern end of the site.   Effective recharge is through the upper alluvium, where the groundwater table is normally between 1 and 2 m below the ground level.

The US Geological Survey computer program MODFLOW, a three-dimensional finite-difference groundwater model [3], has been used in this study with the river package being used to simulate the river-aquifer interactions. A traditional trial and error approach and MODFLOWP [4], a nonlinear regression method, were jointly employed during model calibration and sensitivity analysis.

Detailed model setup, grid design, model calibration and sensitivity analysis have been presented elsewhere and are thus not repeated here [5].

Delienation of Groundwater Protection Zones (GPZs)

After the groundwater flow simulation, the reverse particle tracking technique, using MODPATH [6], was applied to compute flow paths and delineate   travel time-related capture zones at the projected maximum pumping rate of 27000 m^3/d. An effective porosity of 0.40 was adopted to calculate the average linear velocity fields from the simulated head distribution.

Three capture zones (or groundwater protection zones) have been   defined using the 50-day and 400-day travel isochrons for layer 1. The size and shape of each zone are determined mainly by the hydrogeological characteristics of the river-aquifer system and the direction of groundwater flow.

·   50-day capture zone (Zone 1)

The 50-day capture zones in both model layers are nearly identical in geometry except that the Eastern limit in layer 1 is bounded by the river itself. Because of the high water-transmitting properties of the alluvium and the complex hydrogeological systems that the model represents, the 50-day capture zones take highly irregular shapes. These consist of several discrete regions, which are attributed mainly to the distance and the pumping rates of two neighbouring wells. The extent of this zone is generally situated within 200 m of the river and is centered on the wellfield (Fig. 3). The cumulative size of the 50-day capture zones is small for both model layers, being approximately 0.5 km^2, or one-twelfth of the modelled area.

·   400-day capture zone (Zone 2)

Like the 50-day capture zones, the 400-day capture zones (*zone* 2) exhibit highly irregular geometry around the wellfield. They extend further westward for both the model layers and eastward for the confined alluvium. The eastern limit of the 400-day capture zone in the layer 1 is bounded by the river (Fig. 3). *Zone 2* of layer 1 covers approximately 0.75 km^2.

· Catchment Source Zone (Zone 3)

Zone 3 covers the rest of the modelled catchment beyond the 400-day travel isochron. Any contaminant emanating from *zone 3* would eventually discharge into the wellfield with the varying time scales all exceeding 400 days. The time required for contaminants to reach the wellfield depends on many factors, such as their initial positions, hydrogeological properties of the alluvium and the distance to the pumping wells.

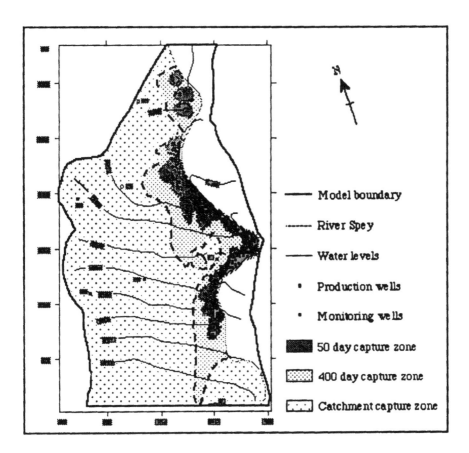

Fig. 3 Simulated potentiometric surface and groundwater protection zones of layer 1 under the projected pumping rate of 27000 m³/d.

Implications in Aquifer Protection

The implementation of groundwater protection zones for the Spey Abstraction Scheme demonstrated the potential of numerical groundwater models as management tools in groundwater protection.

The capture zone delineation suggested that the primary potential source of pollution for the wellfield is the river Spey itself. However, the modelling also indicated that significant potential threats to groundwater integrity could come from pollution sources within 50-day (*Zone 1*) and 400-day (*Zone 2*) capture zones [5]. Thus, there is a need to carefully control land use and other activities in the two zones in accordance with the Groundwater Protection Strategy for Scotland [2].

The critical components for a comprehensive monitoring programme for groundwater protection are a network of monitoring wells around the perimeter of groundwater protection zones and a continuous pollution monitor in the river upstream. The purpose of the these monitoring measures is to allow   early detection of possible groundwater and surface water contamination and to take any remedial action, such as   pump shutdown and backflushing the wellfield during any period of river pollution.

This modelling study suggests that the current plan for water quality monitoring perhaps needs to be enhanced. First, the planned pollution monitoring at the Boat O'Brig, some 6 km upstream of the wellfield (Fig. 1a) should be installed as soon as possible and this work is currently underway.

For the purpose of land-use control, the Scheme operators have purchased the two farms within the wellfield and have already imposed restrictions on potentially polluting farm practices. An irrigation system has also been introduced to replenish soil moisture deficits during droughts that may be exacerbated by   drawdown resulting from wellfield pumping. However, current and simulated drawdown indicate that the magnitude and extent of   drawdown caused by pumping more than 20% of the water from the alluvium itself may previously have been underestimated. This, together with strict restrictions in farm practices (eg limiting fertilizer applications *etc*) could reduce crop production. At present,   restrictions on farm practices have been imposed over a larger area than the capture zones indicate is necessary. However, this strategy may be considered favourable as it adopts a precautionary approach to groundwater protection. Clearly, economic estimates of the costs of land-use restrictions and irrigation supplies should be evaluated in relation to the benefits of groundwater protection if the operating costs of the scheme are to be properly appreciated.

CONCLUSIONS

Groundwater protection zones are useful tools in the protection of groundwater resources from pollution, but these concepts may be difficult to implement in reality. These difficulties often result from the lack of public awareness and publicity about the risks of groundwater protection. In Scotland, the Scottish Environmental Protection Agency can only implement its Groundwater Protection Strategy by education and persuasion as land owners have a overall control on land use. Thus there is a urgent need for public education and possible legal changes in Scotland regarding the implementation of these groundwater protection zones.

Delineating groundwater protection zones for the Spey Abstraction Scheme demonstrated their potential importance in aquifer management. It is recommended that rigorous programs for monitoring water quality both in the river upstream, and along the perimeter of zones 1 and 2 within the modelled catchment are required to protect the Spey Abstraction Scheme.

Acknowledgments

We would like to thank the Bank of Scotland and Aberdeen University Research Committee for funding this study.

REFERENCES

1. NRA. Policy and practice for the protection of groundwater. London, HMSO (1992).

2. ADRIS. Groundwater protection strategy for Scotland. Association of Directors and River Inspectors of Scotland (1995).

3. M.G. McDonald and A.W. Harbaugh. A modular three-dimensional finite-difference ground water model. *U.S. Geological Survey, Techniques of Water-Resources Investigations*, Book 6,   Chapter A1 (1988).

4. M.C. Hill. A computer program (MODFLOWP) for estimating parameters of a transient, three-dimensional, groundwater-water flow model using nonlinear regression. *U.S. Geological Survey*, Openfile report 91-484, 358pp (1992).

5. M. Chen, C. Soulsby and B. Willetts. Modelling river-aquifer interactions at the Spey Abstraction Scheme, North East Scotland: implications for aquifer protection, *Quarterly Journal of Engineering Geology* (in press).

6. D.W. Pollock. Documentation of computer programs to complete and display pathlines using results from the U.S. Geological Survey modular three-dimensional finite-difference ground-water model. *US Geological Survey Open File Report* 89-381, 81pp (1989).

7. T. Franz and N. Guiguer. FLOWPATH, two-dimensional horizontal aquifer simulation model. Waterloo Hydrogeologic Software, Waterloo, Ontario, 74pp (1990).

8. M.P. Anderson and W.W. Woessner. Applied groundwater modelling: simulation of flow and advective transport. Academic Press, San Diego, 381pp (1992).

Proc. 30th Int'l Geol. Congr., Vol.22, pp. 308-315
Fei Jin and N.C. Krothe (Eds.)
© VSP 1997

An Inverstigation of Salt Contamination of Groundwater in Inner Mongolia

YUN-SHENG YU
Department of Civil and Environmental Engineering, University of Kansas, Lawrence, KS 66045, U.S.A.
B. C. WANG, W. F.SUN, C. H. DONG, M. Z. WANG
CIGIS , Ministry of Construction, Beijing, China
MING-SHUTSOU
Department of Civil and Environmental Engineering, University of Kansas, Lawrence, KS 66045, U. S. A.

Abstract

This case study concerns the salt intrusion into the fresh groundwater in an Inner Mongolia village called Yaobata located approximately 1300 km west of Beijing, China. Yaobata borders the Dengali desert on the west and the Helansan mountain on the east with a reclaimed desert land of 64 sq. km. for agricultural production and a population over 8000 people. Over the years, continual excessive withdrawals of the groundwater with little recharge have resulted in drop in groundwater level by 4 meters and increases in salinity is exceeding 3 g/l at some pumped wells. Extensive field data collected show that the source of salt is from the Dowsu Lake located in the southwest corner of Yaobata A deterministic numerical model was developed and used to estimate future piezometric head and salinity distributions under the present scheme of groundwater usage. In addition, model parameters are considered random ariables in space. Conditional simulation based on simulated annealing is used to generate 100 realizations of model parameters in the numerical model. The statistics of the resulting realizations of the piezometric heads and salinity re obtained. Based on these results management plans are proposed to insure sustainable groundwater supply for the community.

Keyords : groundwater, saltwater intrusion, numerical model, geostatistical simulation

INTRODUCTION

The discovery of groundwater of good quality in the 1960s has so far sustained the growth of the agricultural community, Yaobata, in Inner Mongolia to a population of 8000 people. Since groundwater is the only source of water supply for the community, it is imperative that good-quality groundwater supply should be maintained for the livelihood of the community. Groundwater levels in some irrigation wells have been lowered by as much as four meters from 1984 to 1990 and the salinity of some groundwater has exceeded 3 g/l resulting in detrimental effects on crop yield. The sustainability of groundwater supply becomes a critical and urgent issue for the community.

In 1987, an interdisciplinary research team of geographers, geologists, hydrogeologists and engineers was organized to study the groundwater system with the main objectives being: (1) to assess the availability of groundwater, (2) to determine the source and migration of salt contaminant, and (3) to find management solutions to insure sustainable groundwater supply for the community. The methods of approach to these objectives include extensive field experimentation and development and validation of mathematical models for the aquifer system.

FIELD EXPERIMENTATION

A network of 30 observation wells was established for monitoring the groundwater levels and the various water quality variables. Pumping tests and tracer tests were carried out to determine the hydraulic conductivity, dispersivity, and flow velocity. Radio,, isotope tracer technique and remote sensing were used to identify groundwater sources. Detailed descriptions of field experimentation are presented in a report published in 1992 (4). The extensive field data clearly indicate that the confined aquifer is a layer extending generally from 15 m to about 100 m below the ground surface and is sandwiched between two parallel relatively impermeable layers. The results of the radio,,isotope tracer tests suggest that there is little groundwater recharge from precipitation. The source of groundwater may be from the connected aquifer near the Helansan mountain.

Soil samples taken from the Dowsu Lake located in the lowest region of the southwest corner of Yaobata show that the salinity of the soil varies from a maximum of 16.0 g/l at the center of the lake to 3g /l near the lake shore. The salinity in the soil at the lake center decreases as the depth from the lake surface increases. In addition, the observed salinity of water in different observation wells decreases as the distance from the Dowsu lake increases. The piezometric heads in wells close to the lake are generally higher than those in the irrigation wells by as much as 10m. These observations would serve to identify the Dowsu Lake area as the main source of salt intrusion.

The field experimentation showed that values of hydraulic conductivity vary between 200 m^2/d and 2200 m^2 /d storativity ranges from 0.001 to 0.08 and that longitudinal dispersivity, a$_L$ varies between 48 and 68 while transverse dispersivity ,a$_L$ remains essentially constant at 40.

NUMERICAL MODEL

The confined aquifer of interest is a layer sandwiched between two relatively impermeable layers and the well water is generally well-mixed. For these reasons, a two-dimensional flow and contaminant transport model governed by the following equations was developed.

The law of conservation of mass expressed in Cartesian coordinates is

$$S\frac{\partial h}{\partial t} = div \ [\ T \cdot grad(h) \ ] \tag{1}$$

where S = storativity T = transmissivity vector, h = piezometric head, t = time.
The conservation of mass of the solute is

$$\frac{\partial (\rho \phi c)}{\partial t} = div \ [\ \rho \ D_{ij} \cdot grad(c) \ ] \ - div \ [\ \rho \ c \ \bar{V} \ ] \tag{2}$$

where ρ = fluid density, ϕ = effective porosity, c=salinity, =velocity vector, and

Dij = dispersivity tensor .

Darcy's law in vector form is

$$\bar{V} = -\frac{k}{\mu} grad(h)$$ (3)

where k=intrinsic permeability and μ = fluid viscosity.

Table 1   Measured and computed   piezometric   heads in meters above   MSL   at observation wells for February 1990

Well Numberà	Computed	Measured	Difference
1	1283.69	1283.70	-0.01
2	1283.11	1283.09	0.02
3	1282.98	1282.94	0.04
4	1282.69	1282.68	0.01
5	1282.66	1282.71	-0.05
6	1282.02	1282.10	-0.08
7	1282.66	1282.64	0.02
8	1282.26	1282.27	0.01
9	1282.37	1282.30	0.07
10	1282.52	1282.57	-0.05
11	1282.49	1282.52	-0.03
12	1282.57	1282.55	0.02
13	1282.56	1282.62	-0.06
14	1282.38	1282.43	-0.05
15	1282.99	1282.03	-0.04
16	1281.86	1281.84	0.02
17	1281.79	1281.75	0.04
18	1282.55	1282.48	0.07
19	1281.90	1281.89	0.01
20	1281.92	1281.97	-0.05
21	1282.56	1282.55	0.01
22	1282.43	1282.37	0.06
23	1282.11	1282.11	0.00
24	1281.89	1281.88	0.01
25	1282.02	1281.99	0.03
26	1282.26	1282.29	-0.03
27	1282.43	1282.41	0.02
28	1282.55	1282.58	-0.03
29	1282.31	1282.30	0.01
30	1282.45	1282.45	0.00

Equations 1 and 2 together with initial and boundary conditions are solved numerically by using the Galerkin   finite,,element method. The study area is covered with 331 triangular elements with 139 interior nodes and 55 boundary nodes. The boundary conditions are all of the first kind.   The observed data of February 1990 were used for model calibration and the April 1992 data for model validation.   The computed and measured   piezometric   heads and their differences for February

1990 are shown in Table 1.

The differences between the computed and measured values as shown in the last column of Table 1 vary from 0.08 m to 0.07 m and are acceptable.For model validation, the piezometric heads for April 1992 are computed and compared with the measured values at the observation wells. The differences are generally larger than those for February 1990 and are between ± 0.5 m. Detailed description of validation model for water quantity and salinity are reported elsewhere (4) and not included in this paper. The results are in generally in fair agreement. Thus, the numerical model is believed to be reasonably accurate as a tool for making management decisions. In the following section, the model will be used to simulate the operation of a well system located near the Dowsu lake to reduce the saltwater intrusion from the Dowsu lake toward the irrigation wells.

METHOD FOR REDUCING SALTWATER INTRUSION

Field experimentation shows that the groundwater under and near the Dowsu lake has very high salinity and the piezometric heads in the area are much higher than those in the irrigated area. Excessive withdrawal of groundwater for irrigation has exacerbated the saltwater intrusion from the Dowsu Lake into the irrigated area. The method proposed for reducing and stopping the saltwater intrusion involves drilling a row of wells located near the Dowsu Lake. These pumped wells would serve to reduce the local piezometric head and consequently would result in less saltwater intrusion. The pumped water with high salinity may be discharged directly into the lake. Due to high evaporation rate in the desert area, water in the lake would evaporate with concentrated salt deposited on the lake bottom. During the year, the lake is usually dry except during occasional summer storms. Since the soil of the lake bottom has very low permeability, the amount of seepage of lake water would be minimal. Thus, simple pumping appears to offer an effective and economical way to reduce saltwater intrusion.

The optimal well system to accomplish this objective has been determined by using both the numerical model to simulate the operation of different configurations of the well system and a nonlinear optimization technique. The configuration of the well system includes the number of wells, pumpage and their locations. Detailed description of the procedure is presented elsewhere (4) and not included in this paper. The final selection of the well system, which is most economical and satisfies the objective, is a row of 6 wells with equal distance of 640 m apart and each well having a pumping capacity of 700 cubic meters per day. This arrangement would create a low piezometric head corridor about 2400 m long to reduce and eventually stop the saltwater intrusion from the Dowsu Lake toward the irrigated area. The results of simulated operations of the well system are very encouraging. After the first 5-year of the simulated operation, the salinity of water in the irrigation well would reduce 15 mg/l every year. The salinity would stop increasing during off -irrigation season. This well system has been recommended for implementation.

CONDITIONAL SIMULATION OF MODEL PARAMETERS BY SIMULATED ANNEALIN

The numerical model described in the previous section is a deterministic model. For a given set of model parameters, namely, transmissivities and dispersivities , and the boundary and initial conditions, the model computes the piezometric heads and salinity at the nodes of the finite -lement grids. In this section, these model parameters are considered as random functions of space. The seventeen sampled values are a partial realization of the random parameters. Realizations of model parameters at unsampled nodes must be estimated using conditional simulation which preserves the geostatistical characteristics of the sampled data by simulated annealing and honors the parameter values at sampled sites.

Simulated annealing is a new simulation method in geostatistics The method was initially formulated to solve the statistical mechanics problem of calculating the variation with temperature of properties of substances composed of interacting molecules (3), such as energy levels in the metallurgical annealing process of prolonged heating and slow cooling of a piece of metal. In the application of annealing to geostatistics , one deals with values of a spatial attribute instead of molecules. The method presumes that the sampling took place at some of the sites to be considered in the characterization through the partial realization of the random function, typically a grid of regular nodes. If that is not the case, the observations are moved to the closest node.   Duplicates are discarded or averaged.

The grid values are assigned in two steps: 1. All nodes coinciding with the sampled sites take the observed value at the node.   2. One assigns values to the remaining nods by drawing at random from a cumulative density function, typically the sampling cumulative density function. The purpose of simulated annealing is to achieve a matching between a function of the original observation and the one for simulated realizations, typically a semivariogram (1). The of squared differences of semivariograms

$$G= [\; \gamma_G(h) - \gamma \; (h)] \;^2$$

where $\gamma \; (h)$ is the model semivariogram for the sampled data and $\gamma_G (h)$ is the semivariogram for the realization.

The minimization of the objective function is achieved by swapping pair of values of the attribute at different locations $Z_a (X_i)$ and $Z_a (X_j)$ chosen at random and observing the dffect on the objective function. The computation of the objective function can be considerably simplified by correcting the effect of the swapping rather than starting the recalculation of G from scratch. A swap is accepted if (I) neither of the locations involved in the swapping coincides whth a sampling site, (ii) there is a retained but the frequency with which those apparently unfavorable swaps are retained decreases with exp $[(G_{old} - G_{new})/t \;]$, where t mimics the temperature parameter in the Boltzmann distribution.

Based on the sampled values of hydraulic conductivity, the best semivariogram model with the minimum AIC (2) is the following spherical model

$$\gamma (h) = 75.36 + 205706 \; \{ (3/2) \, (h/5130) - (1/2) \, (h/5230)^3 \} \; , \text{ for } 0 \le h \le 5130$$
and $\gamma (h) = 280742 \quad \text{for } h \ge 5130$

Using the semivaroigram model and the simulated annealing procedure, one can obtain as many realizations of the random field as required. The values of hydraulic conductivity obtained for each realization are then used in the deterministic model to compute the piezometric heads at various nodes. Similarly, the realizations of salinity distributions are obtained. The results are presented in the following section.

ISOLINES OF MEAN AND STANDARD DEVIATION OF PIEZOMETRIC HEAD AND SALINITY FOR APRIL 2000

One hundred realizations of the model parmeters are obtained from the conditional simulation. The values of each realization at the grid nods are in the numerical model to compute the piezometric heads and the salinity values. At each node, the statistics of the 100 values are computed. A sample piot of the cumulative probability versus the piezometric head on a probability paper as shown in Fig.1. The plotted points generally follow a straight line on the probability paper. Thus, the piezometric head has a normal distribution. The isolines of the mean and the standard deviation computed for April 2000 are show in Figs.2 and 3, respectively. From these two figures, at any given location, the mean and standard deviation of the piezometric head can be found. Given a specific probability of exceedence, the expected piezometric head at that location can then be determined using following the Gaussian distribution. Similar plots of mean salinity and stsndard deviation for April 2000 are shown in Figs.4and 5, respectively.

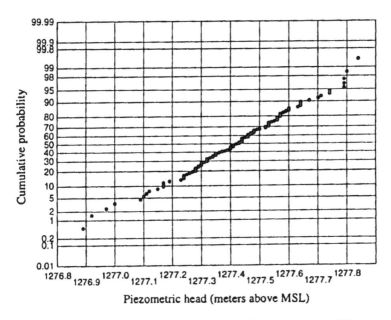

Figure 1   Realizations of computed piezometric head at node #27 plotted on a probability paper

CONCLUSIONS

Extensive field data show convincingly that the source of saltwater intrusion is from the groundwater under and clode to the Dowsu Lake and that there is little groundwater recharge from surface runoff. An effective and economic method for reducing and stopping saltwater intrusion is to use a row of six pumped wells located near the Dowsu lake each with pumping capacity of 700 cubic meters per day with equal spacing of 600 m. The results of 5-year simulated operation of the well system indicate that a reduction of 15 mg/l in salinity per year during the irrigation season would be achievable and that salinity would not increase during the off-irrigation season. The pumped water can be discharged into the Dowsu Lake for repid evaporation with practically no seepage from the lake bottom.

Acombined use of the deterministic numerical model and conditional simulation of model parameters provides a method of stochastic forecasting of the piezometric head and salinity dirtributions for the study area. The distributions mean and standard deviation of piezometric head and salinity for April 2000 can be used to estimate the probability of exceedence for a given piezmetric head or salinity at a given location in the study area to account for the uncertainty of model parameters.

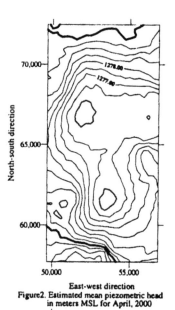

East-west direction
Figure2. Estimated mean piezometric head
in meters MSL for April, 2000

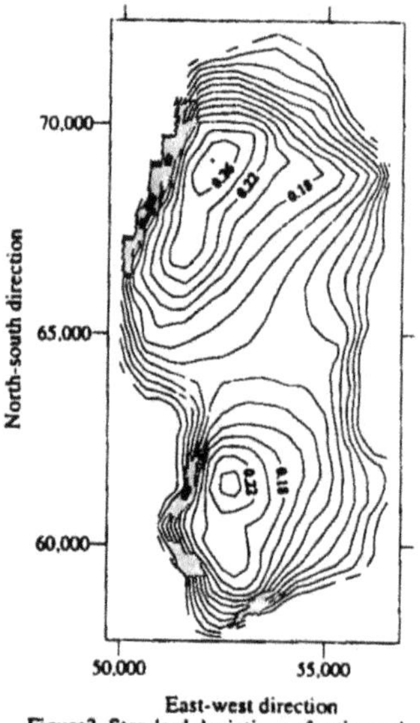

East-west direction
Figure3. Standard deviations of estimated
piezometric head (m) for April, 2000

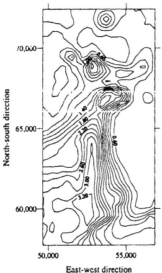

East-west direction

Figure4. Estimated mean salinity (g/l)
for April, 2000

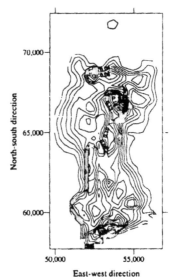

East-west direction

Figure5. Standard deviations of estimated
salinity (g/l) for April, 2000

REFERENCES

1. C. V. Deutsch, and A. G. Journel. GSLIB-Geostatistical software library and user's guide. Oxford Usiversity press, New York (1992).
2. X. D. Jian, R. Olea, and Y. S. Yu. Semivariogram modeling by weighted least squares. Computers & Geosciences, Vol. 22, No.4, 387-397 (1996).
3. N. Metropolis, A. W. Rosenbluth, M. N. Rosenbluth, A. H. Teller, and E. Teller. Equation of state calculations by fast computing machines. The Journal of Chemical Physics, Vol.21, No.6, 1087-1092 (1953).
4. B. C. Wang, W. F. Sun, C. H. Dong, and M. Z. Wang. A research report on management of groundwater resources of Yaobata, Alasan, Inner Mongolia. Published by CIGIS, Ministry of Constroction, China (1992), in Chinese.